"In this terrific book, Ozzie Zehner explains why most current approaches to the world's gathering climate and energy crises are not only misguided but actually counterproductive. We fool ourselves in innumerable ways, and Zehner is especially good at untangling sloppy thinking. Yet *Green Illusions* is not a litany of despair. It's full of hope—which is different from false hope, and which requires readers with open, skeptical minds."—David Owen, author of *Green Metropolis*

"Think the answer to global warming lies in solar panels, wind turbines, and biofuels? Think again. . . . In this thought-provoking and deeply researched critique of popular 'green' solutions, Zehner makes a convincing case that such alternatives won't solve our energy problems; in fact, they could make matters even worse."—Susan Freinkel, author of *Plastic: A Toxic Love Story*

"There is no obvious competing or comparable book. . . . *Green Illusions* has the same potential to sound a wake-up call in the energy arena as was observed with *Silent Spring* in the environment, and *Fast Food Nation* in the food system."—Charles Francis, former director of the Center for Sustainable Agriculture Systems at the University of Nebraska

"This is one of those books that you read with a yellow marker and end up highlighting most of it."—David Ochsner, University of Texas at Austin

Green Illusions

Our Sustainable Future

Series Editors

Charles A. Francis
University of Nebraska–Lincoln

Cornelia Flora
Iowa State University

Paul A. Olson
University of Nebraska–Lincoln

Green Illusions

The Dirty Secrets of Clean Energy and the Future of Environmentalism

Ozzie Zehner

University of Nebraska Press
Lincoln and London

∞

Library of Congress Cataloging-in-Publication Data

Zehner, Ozzie.
Green illusions: the dirty secrets of clean energy and the future of environmentalism / Ozzie Zehner.
p. cm. — (Our sustainable future)
Includes bibliographical references and index.
ISBN 978-0-8032-3775-9 (cloth: alk. paper)
1. Renewable energy sources—United States.
2. Environmentalism—United States. I. Title.
TJ807.9.U6Z44   2012
333.79'40973—dc23   2011042685

Set in Fournier MT.
Designed by Mikah Tacha.

To Mom and Dad,
who gave me a leash,
only to show me how to break it.

§

All of the royalties from this book
will go toward projects supporting
the future of environmentalism.

Contents

Illustrations

Figures

Table

Acknowledgments

To begin, I'd like to extend special thanks to numerous anonymous individuals who risked their standing or job security to connect me with leads, offer guidance, spill dirt, and sneak me into places I perhaps shouldn't have been. These include one World Bank executive, one member of Congress, one engineer at General Motors, two marketing executives, one former teen celebrity, two political strategists in Washington DC, two military contractors, a high school vice principal, one solar industry executive, one solar sales rep, one mining worker, and three especially helpful security guards.

I also extend thanks to those organizations that generously released confidential reports, which I had worked on, so I might draw upon their findings in this very public setting. I appreciate the cooperation from numerous Department of Energy employees, who provided everything from images to insight on internal decisionmaking. This book would not have been possible without brave individuals from the World Bank, U.S. military, academia, and industry who were involved in whistle-blowing, industrial espionage, and leaks exposing wrongdoings. Their courage reminds us there are rules that are provisional and rules that are not.

My enthusiastic and keen agent, Uwe Stender, generously pursued a nonprofit press deal while offering me limitless support and advice. I'd like to thank the editors and staff at the University of Nebraska Press, including Heather Lundine, Bridget Barry, Joeth Zucco, and Cara Pesek, who believed in this work enough to buy the rights, support this book in staff meetings, coordinate expert reviews, considerably improve the manuscript, and put up with this green writer. Special thanks go to Karen Brown, whose copyediting cut a path through my writing for readers to follow.

I'd especially like to mention the generous support of numerous educators, colleagues, and advisers starting with John Grin, Loet Leydesdorff, Stuart Blume, Chunglin Kwa, and, in memory, Olga Amsterdamska from the University of Amsterdam; Steven Epstein (Northwestern University), Naomi Oreskes, Katrina Peterson, and Marisa Brandt from the Department of Science Studies at the University of California–San Diego; Joseph Dumit, Mario Biagioli, Tom Beamish, Tim Choy, Jim Griesemer, Caren Kaplan, Colin Milburn, Chris Kortright, and Michelle Stewart from the Department of Science and Technology Studies at the University of California–Davis; Cori Hayden, Cathryn Carson, Hélène Mialet, Mark Robinson, Mark Fleming, Jerry Zee, Mary Murrell, Diana Wear, and Gene Rochlin from the Science, Technology, and Society Center at the University of California–Berkeley; Reginald Bell at Kettering University; Charles Francis at the University of Nebraska; David Ochsner at the University of Texas at Austin; Brian Steele from the former Kalamazoo Academy; the dedicated staff at Nature Bridge Headlands Institute; and the Science, Technology, and Society faculty at the Massachusetts Institute of Technology.

I am grateful for the editors who dealt with various rough drafts of this manuscript, from proposal to completion, including especially Frieder Christman, Oleg Makariev, and Joe Clement, as well as the acutely helpful input from three anonymous academic reviewers. I'd also like to thank Saad Padela, Jenny Braswell, Harshan Ramadass, Tayo Sowunmi, Nahvae Frost, Myla Mercer, Judy Traub, Sarah Margolis, Cheryl Levy, Garrett Brown, and Karla L. Topper.

I'd like to thank numerous individuals who assisted me with conceptual and theoretical development: Charlie Zuver, D. A. M., Brad Borevitz, Kyla Schuller, Jeffrey Stevens, Jack Markey, L. Chase Smith, Kien Tran, Babs Mondschein, Yao Odamtten, Leonardo Martinez-Diaz, Drika Makariev, Valera Zakharov, Florence Zakharov, Ariel Linet, Ben Wyskida, Daniel Williford, Jesse Burchfield, Jessica Upson, Nathalie Jones, Nicholas Sanchez, Allison Rooney, Rex Norton, Hilda Norton, Jens Maier, Paul Burow, Santani Teng, Thomas Kwong, Stefanie Graeter, Susan Elliott Sim, Tom Waidzunas, Olivia Iordanescu, James Dawson, Maurice van den Dobbelsteen, Thomas Gurney, Jurjen van Rees, and the many other people who helped out along the way.

Most importantly, I am thankful for the boundless support of my family: Patti Zehner, Tom Zehner, Robby Zehner, Aaron Norton, Sabin Blake, and Randy Shannon.

All errors in judgment, content, or otherwise lie with me.

Introduction: Unraveling the Spectacle

The world will not evolve past its current state of crisis by using the same thinking that created the situation. - Albert Einstein

If the title of this book makes you a little suspicious of what I'm up to, then all is well. We'll get along just fine. That's because the dirty secrets ahead aren't the kind you can be *told* (you probably wouldn't believe me anyway), but rather are the kind you must be *shown*. But even then, I don't expect you to accept all of my particular renderings.

Ahead you'll see that this certainly isn't a book *for* alternative energy. Neither is it a book *against* it. In fact, we won't be talking in simplistic terms of for and against, left and right, good and evil. I wouldn't dare bludgeon you with a litany of environmental truths when I suspect you'd rather we consider the far more intriguing questions of how such truths are made. Ultimately, this is a book of shades. This is a book for you and others who like to think.

Ahead, we'll interrogate the very idea of being *for* or *against* energy technologies at all. Many energy debates arise from special interests as they posture to stake flags on the future—flags adorned with the emblems of their favorite pet projects. These iridescent displays have become spectacles in their own right. And oh, how we do delight in a spectacle with our morning coffee. Needless to say, these spectacles influence the answers we get—there is nothing new about this observation—but these energy spectacles do much more. They narrow our focus. They misdirect our attention. They sidetrack our most noble intentions. *They limit the very questions we even think to ask.*

Consider, for instance, America's extensive automotive transportation system that, alongside impressive benefits, yields a host of negative side effects such as smog, particulates, CO_2, and deadly accidents. America's overwhelming response has been to adjust the technology, the automobile itself. Our politicians, corporations, universities, and the media open their palms to show us an array of biofuel, electric, and hydrogen vehicles as alternatives. But even though these vehicles might not emit toxic fumes directly, their manufacture, maintenance, and disposal certainly do. Even if we could run our suburbs on batteries and hydrogen fuels cells, these devices wouldn't prevent America's thirty thousand automobile collision fatalities per year.[1] Nor would they slow suburban proliferation or the erosion of civil society that many scholars link to car culture. And it doesn't seem that people enjoy being in cars much in the first place—40 percent of people say they'd be willing to pay twice the rent to cut their commute from forty-five to ten minutes, and a great many more would accept a pay cut if they could live closer to friends.[2]

Might we be better served to question the *structure* and *goals* of our transportation sector rather than focus so narrowly on alternative vehicles? Perhaps. Yet during times of energy distress we Americans tend to gravitate toward technological interventions instead of addressing the underlying conditions from which our energy crises arise.[3] As we shall discover in the chapters to follow, these fancy energy technologies are not without side effects and limitations of their own.

When I speak on energy, the most frequent questions I receive are variants of "What energy technology is best?"—as if there is a straightforward answer. Every energy technology causes aches and pains; shifting to alternative energy represents nothing more than a shift to alternative aches and pains. Still, I find most people are interested in exploring genuine solutions to our energy problems; they're eager to latch on and advocate for one if given the opportunity. As it turns out, there are quite a few solutions that could use some latching onto. But they're not the ones you'll read about in glossy magazines or see on television news—they're far more intriguing, powerful, and rewarding than that.

In the latter part of this book, we'll imagine tangible strategies that cross-examine technological politics. But don't worry, I won't waste your time with dreamy visions that are politically naïve or socially unworkable. The durable first steps we'll discuss are not technologically based, but they stand on the same ground—that of human creativity and imagination. And you don't need to live in any particular location or be trained as an engineer or a scientist (or any other trade for that matter) to take part.

But enough about you.

Who is this author, with the peculiar name? (And if you don't much care, well then skip the next couple of paragraphs.) Your author fittingly grew up in Kalamazoo—home to numerous quirky Midwesterners, a couple of universities, a pharmaceutical company, and an industrial power plant where, it just so happens, he had a job one summer long ago.

At 4:30 a.m. daily, I would awake in time to skip breakfast, drive to the remote facility, and crawl into a full-body suit designed for someone twice my size, complete with a facemask and headlight. Holding a soot-scraping tool high above my head in the position of a diver, my coworkers plunged me head first into a narrow exhaust manifold that twisted down into the dark crypt where I would work out the day. I remember the weight of silence that lay upon my eardrums and how my scraping would chop it into rhythms, then tunes. I learned the length of eight hours down there. I haven't forgotten the lunchtime when we gossiped about our supervisor's affair, or the day my breathing filter didn't seal properly, or the pain of the rosy mucus I coughed up that night. I was tough then. But at the time, I didn't know I'd been breathing in asbestos courtesy of a company that has since gone bankrupt. Nor did I realize that my plant's radiation levels exceeded those inside a nearby nuclear power facility. These were the kind of answers that demanded more questions.

I suspect my summer spent cleaning out the bowels of that beast still informs the questions that attract me, though my work today is much different. Your author is currently an environmental researcher and a visiting scholar at the University of California–Berkeley. As an adviser to organizations, governments, and philanthropists, I deal with the frustration of these groups as they draw upon their resources or notoriety in attempts to create positive change. Sadly, some of them have come to me for assistance after supporting environmental initiatives that actually harmed those they had intended to help. With overwhelming requests pouring in, where can policymakers, professors, business leaders, concerned citizens, voters, and even environmentalists—best direct their energies?

In order to get some answers (and more importantly, find the right questions to ask), I geared up again. But this time I held a pen and notepad above my head as I dove into the underbelly of America's energy infrastructure to perform a long overdue colonoscopy. What I began to uncover haunted me—unsettling realizations that pulled me to investigate further. I pieced together funding for the first year. A year turned into two, then four—it's now been a decade since I began crisscrossing the energy world: arctic glaciers, oil fields in the frigid North Sea, turbine manufacturing facilities in Ireland, wind farms in Northern California, sun-scorched solar installations in Africa, biofuel farms in Iowa, unmarked government facilities in New Mexico, abandoned uranium mines in Colorado, modest dwellings in rural China, bullet trains in Japan, walkable villages in Holland, dusty archives in the Library of Congress, and even the Senate and House chambers in Washington DC.

I aimed to write an accessible yet rigorous briefing—part investigative journalism, part cultural critique, and part academic scholarship. I chose to publish with a nonprofit press and donate all author royalties to the underserved initiatives outlined ahead.

While I present a critique of environmentalism in America, I don't intend to criticize my many colleagues dedicated to working toward positive change. I aim only to scrutinize our creeds and biases. For that reason, you'll notice I occasionally refer to "the mainstream environmental movement," an admittedly vague euphemism for a heterogeneous group. Ultimately, we're all in this together, which means we're all going to be part of the solution. I'd like to offer a constructive critique of those efforts, not a roadblock. I don't take myself too seriously and I don't expect others to either. I ask only for your consideration of an alternate view. And even while I challenge claims to truth making, you'll see I emerge from the murky depths to voice my own claims to truth from time to time. This is the messy business of constructive argumentation, the limits of which are not lost on me.

Producing power is not simply a story of technological possibility, inventors, scientific discoveries, and profits; it is a story of meanings, metaphor, and human experience as well. The story we'll lay bare is far from settled. This book is but a snapshot. It is my hope that you and other readers will help complete the story. If after reading, scanning, or burning this book you'd care to continue the dialogue, I'd be honored to speak at your university, library, community group, or other organization (see GreenIllusions.org or OzzieZehner.com). I also invite you to enjoy a complimentary subscription to an ongoing series of environmental trend briefings at CriticalEnvironmentalism.org.

Part I
Seductive Futures

1. Solar Cells and Other Fairy Tales

To a man with a hammer, everything looks like a nail. —Mark Twain

Once upon a time, a pair of researchers led a group of study participants into a laboratory overlooking the ocean, gave them free unlimited coffee, and assigned them one simple task. The researchers spread out an assortment of magazine clippings and requested that participants assemble them into collages depicting what they thought of energy and its possible future.[1] No cost-benefit analyses, no calculations, no research, just glue sticks and scissors. They went to work. Their resulting collages were telling—not for what they contained, but for what they didn't.

They didn't dwell on energy-efficient lighting, walkable communities, or suburban sprawl. They didn't address population, consumption, or capitalism. They instead pasted together images of wind turbines, solar cells, biofuels, and electric cars. When they couldn't find clippings, they asked to sketch. Dams, tidal and wave-power systems, even animal power. They eagerly cobbled together fantastic totems to a gleaming future of power *production*. As a society, we have done the same.

The seductive tales of wind turbines, solar cells, and biofuels foster the impression that with a few technical upgrades, we might just sustain our current energy trajectories (or close to it) without consequence. Media and political coverage lull us into dreams of a clean energy future juxtaposed against a tumultuous past characterized by evil oil companies and the associated energy woes they propagated. Like most fairy tales, this productivist parable contains a tiny bit of truth. And a whole lot of fantasy.

Act I

I should warn you in advance; this book has a happy ending, but the joust in this first chapter might not. Even so, let's first take a moment to consider the promising allure of solar cells. Throughout the diverse disciplines of business, politics, science, academia, and environmentalism, solar cells stand tall as a valuable technology that everyone agrees is worthy of advancement. We find plenty of support for solar cells voiced by:

politicians,

> If we take on special interests, and make aggressive investments in clean and renewable energy, like Google's done with solar here in Mountain View, then we can end our addiction to oil, create millions of jobs and save the planet in the bargain. –Barack Obama

textbooks,

> Photovoltaic power generation is reliable, involves no moving parts, and the operation and maintenance costs are very low. The operation of a photovoltaic system is silent, and creates no atmospheric pollution. . . . Power can be generated where it is required without the need for transmission lines. . . . Other innovative solutions such as photovoltaic modules integrated in the fabric of buildings reduce the marginal cost of photovoltaic

energy to a minimum. The economic comparison with conventional energy sources is certain to receive a further boost as the environmental and social costs of power generation are included fully in the picture. –From the textbook, *Solar Electricity*

environmentalists,

Solar power is a proven and cost-effective alternative to fossil fuels and an important part of the solution to end global warming. The sun showers the earth with more usable energy in one minute than people use globally in one year. –Greenpeace

and even oil companies,

Solar solutions provide clean, renewable energy that save you money. –BP[2]

We ordinarily encounter the dissimilar views of these groups bound up in a tangle of conflict, but solar energy forms a smooth ground of commonality where environmentalists, corporations, politicians, and scientists can all agree. The notion of solar energy is flexible enough to allow diverse interest groups to take up solar energy for their own uses: corporations crown themselves with halos of solar cells to cast a green hue on their products, politicians evoke solar cells to garner votes, and scientists recognize solar cells as a promising well of research funding. It's in everyone's best interest to broadcast the advantages of solar energy. And they do. Here are the benefits they associate with solar photovoltaic technology:

- *CO_2 reduction*: Even if solar cells are expensive now, they're worth the cost to avoid the more severe dangers of climate change.
- *Simplicity*: Once installed, solar panels are silent, reliable, and virtually maintenance free.
- *Cost*: Solar costs are rapidly decreasing.
- *Economies of scale*: Mass production of solar cells will lead to cheaper panels.

- *Learning by doing*: Experience gained from installing solar systems will lead to further cost reductions.
- *Durability*: Solar cells last an extremely long time.
- *Local energy*: Solar cells reduce the need for expensive power lines, transformers, and related transmission infrastructure.[3]

Where the Internet Ends

All of these benefits seem reasonable, if not downright encouraging; it's difficult to see why anyone would want to argue with them. Over the past half century, journalists, authors, politicians, corporations, environmentalists, scientists, and others have eagerly ushered a fantasmatic array of solar devices into the spotlight, reported on their spectacular journeys into space, featured their dedicated entrepreneurs and inventors, celebrated their triumphs over dirty fossil fuels, and dared to envisage a glorious solar future for humanity.

The sheer magnitude of literature on the subject overwhelms—not just in newspapers, magazines, and books, but also in scientific literature, government documents, corporate materials, and environmental reports—far, far too much to sift through. The various tributes to solar cells could easily fill a library; the critiques would scarcely fill a book bag.

When I searched for critical literature on photovoltaics, Google returned numerous "no results found" errors—an error I'd never seen (or even realized existed) until I began querying for published drawbacks of solar energy. Bumping into the end of the Internet typically requires an especially arduous expedition into the darkest recesses of human knowledge, yet searching for drawbacks of solar cells can deliver you in a click. Few writers dare criticize solar cells, which understandably leads us to presume this sunny resource doesn't present serious limitations and leaves us clueless as to why nations can't seem to deploy solar cells on a grander scale. Though if we put on our detective caps and pull out our flashlights, we might just find some explanations lurking in the shadows—perhaps in the most unlikely of places.

Photovoltaics in Sixty Seconds or Less

Historians of technology track solar cells back to 1839 and credit Alexandre-Edmond Becquerel for discovering that certain light-induced chemical reactions produce electrical currents. This remained primarily an intellectual curiosity until 1940, when solid-state diodes emerged to form a foundation for modern silicon solar cells. The first solar cells premiered just eighteen years later, aboard the U.S. Navy's Vanguard 1 satellite.[4]

Today manufacturers construct solar cells using techniques and materials from the microelectronics industry. They spread layers of p-type silicon and n-type silicon onto substrates. When sunlight hits this silicon sandwich, electricity flows. Brighter sunlight yields more electrical output, so engineers sometimes incorporate mirrors into the design, which capture and direct more light toward the panels. Newer thin-film technologies employ less of the expensive silicon materials. Researchers are advancing organic, polymer, nanodot, and many other solar cell technologies.[5] Patent activity in these fields is rising.

Despite being around us for so long, solar technologies have largely managed to evade criticism. Nevertheless, there is now more revealing research to draw upon—not from Big Oil and climate change skeptics—but from the very government offices, environmentalists, and scientists promoting solar photovoltaics. I'll draw primarily from this body of research as we move on.

Powering the Planet with Photovoltaics

When I give presentations on alternative energy, among the most common questions philanthropists, students, and environmentalists ask is, "Why can't we get our act together and invest in solar cells on a scale that could really create an impact?" It is a reasonable question, and it deserves a reasonable explanation.

Countless articles and books contain a statistic reading something like this: Just a fraction of some-part-of-the-planet would provide all of the earth's power if we simply installed solar cells there. For instance, environmentalist Lester Brown, president of the Earth Policy Institute, indicates that it is "widely known within the energy community that there is enough solar energy reaching the earth each hour to power the world economy for one year."[6] Even Brown's nemesis, skeptical environmentalist Bjorn Lomborg claims that "we could produce the entire energy consumption of the world with present-day solar cell technology placed on just 2.6 percent of the Sahara Desert."[7] Journalists, CEOs, and environmental leaders widely disseminate variations of this statistic by repeating it almost ritualistically in a mantra honoring the monumental promise of solar photovoltaic technologies. The problem with this statistic is not that it is flatly false, but that it is somewhat true.

"Somewhat true" might not seem adequate for making public policy decisions, but it has been enough to propel this statistic, shiny teeth and all, into the limelight of government studies, textbooks, official reports, environmental statements, and into the psyches of millions of people. It has become an especially powerful rhetorical device despite its misleading flaw. While it's certainly accurate to state that the quantity of solar energy hitting that small part of the desert is equivalent to the amount of energy we consume, it does not follow that we *can harness it*, an extension many solar promoters explicitly or implicitly assume when they repeat the statistic. Similarly, any physicist can explain how a single twenty-five-cent quarter contains enough energy bound up in its atoms to power the entire earth, but since we have no way of accessing these forces, the quarter remains a humble coin rather than a solution to our energy needs. The same limitation holds when considering solar energy.

Skeptical? I was too. And we'll come to that. But first, let's establish how much it might actually cost to build a solar array capable of powering the planet with *today's* technology (saying nothing yet about the potential for future cost reductions). By comparing global energy consumption with the most rosy photovoltaic cost estimates, courtesy of solar proponents themselves, we can roughly sketch a total expense. The solar cells would cost about $59 trillion; the mining, processing, and manufacturing facilities to build them would cost about $44 trillion; and the batteries to store power for evening use would cost $20 trillion; bringing the total to about $123 trillion plus about $694 billion per year for maintenance.[8] Keep in mind that the entire gross domestic product (GDP) of the United States, which includes all food, rent, industrial investments, government expenditures, military purchasing, exports, and so on, is only about $14 trillion. This means that if every American were to go without food, shelter, protection, and everything else while working hard every day, naked, we might just be able to build a photovoltaic array to power the planet in about a decade. But, unfortunately, these estimations are optimistic.

If *actual* installed costs for solar projects in California are any guide, a global solar program would cost roughly $1.4 quadrillion, about one hundred times the United States GDP.[9] Mining, smelting, processing, shipping, and fabricating the panels and their associated hardware would yield about 149,100 megatons of CO_2.[10] And everyone would have to move to the desert, otherwise transmission losses would make the plan unworkable.

That said, few solar cell proponents believe that nations ought to rely exclusively on solar cells. They typically envision an alternative energy future with an assortment of energy sources— wind, biofuels, tidal and wave power, and others. Still, calculating the total bill for solar brings up some critical questions. Could manufacturing and installing photovoltaic arrays with today's technology on *any scale* be equally absurd? Does it just not seem as bad when we are throwing away a few billion dollars at a time? Perhaps. Or perhaps none of this will really matter since photovoltaic costs are dropping so quickly.

Price Check

Kathy Loftus, the executive in charge of energy initiatives at Whole Foods Market, can appreciate the high costs of solar cells today, but she is optimistic about the future: "We're hoping that our purchases along with some other retailers will help bring the technology costs down."[11] Solar proponents share her enthusiasm. The Earth Policy Institute claims solar electricity costs are "falling fast due to economies of scale as rising demand drives industry expansion."[12] The Worldwatch Institute agrees, claiming that "analysts and industry leaders alike expect continued price reductions in the near future through further economies of scale and increased optimization in assembly and installation."[13]

At first glance, this is great news; if solar cell costs are dropping so quickly then it may not be long before we can actually afford to clad the planet with them. There is little disagreement among economists that manufacturing ever larger quantities of solar cells results in noticeable economies of scale. Although it's not as apparent whether they believe these cost reductions are particularly significant in the larger scheme of things. They cite several reasons.

First, it is precarious to assume that the solar industry will realize substantial quantities of scale before solar cells become cost competitive with other forms of energy production. Solar photovoltaic investments have historically been tossed about indiscriminately like a small raft in the larger sea of the general economy. Expensive solar photovoltaic installations gain popularity during periods of high oil costs, but are often the first line items legislators cut when oil becomes cheaper again. For instance, during the oil shock of the 1970s, politicians held up solar cells as a solution, only to toss them aside once the oil price

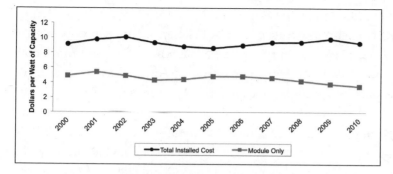

Figure 1: California solar system costs Installed photovoltaic system costs in California remain high due to a variety of expenses that are not technically determined. (Data from California Energy Commission and Solarbuzz)

tide subsided. More recent economic turmoil forced Duke Energy to slash $50 million from its solar budget, BP cut its photovoltaic production capacity, and Solyndra filed for Chapter 11 bankruptcy.[14] Economists argue that it's difficult to achieve significant economies of scale in an industry with such violent swings between investment and divestment.

Second, solar advocates underscore dramatic photovoltaic cost reductions since the 1960s, leaving an impression that the chart of solar cell prices is shaped like a sharply downward-tilted arrow. But according to the solar industry, prices from the most recent decade have flattened out. Between 2004 and 2009, the installed cost of solar photovoltaic modules actually increased—only when the financial crisis swung into full motion over subsequent years did prices soften. So is this just a bump in the downward-pointing arrow? Probably. However, even if solar cells become markedly cheaper, the drop may not generate much impact since photovoltaic panels themselves account for less than half the cost of an installed solar system, according to the industry.[15] Based on research by solar energy proponents and field data from the California Energy Commission (one of the largest clearinghouses of experience-based solar cell data), cheaper photovoltaics won't offset escalating expenditures for insurance, warranty expenses, materials, transportation, labor, and other requirements.[16] Low-tech costs are claiming a larger share of the high-tech solar system price tag.

Finally, unforeseen limitations are blindsiding the solar industry as it grows.[17] Fire departments restrict solar roof installations and homeowner associations complain about the ugly arrays. Repair and maintenance costs remain stubbornly high. Adding to the burden, solar arrays now often require elaborate alarm systems and locking fasteners; without such protection, thieves regularly steal the valuable panels. Police departments throughout the country are increasingly reporting photovoltaic pilfering, which is incidentally inflating home insurance premiums. For instance, California resident Glenda Hoffman woke up one morning to discover thieves stole sixteen solar panels from her roof as she slept. The cost to replace the system chimed in at $95,000, an expense her insurance company covered. Nevertheless, she intends to protect the new panels herself, warning, "I have a shotgun right next to the bed and a .22 under my pillow."[18]

Disconnected: Transmission and Timing

Solar cells offer transmission benefits in niche applications when they supplant disposable batteries or other expensive energy supply options. For example, road crews frequently use solar cells in tandem with rechargeable battery packs to power warning lights and monitoring equipment along highways. In remote and poor equatorial regions of the world, tiny amounts of expensive solar energy can generate a sizable impact on families and their communities. Here, solar cells provide a viable alternative to candles, disposable batteries, and kerosene lanterns, which are expensive, dirty, unreliable, and dangerous.

Given the appropriate socioeconomic context, solar energy can help villages raise their standards of living. Radios enable farmers to monitor the weather and connect families with news and cultural events. Youth who grow up with evening lighting, and thus a better chance for education, are more likely to wait before getting married and have fewer, healthier children if they become parents.[19] This allows the next generation of the village to grow up in more economically stable households with extra attention and resources allotted to them.

Could rich nations realize similar transmission-related benefits? Coal power plants require an expensive network of power lines and transformers to deliver their power. Locally produced solar energy may still require a transformer but it bypasses the long-distance transmission step. Evading transmission lines during high midday demand is presumably beneficial since this is precisely when fully loaded transmission lines heat up, which increases their resistance and thus wastes energy to heat production. Solar cells also generate their peak output right when users need it most, at midday on hot sunny days as air conditioners run full tilt. Electricity is worth more at these times because it is in short supply. During these periods, otherwise dormant power facilities, called peaker plants, fire up to fulfill spikes in electrical demand. Peaker plants are more expensive and less efficient than base-load plants, so avoiding their use is especially valuable. Yet analysts often evaluate and compare solar power costs against *average* utility rates.[20] This undervalues solar's midday advantage. Taken into account, timing benefits increase the value of solar cell output by up to 20 percent.

Transmission and timing advantages of solar electricity led the director of the University of California Energy Institute, Severin Borenstein, to find out how large these benefits are in practice. His conclusions are disheartening.

Borenstein's research suggests that "actual installation of solar PV [photovoltaic] systems in California has not significantly reduced the cost of transmission and distribution infrastructure, and is unlikely to do so in other regions." Why? First, most transmission infrastructure has already been built, and localized solar-generation effects are not enough to reduce that infrastructure. Even if they were, the savings would be small since solar cells alone would not shrink the breadth of the distribution network. Furthermore, California and the other thirty states with solar subsidies have not targeted investments toward easing tensions in transmission-constrained areas. Dr. Borenstein took into account the advantageous timing of solar cell output but he ultimately concludes: "The market benefits of installing the current solar PV technology, even after adjusting for its timing and transmission advantages, are calculated to be much smaller than the costs. The difference is so large that including current plausible estimates of the value of reducing greenhouse gases still does not come close to making the net social return on installing solar PV today positive."[21] In a world with limited funds, these findings don't position solar cells well. Still, solar advocates insist the expensive panels are a necessary investment if we intend to place a stake in the future of energy.

Learning by Doing: Staking Claims on the Future

In the 1980s Ford Motor Company executives noticed something peculiar in their sales figures. Customers were requesting cars with transmissions built in their Japanese plant instead of the American one. This puzzled engineers since both the U.S. and Japanese transmission plants built to the same blueprints and same tolerances; the transmissions should have been identical. They weren't. When Ford engineers disassembled and analyzed the transmissions, they discovered that even though the American parts met allowable tolerances, the Japanese parts fell within an even tighter tolerance, resulting in transmissions that ran more smoothly and yielded fewer defects—an effect researchers attribute to the prevalent Japanese philosophy of Kaizen. Kaizen is a model of continuous improvement achieved through hands-on experience with a technology. After World War II, Kaizen grew in popularity, structured largely by U.S. military innovation strategies developed by W. Edwards Deming. The day Ford engineers shipped their blueprints to Japan marked the beginning of this design process, not the end. Historians of technological development point to such learning-by-doing effects when explaining numerous technological success stories. We might expect such effects to benefit the solar photovoltaic industry as well.

Indeed, there are many cases where this kind of learning by doing aids the solar industry. For instance, the California Solar Initiative solved numerous unforeseen challenges during a multiyear installation of solar systems throughout the state—unexpected and burdensome administration requirements, lengthened application processing periods, extended payment times, interconnection delays, extra warranty expenses, and challenges in metering and monitoring the systems. Taken together, these challenges spurred learning that would not have been possible without the hands-on experience of running a large-scale solar initiative.[22] Solar proponents claim this kind of learning is bringing down the cost of solar cells.[23] But what portion of photovoltaic price drops over the last half century resulted from learning-by-doing effects and what portion evolved from other factors?

When Gregory Nemet from the Energy and Resources Group at the University of California disentangled these factors, he found learning-by-doing innovations contributed only slightly to solar cell cost reductions over the last thirty years. His results indicate that learning from experience "only weakly explains change in the most important factors—plant size, module efficiency, and the cost of silicon."[24] In other words, while learning-by-doing effects do influence the photovoltaic manufacturing industry, they don't appear to justify massive investments in a fabrication and distribution system just for the sake of experience.

Nevertheless, there is a link that Dr. Nemet didn't study: silicon's association with rapid advancements in the microelectronics industry. Microchips and solar cells are both crafted from silicon, so perhaps they are both subject to Moore's law, the expectation that the number of transistors on a microchip will double every twenty-four months. The chief executive of Nanosolar points out, "The solar industry today is like the late 1970s when mainframe computers dominated, and then Steve Jobs and IBM came out with personal computers." The author of a *New York Times* article corroborates the high-tech comparison: "A link between Moore's law and solar technology reflects the engineering reality that computer chips and solar cells have a lot in common." You'll find plenty of other solar proponents industriously evoking the link.[25]

You'll have a difficult time finding a single physicist to agree.

Squeezing more transistors onto a microchip brings better performance and subsequently lower costs, but miniaturizing and packing solar cells tightly together simply reduces their surface area exposed to the sun's energy. Smaller is worse, not better. But size comparisons are a literal interpretation of Moore's law. Do solar technologies follow Moore's law in terms of cost or performance?

No and no.

Proponents don't offer data, statistics, figures, or any other explanation beyond the comparison itself—a hit and run. Microchips, solar cells, and Long Beach all contain silicon, but their similarities end there. Certainly solar technologies will improve—there is little argument on that—but expecting them to advance at a pace even approaching that of the computer industry, as we shall see, becomes far more problematic.

Solar Energy and Greenhouse Gases

Perhaps no single benefit of solar cells is more cherished than their ability to reduce CO_2 emissions. And perhaps no other pur-

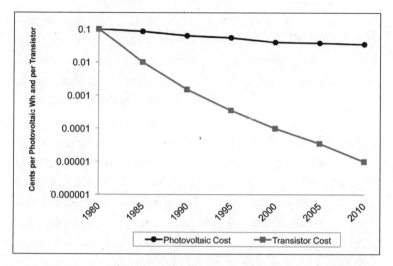

Figure 2: Solar module costs do not follow Moore's law Despite the common reference to Moore's law by solar proponents, three decades of data show that photovoltaic module
cost reductions do not mirror cost reductions in the microelectronics industry. Note the logarithmic scale. (Data from
Solarbuzz and Intel)

ported benefit stands on softer ground. To start, a group of Columbia University scholars calculated a solar cell's lifecycle carbon
footprint at twenty-two to forty-nine grams of CO_2 per kilowatt-
hour (kWh) of solar energy produced.[26] This carbon impact is
much lower than that of fossil fuels.[27] Does this offer justification for subsidizing solar panels?

We can begin by considering the market price of greenhouse
gases like CO_2. In Europe companies must buy vouchers to emit
CO_2, which trade at around twenty to forty dollars per ton. Most
analysts expect American permits to stabilize on the open market
somewhere below thirty dollars per ton.[28] Today's solar technologies would compete with coal only if carbon credits rose
to three hundred dollars per ton. Photovoltaics could nomi-
nally compete with natural gas only if carbon offsets skyrocketed to six hundred dollars per ton.[29] It is difficult to conceive of

conditions that would thrust CO_2 prices to such stratospheric levels in real terms. Even some of the most expensive options for dealing with CO_2 would become cost competitive long before today's solar cell technologies. If limiting CO_2 is our goal, we might be better off directing our time and resources to those options first; solar cells seem a wasteful and pricey strategy.

Unfortunately, there's more. Not only are solar cells an overpriced tool for reducing CO_2 emissions, but their manufacturing process is also one of the largest emitters of hexafluoroethane (C_2F_6), nitrogen trifluoride (NF_3), and sulfur hexafluoride (SF_6). Used for cleaning plasma production equipment, these three gruesome greenhouse gases make CO_2 seem harmless. As a greenhouse gas, C_2F_6 is twelve thousand times more potent than CO_2, is 100 percent manufactured by humans, and survives ten thousand years once released into the atmosphere.[30] NF_3 is seventeen thousand times more virulent than CO_2, and SF_6, the most treacherous greenhouse gas, according to the Intergovernmental Panel on Climate Change, is twenty-five thousand times more threatening.[31] The solar photovoltaic industry is one of the leading and fastest growing emitters of these gases, which are now measurably accumulating within the earth's atmosphere. A recent study on NF_3 reports that atmospheric concentrations of the gas have been rising an alarming 11 percent per year.[32]

Check the Ingredients: Toxins and Waste

In the central plains of China's Henan Province, local residents grew suspicious of trucks that routinely pulled in behind the playground of their primary school and dumped a bubbling white liquid onto the ground. Their concerns were justified. According to a *Washington Post* investigative article, the mysterious waste was silicon tetrachloride, a highly toxic chemical that burns human skin on contact, destroys all plant life it comes near, and violently reacts with water.[33] The toxic waste was too expensive to recycle, so it was simply dumped behind the playground—daily—for over nine months by Luoyang Zhonggui High-Technology Company, a manufacturer of polysilicon for solar cells. Such cases are far from rare. A report by the Silicon Valley Toxics Coalition claims that as the solar photovoltaic industry expands,

> little attention is being paid to the potential environmental and health costs of that rapid expansion. The most widely used solar PV panels have the potential to create a huge new wave of electronic waste (e-waste) at the end of their useful lives, which is estimated to be 20 to 25 years. New solar PV technologies are increasing cell efficiency and lowering costs, but many of these use extremely toxic materials or materials with unknown health and environmental risks (including new nanomaterials and processes).[34]

For example, sawing silicon wafers releases a dangerous dust as well as large amounts of sodium hydroxide and potassium hydroxide. Crystalline-silicon solar cell processing involves the use or release of chemicals such as phosphine, arsenic, arsine, trichloroethane, phosphorous oxychloride, ethyl vinyl acetate, silicon trioxide, stannic chloride, tantalum pentoxide, lead, hexavalent chromium, and numerous other chemical compounds. Perhaps the most dangerous chemical employed is silane, a highly explosive gas involved in hazardous incidents on a routine basis according to the industry.[35] Even newer thin-film technologies employ numerous toxic substances, including cadmium, which is categorized as an extreme toxin by the U.S. Environmental Protection Agency and a Group 1 carcinogen by the International Agency for Research on Cancer. At the end of a solar panel's usable life, its embedded chemicals and compounds can either seep into groundwater supplies if tossed in a landfill or contaminate air and waterways if incinerated.[36]

Are the photovoltaic industry's secretions of heavy metals, hazardous chemical leaks, mining operation risks, and toxic wastes especially problematic today? If you ask residents of Henan

Province, the answer will likely be yes. Nevertheless, when pitted against the more dangerous particulate matter and pollution from the fossil-fuel industry, the negative consequences of solar photovoltaic production don't seem significant at all. Compared to the fossil-fuel giants, the photovoltaic industry is tiny, supplying less than a hundredth of 1 percent of America's electricity.[37] (If the text on this page represented total U.S. power supply, the photovoltaic portion would fit inside the period at the end of this sentence.) If photovoltaic production grows, so will the associated side effects.

Further, as we'll explore in future chapters, even if the United States expands solar energy capacity, this may *increase* coal use rather than replace it. There are far more effective ways to invest our resources, ways that *will* displace coal consumption—strategies that will lessen, not multiply, the various ecological consequences of energy production. Yet we have much to discuss before coming to those—most immediately, a series of surprises.

Photovoltaic Durability: A Surprise Inside Every Panel

The United Arab Emirates recently commissioned the largest cross-comparison test of photovoltaic modules to date in preparation for building an ecometropolis called Masdar City. The project's technicians installed forty-one solar panel systems from thirty-three different manufacturers in the desert near Abu Dhabi's international airport.[38] They designed the test to differentiate between cells from various manufacturers, but once the project was initiated, it quickly drew attention to something else—the drawbacks that all of the cells shared, regardless of their manufacturer.

Solar cell firms generally test their panels in the most ideal of conditions—a Club Med of controlled environments. The real-world desert outside Masdar City proved less accommodating. Atmospheric humidity and haze reflected and dispersed the sun's rays. Even more problematic was the dust, which technicians had to scrub off almost daily. Soiling is not always so easy to remove. Unlike Masdar's panels, hovering just a few feet above the desert sands, many solar installations perch high atop steep roofs. Owners must tango with gravity to clean their panels or hire a stand-in to dance for them. Researchers discovered that soiling routinely cut electrical output of a San Diego site by 20 percent during the dusty summer months. In fact, according to researchers from the photovoltaic industry, soiling effects are "magnified where rainfall is absent in the peak-solar summer months, such as in California and the Southwest region of the United States," or in other words, right where the prime real estate for solar energy lies.[39]

When it comes to cleanliness, solar cells are prone to the same vulnerability as clean, white dress shirts; small blotches reduce their value dramatically. Due to wiring characteristics, solar output can drop disproportionately if even tiny fragments of the array are blocked, making it essential to keep the entire surface clear of the smallest obstructions, according to manufacturers. Bird droppings, shade, leaves, traffic dust, pollution, hail, ice, and snow all induce headaches for solar cell owners as they attempt to keep the entirety of their arrays in constant contact with the sunlight that powers them. Under unfavorable circumstances, these soiling losses can climb to 80 percent in the field.[40]

When journalists toured Masdar's test site, they visited the control room that provided instant energy readouts from each company's solar array. On that late afternoon, the journalists noted that the most productive unit was pumping out four hundred watts and the least productive under two hundred. All of the units were rated at one thousand watts maximum. This peak output, however, can only theoretically occur briefly at midday, when the sun is at its brightest, and only if the panels are located within an ideal latitude strip and tilted in perfect alignment with the sun (and all other conditions are near perfect as well). The desert outside Masdar City seems like one of the few ideal locations on the planet for such perfection. Unfortunately, during the midday hours of the summer, all of the test cells became extremely hot, up to 176 degrees Fahrenheit (80°C), as they baked in the desert sun. Due to the temperature sensitivity of the photovoltaic cells, their output was markedly hobbled across the board, right at the time they should have been producing their highest output.[41] So who won the solar competition in Masdar City? Perhaps nobody.

In addition to haze, humidity, soiling, misalignment, and temperature sensitivity, silicon solar cells suffer an aging effect that decreases their output by about 1 percent or more per year.[42] Newer thin-film, polymer, paint, and organic solar technologies degrade even more rapidly, with some studies recording degradation of up to 50 percent within a short period of time. This limitation is regularly concealed because of the way reporters, corporations, and scientists present these technologies.[43]

For instance, scientists may develop a thin-film panel achieving, say, 13 percent overall efficiency in a laboratory. However, due to production limitations, the company that *commercializes* the panel will typically only achieve a 10 percent overall efficiency in a prototype. Under the best conditions in the field this may drop to 7–8.5 percent overall efficiency due just to degradation effects.[44] Still, the direct current (DC) output is not usable in a household until it is transformed. Electrical inverters transform the DC output of solar cells into the higher voltage and oscillating AC that appliances and lights require. Inverters are 70–95 percent efficient, depending on the model and loading characteristics. As we have seen, other situational factors drag performance down even further. Still, when laboratory scientists and corporate PR teams write press releases, they report the more favorable figure, in this case 13 percent. Journalists at even the most esteemed publications will often simply transpose this figure into their articles. Engineers, policy analysts, economists, and others in turn transpose the figure into their assessments.

Illustration 1: Solar system challenges The J. F. Williams Federal Building in Boston was one of the earliest Million Solar Roofs sites and the largest building-integrated array on the East Coast. As with most integrated systems, the solar cells do not align with the sun, greatly reducing their performance. In 2001 technicians replaced the entire array after a system malfunction involving electrical arcing, water infiltration, and broken glass. The new array has experienced system-wide aging degradation as well as localized corrosion, delamination, water infiltration, and sudden module failures. (Photo by Roman Piaskoski, courtesy of the U.S. Department of Energy)

With such high expectations welling up around solar photo-voltaics, it is no wonder that newbie solar cell owners are often shocked by the underwhelming performance of their solar arrays in the real world. For example, roof jobs may require that they disconnect, remove, and reinstall their rooftop arrays. Yet an even larger surprise awaits them—within about five to ten years, their solar system will abruptly stop producing power. Why? Because a key component of the solar system, the electrical inverter, will eventually fail. While the solar cells themselves can survive for twenty to thirty years, the associated circuitry does not. Inverters for a typical ten-kilowatt solar system last about five to eight years and therefore owners must replace them two to five times during the productive life of a solar photovoltaic system. Fortunately, just about any licensed electrician can easily swap one out. Unfortunately, they cost about eight thousand dollars each.[45]

Free Panels, Anyone?

Among the CEOs and chief scientists in the solar industry, there is surprisingly little argument that solar systems are expensive.[46] Even an extreme drop in the price of polysilicon, the most expensive technical component, would do little to make solar cells more competitive. Peter Nieh, managing director of Lightspeed Venture Partners, a multibillion-dollar venture capital firm in Silicon Valley, contends that cheaper polysilicon won't reduce the overall cost of solar arrays much, even if the price of the expensive material dropped to zero.[47] Why? Because the cost of other materials such as copper, glass, plastics, and aluminum, as well as the costs for fabrication and installation, represent the bulk of a solar system's overall price tag. The technical polysilicon represents only about a fifth of the total.

Furthermore, Keith Barnham, an avid solar proponent and senior researcher at Imperial College London, admits that unless efficiency levels are high, "even a zero cell cost is not competitive."[48] In other words, even if someone were to offer you solar cells *for free*, you might be better off turning the offer down than paying to install, connect, clean, insure, maintain, and eventually dispose of the modules—especially if you live outside the remote, dry, sunny patches of the planet such as the desert extending from southeast California to western Arizona. In fact, the unanticipated costs, performance variables, and maintenance obligations for photovoltaics, too often ignored by giddy proponents of the technology, can swell to unsustainable magnitudes. Occasionally buyers decommission their arrays within the first decade, leaving behind graveyards of toxic panels teetering above their roofs as epitaphs to a fallen dream. Premature decommissioning may help explain why American photovoltaic electrical generation dropped during the last economic crisis even as purported solar capacity expanded.[49] Curiously, while numerous journalists reported on solar infrastructure expansion during this period, I was unable to locate a single article covering the contemporaneous drop in the nation's solar electrical output, which the Department of Energy quietly slid into its annual statistics without a peep.

The Five Harms of Solar Photovoltaics

Are solar cells truly such a waste of money and resources? Is it really possible that today's solar technologies could be so ineffectual? Could they even be harmful to society and the environment?

It would be egotistically convenient if we could dismiss the costs, side effects, and limitations of solar photovoltaics as the blog-tastic hyperbole of a few shifty hacks. But we can't. These are the limitations of solar cells as directly reported from the very CEOs, investors, and researchers most closely involved with their real-world application. The side effects and limitations collected here, while cataclysmically shocking to students, activists, business people, and many other individuals I meet, scarcely raise an eyebrow among those in the solar industry who are most intimately familiar with these technologies.

Few technicians would claim that their cells perform as well in the field as they do under the strict controls of a testing laboratory; few electricians would deny that inverters need to be replaced regularly; few energy statisticians would argue that solar arrays have much of an impact on global fossil-fuel consumption; few economists would insist we could afford to power the planet with existing solar technologies. In fact, the shortcomings reported in these pages have become remarkably stable facts within the communities of scientists, engineers, and other experts who work on solar energy. But because of specialization and occupational silo effects, few of these professionals capture the entire picture at once, leaving us too often with disjointed accounts rather than an all-inclusive rendering of the solar landscape.

Collected and assembled into one narrative, the costs, side effects, and limitations of solar photovoltaics become particularly worrisome, especially within the context of our current national finances and limited resources for environmental investments. The point is not to label competing claims about solar cells as simply true or false (we have seen they are a bit of both), but to determine if these claims have manifested themselves in ways and to degrees that validate solar photovoltaics as an appropriate means to achieve our environmental goals.

I'd like to consider some alternate readings of solar cells that are a bit provocative, perhaps even strident. What if we can't simply roll our eyes at solar cells, shrugging them off as the harmless fascination of a few silver-haired hippies retired to the desert? What if we interpret the powerful symbolism of solar cells as metastasizing in the minds of thoughtful people into a form that is disruptive and detrimental?

First, we could read these technologies as lucrative forms of misdirection—shiny sleights of hand that allow oil companies, for example, to convince motorists that the sparkling arrays atop filling stations somehow make the liquid they pump into their cars less toxic. The fact that some people now see oil companies that also produce solar cells as "cleaner" and "greener" is a testament to a magic trick that has been well performed. Politicians have proven equally adept in such trickery, holding up solar cells to boost their poll numbers in one hand while using the other to palm legislation protecting the interests of status quo industries.

Second, could the glare from solar arrays blind us to better alternatives? If solar cells are seen as the answer, why bother with less sexy options? Many homeowners are keen on upgrading to solar, but because the panels require large swaths of unobstructed exposure to sunlight, solar cells often end up atop large homes sitting on widely spaced lots cleared of surrounding trees, which could have offered considerable passive solar benefits. In this respect, solar cells act to "greenwash" a mode of suburban residential construction, with its car-dependent character, that is hardly worthy of our explicit praise. Sadly, as we shall see ahead, streams of money, resources, and time are diverted away from less visible but more durable solutions in order to irrigate the infertile fields of solar photovoltaics.

Third, might the promise of solar cells act to prop up a productivist mentality, one that insists that we can simply generate more and more power to satisfy our escalating cravings for energy? If clean energy is in the pipeline, there is less motivation to use energy more effectively and responsibly.[50]

Fourth, we could view solar photovoltaic subsidies as perverse misallocations of taxpayer dollars. For instance, the swanky Honig Winery in Napa Valley chopped down some of its vines to install $1.2 million worth of solar cells. The region's utility customers paid $400,000 of the tab. The rest of us paid another 30 percent through federal rebates. The 2005 federal energy bill delivered tax write-offs of another 30 percent. Luckily, we can at least visit the winery to taste the resulting vintages, but even though you're helping pay their electric bill, they'll still charge you ten dollars for the privilege. Honig is just one of several dozen wineries to take advantage of these government handouts of your money, and wineries represent just a handful of the thousands of industries and mostly wealthy households that have done the same.

Fifth, photovoltaic processes—from mineral exploration to fabrication, delivery, maintenance, and disposal—generate their own environmental side effects. Throughout the photovoltaic lifecycle, as we have reviewed, scientists are discovering the same types of short- and long-term harms that environmentalists have historically rallied against.

Finally, it is worth acknowledging that there are a few places where solar cells can generate an impact today, but for the most part, it's not here. Plopping solar cells atop houses in the well-trimmed suburbs of America and under the cloudy skies of Germany seems an embarrassing waste of human energy, money, and time. If we're searching for meaningful solar cell applications today, we'd better look to empowering remote communities whose only other options are sooty candles or expensive disposable batteries—not toward haplessly supplementing power for the wine chillers, air conditioners, and clothes dryers of industrialized nations. We environmentalists have to consider whether it's reasonable to spend escalating sums of cash to install primitive solar technologies when we could instead fund *preconditions* that might someday make them viable solutions for a greater proportion of the populace.

Charting a New Solar Strategy

Current solar photovoltaic technologies are ineffective at preventing greenhouse gas emissions and pollution, which is especially disconcerting considering how rapidly their high costs suck money away from more potent alternatives. To put this extreme waste of money into perspective, it would be terrifically more cost effective to patch the leaky windows of a house with *gold leaf* rather than install solar cells on its roof. Would we support a government program to insulate people's homes using gold? Of course not. Anyone could identify the absurd profligacy of such a scheme (in part, because we have not been repeatedly instructed since childhood that it is a virtuous undertaking).

Any number of conventional energy strategies promise higher dividends than solar cell investments. If utilities care to reduce CO_2, then for just 10 percent of the Million Solar Roofs Program cost, they could avoid twice the greenhouse gas emissions by simply converting one large coal-burning power plant over to natural gas. If toxicity is a concern, legislators could direct the money toward low-tech solar strategies such as solar water heating, which has a proven track record of success. Or for no net cost at all, we could support strategies to bring our homes and commercial buildings into sync with the sun's energy rather than working against it. A house with windows, rooflines, and walls designed to soak up or deflect the sun's energy in a passive way, will continue to do so unassumingly for generations, even centuries. Fragile solar photovoltaic arrays, on the other hand, are sensitive to high temperatures, oblige owners to perform constant maintenance, and require extremely expensive components to keep them going for their comparably short lifespan.

Given the more potent and viable alternatives, it's difficult to see why federal, state, and local photovoltaic subsidies, including the $3.3 billion solar roof initiative in California, should not be quickly scaled down and eliminated as soon as practicable. It is hard to conceive of a justification for extracting taxes from the working class to fund installations of Stone Age photovoltaic technologies high in the gold-rimmed suburbs of Arizona and California.

The solar establishment will most certainly balk at these observations, quibble about the particulars, and reiterate the benefits of learning by doing and economies of scale. These, however, are tired arguments. Based on experiences in California,

Japan, and Europe, we now have solid field data indicating that (1) the benefits of solar cells are insignificant compared to the expense of realizing them, (2) the risks and limitations are substantial, and (3) the solar forecast isn't as sunny as we've been led to believe.

Considering the extreme risks and limitations of today's solar technologies, the notion that they could create any sort of challenge to the fossil-fuel establishment starts to appear not merely optimistic, but delusional. It's like believing that new parasail designs could form a challenge to the commercial airline industry. Perhaps the only way we could believe such an outlandish thought is if we are told it over, and over, and over again. In part, this is what has happened. Since we were children, we've been promised by educators, parents, environmental groups, journalists, and television reporters that solar photovoltaics will have a meaningful impact on our energy system. The only difference today is that these fairy tales come funded through high-priced political campaigns and the advertising budgets of BP, Shell, Walmart, Whole Foods, and numerous other corporations.

Solar cells shine brightly within the idealism of textbooks and the glossy pages of environmental magazines, but real-world experiences reveal a scattered collection of side effects and limitations that rarely mature into attractive realities. There are many routes to a more durable, just, and prosperous energy system, but the glitzy path carved out by today's archaic solar cells doesn't appear to be one of them.

2. Wind Power's Flurry of Limitations

Evidence conforms to conceptions just as often as conceptions conform to evidence. –Ludwik Fleck, *Genesis and Development of a Scientific Fact*

By the end of grade school, my mother maintains, I had attempted to deconstruct everything in the house at least once (including a squirrel that fell to its death on the front walk). Somewhere in the fog of my childhood, I shifted from deconstruction to construction, and one of my earliest machinations was a windmill, inspired by a dusty three-foot-diameter turbine blade laying idle in the garage thanks to my father's job at a fan-and-turbine manufacturer. Fortunately, the turbine's hub screws fit snugly around a found steel pipe, which formed a relatively solid, if rusty, axle for the contraption. I mounted the axle in wood rather than steel, since my parents had neglected to teach me to weld. There were no bearings, but I dusted the naked holes with powdered graphite for lubrication; I was serious. Lacking the resources to design a tower, a wood picnic table in the backyard proved sufficient.

Some subsequent day, as cool winds ripped leaves from surrounding oak trees and threw them at passersby, I hauled the rickety contraption from the garage to the picnic table, exposed nails and all. I first pulled the wooden mount up onto the table, weighing it down with bricks and other heavy objects. I then inserted the axle-and-turbine assembly. The already rotating blades hovered out over the table's edge, but there was little time to appreciate my work. Before the lock pin was properly secured, the heavy blade had already begun to spin uncomfortably fast. Only at that moment did it become apparent that I had neglected to install a braking mechanism, but it was too late.

I removed a brick from the base and pressed it against the rotating axle to slow it down, pushing with all my might. The axle hissed as the blades effortlessly accumulated greater speed. I jumped back when the axle's partially engaged lockpin flew out. The picnic table vibrated as the dull black blades melted into a grayish blur. The steel sails thumped through the air with a quickening rhythm of what in essence had become an upended lawnmower shrieking the song of a helicopter carrying a hundred cats in heat. What happened thereafter can only be deduced, because by the time the howling and clamor came to an abrupt end, my adrenaline-filled legs had already carried me well beyond the far side of the house.

I returned to find an empty picnic table in flames.

Now, if you can imagine a force ten thousand times as strong, you'll begin to appreciate the power of modern wind turbines, weighing in at 750 tons and with blade sweeps wider than eleven full-size school buses parked end-to-end.[1]

Like solar cells, wind turbines run on a freely available resource that is exhibiting no signs of depletion. Unlike solar cells, though, wind turbines are economical—just a sixth the cost of photovoltaics, according to an HSBC bank study. Proponents insist that wind power's costs have reached parity with natural-gas electrical generation. Coal-fired electricity is still less expensive, but if a carbon tax of about thirty dollars per ton is figured into

Illustration 2: An imposing scale Raising a blade assembly at night outside Brunsbüttel, Germany, with a second tower in the background. The turbine sits on 1,700 cubic yards of concrete with forty anchors each driven eighty feet into the earth. (Photo by Jan Oelker, courtesy of Repower Systems AG)

the equation, proponents insist that wind achieves parity with coal as well.[2] Either way, wind turbines seem far more pleasant as they sit in fields and simply whirl away.

Today's wind turbines are specially designed for their task and as a result are far more technologically advanced than even those built a decade ago. New composites enable the spinning arms to reach farther and grab more wind while remaining flexible enough to survive forceful gusts. New turbines are also more reliable. In 2002, about 15 percent of turbines were out of commission at any given time for maintenance or repair; now downtime has dropped below 3 percent. Whereas a coal or nuclear plant mishap could slash output dramatically or even completely, wind farms can still pump out electricity even as individual turbines cycle through maintenance. Similarly, new wind farms start to produce power long before they are complete. A half-finished nuclear plant might be an economic boondoggle, but a half-finished wind farm is merely one that produces half the power. Adding capacity later is as simple as adding more turbines. Farmers who are willing to give up a quarter of an acre to mount a large turbine in their fields can expect to make about ten thousand dollars per year in profit without interrupting cultivation of the surrounding land. That's not bad considering the same plot seeded with corn would net just three hundred dollars' worth of bioethanol.[3]

At first glance, deploying wind turbines on a global scale does not apparently pose much of a challenge, at least not an insurmountable one. It seems that no matter what yardstick we use, wind power is simply the perfect solution.

If only it were that simple.

Wind Power in Sixty Seconds or Less

As our sun heats the earth's lower atmosphere, pockets of hot air rise and cooler air rushes in to fill the void. This creates wind. For over two thousand years humans harnessed wind for pumping water, grinding grain, and even transatlantic travel. In fact, wind power was once a primary component of the global energy supply. No more. The Industrial Revolution (which could just as easily have been dubbed the Coal Revolution) toppled wind power's reign. Shipbuilders replaced masts with coal-fired steam engines. Farmers abandoned windmills for pumps that ran on convenient fossil fuels. Eventually, industrialists led the frail wind-power movement to its grave, and gave it a shove. There it would lie, dead and forgotten, for well over a hundred years, until one crisp fall day when something most unexpected occurred.

A hundred years is a short beat in the history of humans but a rather lengthy period in their history of industrialization. And when wind power was eventually exhumed, it found itself in a much-altered world, one that was almost entirely powered by fossil fuels. There were many more humans living at much higher standards of living. A group of them was rather panicked over the actions of an association called the Organization of Arab Petroleum Exporting Countries (OPEC). The scoundrels had decided to turn off their fossil-fuel spigot.

The oil embargo of 1973 marked the resurrection of wind power. Politicians dusted off wind power, dressed it in a green-collared shirt, and shoved it into the limelight as the propitious savior of energy independence. Wind power was worshiped everywhere, but nowhere more than in California. During the great wind rush of the early 1980s, California housed nearly 90 percent of global wind-generation capacity, fueled by tax subsidies and a wealthy dose of sunny optimism.[4] And since the windmill industry had vanished long ago, fabricators cobbled together the new turbines much like the one of my youth, with an existing hodgepodge of parts already available from shipbuilders and other industries. Perhaps predictably, when the oil started to flow again, political support for wind energy subsidies waned. Eventually they vanished altogether. But now, with so many humans using so much energy, it wouldn't be another hundred years before they would call on wind power again.

During the first decade of the twenty-first century, oil prices skyrocketed. But another phenomenon shot up faster: media and political reporting on wind energy.[5] For every doubling of oil prices, media coverage of wind power *tripled*. Capacity grew too—as much as 30 percent annually. But at the end of the decade, an economic crisis smacked wind down again. Wind projects across the planet were cancelled, signaled most prominently by the flapping coat tails of energy tycoon T. Boone Pickens, as he fled from his promise to build massive wind farms in Texas. Financial turmoil further embrittled the fragile balance sheets of turbine manufacturers until orders began to stabilize again around 2011.

By 2012, worldwide wind-power generation capacity had surpassed two hundred gigawatts—many times the capacity of solar photovoltaics but not enough to fulfill even a single percent of global energy demand. We have thrice witnessed the fortunes of wind shifting in the industry's sail and we may find the future of wind power to be similarly constrained, as its detractors are raring to point out.

The Detractors

A boot tumbling around in a clothes dryer—that's how residents of Cape Cod describe the wind turbine whining and thumping that keeps them awake at night and gives them headaches during the day. One wind turbine engineering manual confirms that this noise, produced when blades swoop by the tower, can reach one hundred decibels, or about as loud as a car alarm. Multiple turbines can orchestrate an additive effect that is especially maddening to nearby residents. The fact that there is already a condition recognized as "wind turbine syndrome" testifies to the seriousness of their protest. In addition to noise, detractors point to various other grievances. For instance, turbine blades occasionally ice up, dropping or throwing ice at up to two hundred miles per hour. They may also toss a blade or two, creating a danger zone within a radius of half a mile.⁶ Beyond this zone, residents are relatively safe from harm, and outside a one-mile radius the racket of wind turbines diminishes to the level of a quiet conversation. Ideally energy firms would not build wind turbines near homes and businesses but many of the other prime windy locations are already taken, geologically unstable, inaccessible, or lie within protected lands such as national parks. As a result, desperate wind power developers are already pushing their turbines both closer to communities and out into the sea, a hint as to limitations ahead.

Wind farm opponents tend to arise from one of two groupings, which are not always so easily distinguished from one another. The first are the hundreds of NIMBY (Not in My Backyard) organizations. NIMBY activists live near beautiful pastures, mountain ridges, and other sights they'd prefer to pass on to their children untarnished. They rarely have an economic interest (or anything else to gain) by erecting lines of wind turbines across their landscapes, each taller than the statue of liberty. Can we really blame them for being upset? Generating the power of a single coal plant would require a line of turbines over one hundred miles long. In a *New York Times* editorial, Robert F. Kennedy Jr. declared,

> I wouldn't build a wind farm in Yosemite Park. Nor would I build one on Nantucket Sound. . . . Hundreds of flashing lights to warn airplanes away from the turbines will steal the stars and nighttime views. The noise of the turbines will be audible onshore. A transformer substation rising 100 feet above the sound would house giant helicopter pads and 40,000 gallons of potentially hazardous oil.⁷

Kennedy and other politically well-connected residents of the Sound echo concerns voiced around the world. Even in Europe, where residents generally support wind power, locals often squash plans to build the rotating giants. In the Netherlands, local planning departments have denied up to 75 percent of wind project proposals.[8]

The second group of wind detractors is an unofficial assemblage of coal, nuclear power, and utility companies happy to keep things just as they are. Contrary to public opinion, they aren't too concerned about wind turbines eroding their market share. They're far more concerned that legislators will hand over their subsidies to wind-farm developers or institute associated regulations. These mainstay interests occasionally speak through their CEOs or public relations departments but their views more frequently flow to the media via a less transparent route interceded by think tanks and interest groups. The Cato Institute has taken aim at wind power for over a decade, and their criticisms have been published in *The National Review*, *Marketplace*, the *Washington Times*, and *USA Today*. The Centre for Policy Studies, founded in part by Margaret Thatcher, has done the same. A keen eye can identify these corporate perspectives, which emanate in the form of white papers, newspaper articles, research reports, letters to the editor, and op-eds because they all have one distinct marking in common. They invariably conclude with policy recommendations calling on public and legislative support for our friends in the fossil-fuel and nuclear industries.

NIMBY groups have found a strange bedfellow in these corporate energy giants. Each faction is more than willing to evoke wind power drawbacks that the other develops. Environmentalists sometimes find themselves caught in the mix. For instance, during the 1980s, the Sierra Club rose in opposition to a wind farm proposed for California's Tejon Pass, citing risks to the California condor, an extinct bird in the wild that biologists were planning to reestablish from a small captive population. A Sierra Club representative quipped that the turbines were "Cuisinarts of the sky," and the label stuck. Our detractors passionately cite the dangers to birds and bats as giant blades weighing several tons, their tips moving at two hundred miles per hour, spin within flight paths. However, newer turbine models spin more slowly, making them less a threat. Their smooth towers are less appealing for nesting than the latticed towers of earlier designs. According to one study, each turbine kills about 2.3 birds per year, which, even when multiplied by ten thousand turbines, is a relatively small number compared to the four million birds that crash into communication towers annually, or the hundreds of millions killed by house cats and windows every year.[9] Even the Sierra Club no longer seems overly concerned, pointing out that progress is being made to protect many bird habitats and that turbine-related death "pales in comparison to the number of birds and other creatures that would be killed by catastrophic global warming."[10] The Sierra Club's new positive spin on wind turbines is indicative of a shift in focus within the mainstream environmental movement—toward a notion that technologies such as wind turbines will mitigate climate change and related environmental threats posed by coal-fired power plants. Ahead, we'll consider why this is a frightfully careless assumption to make.

Detractors also cite wind turbines' less-well-known propensity to chop and distort radio, television, radar, and aviation signals in the same way a fan blade can chop up a voice. The United Kingdom has blocked several proposals for offshore wind farms, citing concerns about electromagnetic interference.[11] The 130-turbine Nantucket Sound project (known as the Cape Wind Project) stumbled in 2009 when the Federal Aviation Administration (FAA) claimed the offshore wind farm would interrupt navigation signals. FAA regulators insisted the developer pay $1.5 million to upgrade the radar system at Massachusetts Military Reservation or, if the upgrade could not solve the interference problem, pay $12 million to $15 million to construct an entirely new radar facility elsewhere.[12] A large expense to be sure, but not an insurmountable cost for a large wind-farm developer. Other wind-farm risks are not so easily reconciled.

Illustration 3: Road infiltrates a rainforest Roads offer easy access to loggers and poachers. Here a roadway backbone supports emerging ribs of access roads, which are dissolving this rainforest from the inside. (Image courtesy of Jacques Descloitres, MODIS Land Rapid Response Team, NASA/GSFC)

For instance, if you view satellite images of the Brazilian state of Pará, you'll see strange brownish formations of barren land that look like gargantuan fish skeletons stretching into the lush rainforest. These are roads. A full 80 percent of deforestation occurs within thirty miles of a road. Many of the planet's strongest winds rip across forested ridges. In order to transport fifty-ton generator modules and 160-foot blades to these sites, wind developers cut new roads. They also clear strips of land, often stretching over great distances, for power lines and transformers.[13] These provide easy access to poachers as well as loggers, legal and illegal alike. Since deforestation degrades biodiversity, threatens local livelihoods, jeopardizes environmental services, and represents about 20 percent of greenhouse gas emissions, this is no small concern.

Considering Carbon

The presumed carbon benefits of a remote wind farm, if thoughtlessly situated, could be entirely wiped out by the destructive impact of the deforestation surrounding it—a humbling reminder that *the technologies we create are only as durable as the contexts we create for them.*

Wind proponents are keen to proclaim that their turbines don't spew carbon dioxide. This is correct, but it is the answer to the wrong question. We'll consider some more revealing questions soon, but let's begin with a basic one: turbines may not exhaust CO_2 but what about the total carbon footprint of the mining, building, transporting, installing, clearing, maintaining, and decommissioning activities supporting them? Fossil fuels (including, especially, toxic bunker fuels) supply the power behind these operations. The largest and most efficient turbines rest upon massive carbon-intensive concrete bases, which support the hulking towers and (usually) prevent them from toppling in heavy winds. Any thoughtful consideration of the carbon implications of wind turbines should acknowledge these activities.

Nevertheless, carbon footprint calculations can be rather shifty, even silly at times, despite their distinguished columns of numerical support. They hinge on human assumptions and simplifications. They ignore the numerous other harms of energy production, use, and distribution. They say nothing of political, economic, and social contexts. They offer only the most rudimentary place to *start*.

Former UK leader of Parliament David Cameron installed a wind turbine on his London home, winning him positive reviews from econnoisseurs. However symbolically valuable, it was likely a waste of time, money, and energy according to carbon hawks. That's because homes, trees, towers, and other structures in cities choke airflow, which too often leaves the turbines unmotivated to spin. A British study claims that a third of small wind turbine locations in the windy coastal city of Portsmouth will never work off the carbon footprint invested to build and install them. A full two-thirds of Manchester's wind turbines leave their homes with a higher carbon footprint, not a lower one.[14]

Forceful gusts can whip wind shears up and around buildings, resulting in cracked blades or even catastrophic system failure.[15] The unexpected disintegration of a turbine with blades approaching the size and rotational velocity of a helicopter rotor could understandably produce significant damage anywhere, but in a city these harms become especially alarming. A single failure can take down power lines, tear through buildings, and pose obvious risks to residents. In practice, there are so many challenges to installing wind turbines on buildings, such as noise, insurance, and structural issues, that Mike Bergey, founder of a prominent turbine manufacturer, stated he wished people would stop asking him "about mounting turbines on buildings."[16]

Lifecycle calculations reveal that wind power technologies actually rely heavily on fossil fuels (which is partly why their costs have dramatically increased over the last decade). In practice, this leaves so-called renewable wind power as a mere fossil-fuel hybrid. This spurs some questions. First, if fossil-fuel and raw-material prices pull up turbine costs, to what degree can nations rely on wind power as a hedge against resource scarcity? Moreover, where will the power come from to build the next generation of wind turbines as earlier ones retire from service? Alternative-energy productivists would likely point to the obvious—just use the power from the former generation. But if we will presumably be using all of that output for our appliances, lighting, and driving the kids to school, will there be enough excess capacity left over? Probably not—especially given that the most favorable windy spots, which have been largely exploited, are purportedly satisfying less than 1 percent of global power demands. We'll likely have to fall back on fossil fuels.

Wind is renewable. Turbines are not.

Nevertheless, if we were to assume that NIMBY objections could be overcome (many could be), that turbines were built large enough to exceed their carbon footprint of production (as they usually are), and that other safety risks and disturbances could be lessened (certainly plausible), is there really anything to prevent wind energy from supplanting the stranglehold that dirty coal plants have on the world's electricity markets? Wind is a freely available resource around the globe, it doesn't have to be mined, and we don't have to pay to have it imported. There is, however, one little issue—one that is causing headaches on a monumental scale—which will lead us closer to understanding the biggest limitation of wind power.

Occasionally, wind has been known to stop.

A Frustratingly Unpredictable Fuel

Imagine if your home's electrical system were infested by gremlins that would without warning randomly vary your electrical supply—normal power, then half power, then three-quarter power, then off, then on again. Some days you'd be without electricity altogether and on others you'd be overloaded with so much current your appliances would short circuit and perhaps even catch on fire. This is the kind of erratic electrical supply that wind power grid operators deal with on a minute-to-minute basis. Whenever the wind slows, they must fire up expensive and dirty peaker power plants in order to fill the supply gap. Even when the wind is blowing, they often leave the plants on idle, wasting away their fossil fuels so they're ready when the next lull strikes. To make matters worse, grid operators must perform these feats atop a grid of creaky circuitry that was designed decades ago for a far more stable supply.

Traditional coal, natural gas, nuclear, and hydroelectric power stations provide a steady stream of power that operators throttle to match demand. Conversely, wind and solar electrical output varies dramatically. Windy periods are especially difficult to predict. Even when the wind is blowing more consistently, wind turbines encounter minor gusts and lulls that can greatly affect their minute-to-minute output. Over still periods, wind turbines can actually suck energy off the grid since stalled

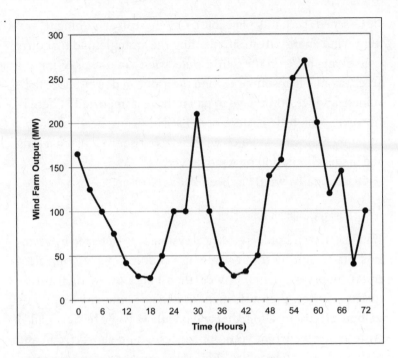

Figure 3: Fussy wind Wind farm output varies unpredictably. This chart shows the output of a large South Australian wind farm (in megawatts) over seventy-two hours. (Data
from Tom Quirk)

turbines require electrical power to operate their massive steer-
ing systems and other idling functions.[17]

Solar radiation is more predictable in frequency but not in intensity, as shown in Figure 4. Even on mostly sunny days, solar
photovoltaic output can vary due to dust, haze, heat, and passing clouds.[18]

Grid operators can handle small solar and wind inputs without much sweat (they manifest as small drops in demand). However, significant unpredictable inputs can endanger the very stability of the grid. Therefore, wind power isn't well suited to
supply base-load power (i.e., the power supplying minimum
demands throughout the day and night). If operators relied on

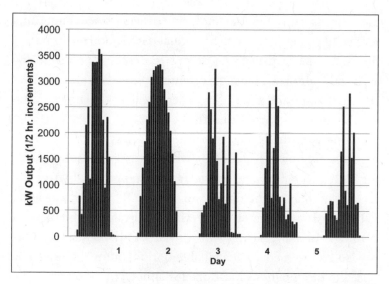

Figure 4: Five days of sun This plot shows the output (in kilo-watts) of a large photovoltaic system in Springerville, Arizona, over five days. Heat, haze, clouds, and other factors affect minute-to-minute solar output unpredictably. (Data from Tucson Electric Power Company)

wind power as a base-load supply, traffic signals, hospitals, and other essential services would be cut whenever the wind stopped. Even though wind power companies employ teams of meteorologists to predict wind speeds on an hour-to-hour basis, they still rely on coal, natural gas, hydroelectric, and nuclear power for backup consistency.

This intermittency is already causing headaches in the country with the highest number of wind turbines per capita, Denmark. Over five thousand turbines produce the equivalent of about 20 percent of the nation's electricity demand but not even half of it can be used or stored within the country.[19] Since the Danes don't suddenly start using more electricity whenever it's windy, the grid verges on excessive supply, and grid operators are forced to dump excess electricity into neighboring Norway, Sweden, and Germany. America's grids appear even more daunting as many cannot handle more than 2 percent intermittent wind power. Even with a national reinvention of the power network, such as the smart grid projects coming online in Hawaii and California, the most optimistic engineers don't expect them to handle any more than 30 percent live wind power, even if more turbines could be erected.

In one way, the Danes are fortunate. They can direct some excess wind power to Norway, where large pumps thrust water high into mountain reservoirs to be tapped by hydroelectric power plants when the wind slows.[20] This is an effective, yet expensive, strategy for buffering the erratic output of wind turbines. In many of the world's flat windy plains, this simply isn't an immediately available option, but turbines can be wired to mountainous locations for about $3 million per mile. Nevertheless, accommodating pumped storage on a large scale would require many more hydropower facilities, which bring their own set of disadvantages, as we will discuss later. Alternately, wind turbines can pressurize air into hermetically sealed underground caverns to be tapped later for power, but the conversion is inefficient and suitable geological sites are rare and often far away from electricity users. Finally, wind energy can be stored in batteries, flywheels, or as hydrogen gas, but these strategies are mind-numbingly pricey, as we shall explore later. Despite all the hype surrounding energy storage, experts debate whether these options could ever become effective large-scale solutions within the next thirty to fifty years, let alone in the more immediate future.

Policymakers, journalists, and wind proponents alike regularly misunderstand or misrepresent these windy realities. Proponents frequently declare that wind power costs the same as natural gas or just a bit more than coal, but this is misleading. Alternative-energy firms aren't required to back up their temperamental products, which makes them seem less pricey than they are in practice. It's during the power conditioning steps that the total costs of wind power start to multiply. The inconsistency of wind power necessitates a dual system, the construction and maintenance of one power supply network for when the wind is blowing and a second network for when it isn't—an incredibly expensive luxury.

Where the Wind Blows

We don't always get wind power *when* we want it, and we less often get it *where* we want it. In the United States, the strongest winds are all offshore. The strongest terrestrial gusts blow within a band stretching from the northern edge of Texas up through the Dakotas—right where almost nobody lives. Getting the wind crop to cities will be both technically knotty and expensive. As the director of North Dakota's Energy and Environmental Resource Center quips, "We produce the crop but we can't get it to the grain elevator."[21] Grid developers will also bump into right-of-way challenges since most residents disapprove of power lines as much as they do of wind turbines. The Sierra Club is actively challenging grid expansion through national forests, noting that the coal industry is ready to pounce on green grids.

Americans cannot count on a comprehensive smart grid any time soon, but the projected cost falls within the bounds of reason and an upgraded grid would bring numerous benefits. Most notably, a comprehensive smart grid would flip the long-held operating rule of power supply. Instead of utilities adjusting their output to meet demand, a smart grid would allow homes and businesses to adjust their electrical use automatically, based on the availability of power. That's because a smart grid coordinates electrical sensors and meters with basic information technology and a communications network akin to the Internet that can transform dumb power lines into a nimble and responsive transmission system. When a wind gust blows, tens of thousands of refrigerators will power up to absorb the added capacity and when the wind lulls, they will immediately shut down again. Of course, not every household and industrial appliance lends itself to be so flexibly controlled—a respirator at a hospital, for example—but a smarter grid will nevertheless minimize the need for expensive peaker power plants and spinning reserve (i.e., idling power plants). Given incentives, consumers could trim peak electricity consumption by 15 percent or more, saving hundreds of billions of dollars in the process.[22]

Smart grids are less vulnerable to power leaks and electricity pilfering—two big holes in the existing national grid. Furthermore, smart grids are less likely to experience power outages, which cost Americans about $150 billion every year and require dirty diesel backup generators to fill gaps in service.[23] By simply plugging leaks and avoiding needless inefficiencies, a nationwide smart grid would save a stream of power equivalent to the raw output of thirty-five thousand large wind turbines. The energy conservation savings that smart grids enable would be greater yet—probably many times greater. And unlike the wildly optimistic conjectures propping up alternative energy policies, smart grid estimations are quite sound; numerous other countries have already rolled out similar upgrades with great success. Sweden, for instance, installed smart meters across the nation quite some time ago.

There is much work to be done if the United States is ever to make similar strides. Regulators will have to coordinate standards and negotiate how the costs and benefits will be shared between the nation's three hundred utilities, five hundred transmission owners, and hundreds of millions of customers. Additionally, a connected smart grid will require a different form of security than comparatively dumb grids of today. Unfortunately, these responsible tasks are all too easy to cast aside when the magical lure of solar cells and wind turbines woos so insistently upon the imaginations of politicians, environmentalists, and the media.

Capacity Versus Production

Do you know the maximum speed of your car? It is safe to venture that most divers don't, save for perhaps German autobahners, since they rarely if ever reach maximum speed. The same holds for power plants—they *can* go faster than they *do*. A plant's maximum output is termed "nameplate capacity," while the actual output over time is called "production." The difference is simple, yet these two measures are confused, conflated, and interchanged by journalists, politicians, and even experts.

A "capacity factor" indicates what percentage of the nameplate maximum capacity a power plant actually produces over time. In traditional plants, operators control production with a throttle. A small one-hundred-megawatt coal plant will only produce 74 percent of that amount on average, or seventy-four megawatts.[24] For wind and solar, as we have already seen, the throttle is monitored by Mother Nature's little gremlins. A large wind farm with a nameplate capacity of one hundred megawatts will produce just twenty-four megawatts on average since the wind blows at varying strengths and sometimes not at all.[25] Every generation mechanism is therefore like a bag of potato chips—only partially full—as shown in Figure 5.

In order to match the production of a large 1,000-megawatt coal-fired power plant with a wind farm, 1,000 megawatts of wind turbines won't be enough. For an even swap, we'd need more than three times the wind capacity, about 3,100 megawatts. Both a 1,000-megawatt coal plant producing on average at 74 percent of capacity and a 3,100-megawatt wind farm producing on average at 24 percent of capacity will yield about the same output over time. Of course, this hypothetical comparison is still inadequate for real-world comparisons given the inconsistency of wind power. Therefore, energy analysts use a reliability factor to measure the minimum percentage of wind power that turbines can deliver 90 percent of the time. Taking this into

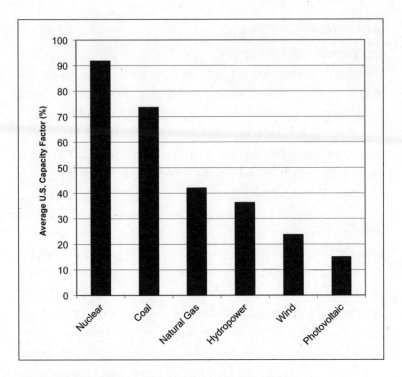

Figure 5: U.S. capacity factors by source A capacity factor is the
percentage of the nameplate maximum capacity that a power
plant actually produces over time. Fossil fuel, hydro, and nuclear plants attain nearly 100 percent of maximum capacity
when fully throttled, but lulls in demand and cost differentials
leave them producing less. Natural gas is more expensive than
coal, so power companies turn off gas plants first when demand drops. Weather variables dictate wind and photovoltaic capacity factors. (Data from U.S. Department of Energy)

account, we would need up to 18,000 megawatts of wind power
to offset a 1,000-megawatt fossil-fuel or nuclear plant 90 percent of the time.[26] As Leigh Glover, a policy fellow at the Center for Energy and Environmental Policy at the University of
Delaware, sums up, "When basic calculations are completed for
the number of wind turbines or PV arrays needed to replace the
world's coal-fired power stations, the resulting scenarios verge
on nothing less than bizarre."[27]

In fact, the rise of wind power in the United States has sadly not shuttered a single coal-powered plant.[28] So why might we think building more turbines will magically serve us any better? Well, it's likely because the story lines surrounding wind power are so compelling. And it just so happens that part of that magic was manufactured.

Manufacturing the Magic

When President Obama premiered his clean energy initiative in Newton, Iowa, he cited a prominent U.S. Department of Energy (DOE) report showing that the nation could easily obtain 20 percent of its electricity from wind turbines by 2030—he may have been completely unaware that the report's key dataset wasn't from the DOE at all. In fact, if genuine DOE cost and performance figures had been used, the report's authors would likely have come to the opposite conclusion—20 percent wind by 2030 will be logistically complex, enormously expensive, and perhaps ultimately unachievable.

Much of the enthusiasm surrounding wind power in recent years has grown out of this prominent Bush-era report entitled *20% Wind Energy by 2030*, which concludes that filling 20 percent of the nation's grid with wind power is achievable and will come at a cost described as "modest." The authoritative DOE report has been held up as a model for charting a course for wind energy funding; it has been covered by media sources across the globe, presented to congressional leaders, evoked by two presidents, and supported by the Sierra Club, the Worldwatch Institute, the Natural Resources Defense Council, and dozens of other organizations.[29] In fact, during my investigative research on the study, I didn't come across a single critical review of its findings. It is therefore particularly intriguing to note that the report is based on key assumptions, hidden within a second appendix, which are so explicitly incongruent with bona fide DOE data that many people might have considered them to be outright fraudulent had they not been produced within the protective halo surrounding alternative-energy research. This DOE report, which probably seemed ecologically progressive to its unwitting list of environmentalist cosponsors, may ultimately prove a tremendous disservice to their cause.

The report's most remarkable conclusion is simple. Filling 20 percent of the grid with wind power over the next twenty years will cost just 2 percent more than a scenario without wind power.[30] The conclusion teeters atop a conspicuous pile of cost and performance figures developed by industry consultants, despite the fact that the DOE already spends millions of dollars tabulating the same sorts of data on a routine basis. The report cites four "major" contributors outside the Department of Energy: a trade organization called American Wind Energy Association (AWEA) and three consulting firms—Black and Veatch, Energetics Incorporated, and Renewable Energy Consulting Services. Would perhaps any one of these groups have something to gain from painting an optimistic rendering of wind's future? It turns out they all do. And that potential gain can be measured in billions.

When the report was written, the AWEA's board of directors included executives from General Electric, JP Morgan, Shell, John Deere, and a handful of wind power companies including T. Boone Pickens's company Mesa Power. As an industry group, the AWEA was interested in orchestrating a positive spin on anything wind. The AWEA salivated in anticipation of preparing a pro-wind report enshrouded by the credibility of the Department of Energy.

But, there was a problem.

The DOE's field data on wind turbine performance was too grim—too realistic—for a report destined to pump up the future of wind power. Far more favorable statistics would be required. And the consultant employed to produce the stand-in datasets would not disappoint.

The authors retained Black and Veatch—a consultancy that designs both wind farms and natural-gas generation plants—to develop cost projections as well as key capacity factors for the analysis.[31] Remember, a capacity factor is simply the percentage of a wind turbine's nameplate capacity that is actually produced under real-world conditions—the difference of a percent or two can make or break a wind farm. According to DOE data, when countries or regions start to install wind turbines, the average capacity factor goes up at first, then levels off or declines as additional turbines are sited in less-ideal locations.[32] For instance, between 1985 and 2001, the average capacity factor in California rose impressively from 13 percent to 24 percent, but has since retreated to around 22 percent. Over recent years, Europe's maturing wind farms have stabilized below 21 percent.[33] The U.S. average is under 26 percent, according to field readings from the DOE. That's why Black and Veatch's capacity-factor assumptions, starting at 35 percent to 52 percent in 2010, and continuing to increase 15 percent by 2030, are particularly shocking.

Black and Veatch's average capacity-factor estimations rank among the highest ever published anywhere, let alone in a formal government report. If Black and Veatch knows how to run the nation's turbines at such high capacity, then they know something that nobody else does. Even the pro-wind AWEA caps realistic capacity factors at a terribly optimistic 40 percent—so, incidentally, does the Department of Energy.[34] In fact, Black and Veatch's expectation that capacity factors for wind turbines will *increase* over the next twenty years conflicts with other DOE reports, which forecast turbulence as future wind farms are forced into subprime locations.

The knowledgeable public servants at the DOE might have laughed Black and Veatch out of Washington. But they didn't. They got them published.

The justifications for employing such extraordinary assumptions are not entirely clear. During my investigation, a DOE official assured me that the Black and Veatch figures "were extensively critiqued and adjusted by experts in the wind and general energy communities." Though when I asked a director at Black and Veatch why their figures differed so dramatically from DOE assumptions, he was rather tight-lipped, insisting only that they stood by the methodology as outlined in the report.[35] That's particularly disconcerting.

The report's methodology section states simply, "Black and Veatch used historical capacity factor data to create a logarithmic best-fit line, which is then applied to each wind power class to project future performance improvements." It seems the consultancy assumed that the wind turbine learning curve (i.e., the idea that past experience with a technology helps to improve the technology and reduce its costs) would continue to produce gains well into the future. While it is well accepted that this occurred through the 1980s and 1990s, the learning curve has since flattened, as the DOE has documented. Therefore, extrapolating a select few years of data into the future without acknowledging the industry's maturation is as problematic as extrapolating the growth of high school students to show that by college they will stand taller than giraffes.

In addition to the optimistic capacity-factor projections, the report's analysis includes mysterious historical data. Black and Veatch "estimated" capacity factors ranging from 32 percent to 47 percent in 2005.[36] The report fails to mention that DOE fieldwork from that year placed the actual nationwide capacity factor closer to 20 percent.[37] (When I asked Black and Veatch about the discrepancy, they offered no further comment.) These discrepancies aren't the only surprises lurking in the report's appendices.

Black and Veatch assumed that the costs for building, installing, and maintaining future wind turbines will not increase, as other DOE reports predict, but will actually decrease, due to what it black-boxes as "technology development." But since today's turbine designs are already close to their theoretical maximum efficiency, the future success of wind power may be less influenced by technological development than by social and environmental variables. Many of the windiest sites present high barriers to entry. Since turbines must be spaced at least five rotor-diameters apart side-to-side and at least ten rotor diameters front-to-back in order to avoid a wind "shading" effect, vast stretches of land rights must be secured in order to create even a modestly scaled wind farm. Offshore sites are easier to procure and have strong, consistent winds, but they are expensive to develop, connect, and maintain for obvious reasons— inaccessibility, deep sea beds, high waves, corrosive salt water, hurricanes, and so on. The Department of Energy expects that suboptimal environments—with greater wind turbulence, wind variability, and unfavorable site factors such as steep slopes, terrain roughness, and reduced accessibility—will push up the cost of most of the remaining wind farm sites by some 200 percent.[38]

When Black and Veatch's capacity-factor assumptions are compounded by their cost assumptions, readers are left with an impression of wind power that is up to *six times* more impressive than if the analysis were run using the DOE's own figures.[39] This raises the question, Why did the Department of Energy base its pivotal wind energy report on numbers conjured up by an engineering firm, with a vested interest in advancing energy production interests, rather than its own data? This is the question I posed to the DOE.

Their response was telling. They made it apparent that even though the report claims to contain "influential scientific information," its analyses might not be recognized as such by the greater scientific community.[40] One of the report's lead editors told me, "The *20% Wind* work was carried out to develop a picture of a future in which 20 percent of the nation's electricity is provided from the wind, and to assess the feasibility of that picture. The work was based on the assumption that reasonable orderly advancement of the technology would continue, and that key issues needing resolution would be addressed and favorably resolved. Hence the work used input information and assumptions that were forward-looking rather than constrained by recent history."[41]

Indeed, the authors did not allow recent history to stand in their way. In fact, some might argue that their answer echoes the rhetoric used to defend the fabrication of data for which no historical justification or cultural context exists. Energy players employed such lines of reasoning to suggest that by the 1960s, nuclear energy would produce abundant clean energy for all, that by the 1970s, fusion power would be too cheap to meter, and that solar cells would be fueling the world's economies by 1986.[42] With the advantage of hindsight, historians of science romp in the particulars of how such declarations rose to prominence. They show how genuine inquiry was often pushed aside to make room for the interests of industrial elites in their attempts to pry open taxpayer coffers for subsidies. Will future historians judge the *20% Wind Energy by 2030* report similarly?

Yes, reasons Nicolas Boccard, author of two academic papers recently published in *Energy Policy*.[43] In his opinion, the kind of tomfoolery going on at the DOE is nothing particularly shocking. Boccard, who studies the phenomenon of capacity-factor exaggerations in Europe, found that when solid data do not exist, wind proponents are all too willing to make "unsubstantiated guesses." They get away with it because the public, politicians, journalists, and even many energy experts don't understand how capacity factors are involved in influencing prospects for wind power development. Or, perhaps caught up in the excitement surrounding wind energy, proponents may simply not care, due to a psychological phenomenon called selection bias, whereby people tend to overvalue information that reinforces their ideology and undervalue that which contradicts it. Boccard insists, "We cannot fail to observe that academic outlets geared at renewable energy sources naturally attract the authors themselves supportive of renewable energy sources, as their writing style clearly indicates. As a consequence, this community has (unconsciously) turned a blind eye to the capacity factor issue." He compared wind farm data across many European countries, where wind power penetration is many times higher than in the United States. He uncovered a worrisome gap between the anticipated and realized output of wind turbines. In fact, Boccard maintains, the difference was so large that wind power ended up being on average 67 percent more expensive and 40 percent less effective than researchers had predicted. As a rule of thumb, he maintains that any country-level assumptions of capacity factors exceeding 30 percent should be regarded as "mere leaps of faith."[44]

It might seem counterproductive for wind firms to risk over-inflating expectations, but only if we assume that real-life turbine performance will impact their profit potential. It won't. Consulting firms such as Black and Veatch stand to lock in profits during the study and design phase, long before the turbines are even brought online. The AWEA manufacturers stand to gain from the sale of wind turbines, regardless of the side effects they produce or the limitations they encounter during operation. And by placing bets on both sides of the line, with both wind turbines and natural gas, Pickens was positioned to gain regardless of the wind's motivations. If the turbines don't return on the promise, it's no big deal for those in the money. The real trick is convincing the government, and ultimately taxpayers, into flipping for as much of the bill as possible. And one of the best tools for achieving that objective? A report that can be summarized in a sound bite struts with an air of authority, and can glide off the president's tongue with ease. *20% Wind Energy by 2030.*

It may be tempting to characterize this whole charade as some sort of cover-up. But the Department of Energy officials I interviewed were certainly open (if nervous) to my questions; anyone with an Internet connection can access the report and its suspect methodologies; and the DOE regularly publishes its field measurements in a report called the *Annual Energy Outlook*. There's no secret. Energy corporations develop "forward-looking" datasets favorable to their cause, government employees slide those datasets into formal reports, the Department of Energy stamps its seal on the reports, and the Government Printing Office publishes them. Then legislators hold up the reports to argue for legislation, the legislation guides the money, and the money gets translated into actions—usually actions with productivist leanings. It isn't a cover-up. It's standard operating procedure. This may be good or bad, depending on your political persuasion. This well-oiled system has operated for years, with all actors performing their assigned duties. As a result, Americans enjoy access to ample and inexpensive energy services and we have a high standard of living to show for it. But this process nevertheless leads to a certain type of policy development—one that is intrinsically predisposed to favor energy *production* over energy *reduction*. As we shall see, this sort of policy bent—while magnificently efficient at creating wealth for those involved—does not so clearly lead to long-term well-being for everyone else.

Step *Away* from the Pom-Poms

When Big Oil leverages questionable science to their benefit, environmentalists fight back en masse. As they should. But when it comes to the mesmerizing power of wind, they acquiesce. No op-eds. No investigative reports. No magazine covers.

Nothing.

If environmentalists suspected anything funny about the *20% Wind Energy by 2030* report, they didn't say anything about it in public. Instead, fifty environmental groups and research institutes, including the Natural Resources Defense Council, Sierra Club, and Lawrence Berkeley National Laboratory opted to double-down their windy bets by formally backing the study. When the nation's smartest and most dedicated research scientists, physicists, and environmentalists roll over to look up googly-eyed at *any* corporate energy production report, it's worthy of our attention. This love affair, however, is harmful to the environmentalists' cause for a number of reasons.

First, fetishizing overly optimistic expectations for wind power takes attention away from another grave concern of environmental groups—reducing dirty coal use. Even if the United States could attain 20 percent wind energy by 2030, the achievement alone might not remove a single fossil-fuel plant from the grid. There is a common misconception that building additional alternative-energy capacity will displace fossil-fuel use; however, over past years, this hasn't been the case. Producing more energy simply increases supply, lowers cost, and stimulates additional energy consumption. Incidentally, some analysts argue that the mass deployment of wind turbines in Europe has not decreased the region's carbon footprint by even a single gram. They point to Spain, which prided itself on being a solar and wind power leader over the last two decades only to see its greenhouse gas emissions rise 40 percent over the same period.

Second, the pomp and circumstance around wind diverts attention from competing solutions that possess promising social and ecological value. In a cash-strapped economy, we have to consider the trade-offs. As journalist Anselm Waldermann points out, "when it comes to climate change, investments in wind and solar energy are not very efficient. Preventing one ton of CO_2 emissions requires a relatively large amount of money. Other measures, especially building renovations, cost much less—and have the same effect."[45]

The third problem is the problem with all myths. When they

don't come true, people grow cynical. Inflated projections to-day endanger the very legitimacy of the environmental move-ment tomorrow.

Every energy-production technology carries its own yoke of drawbacks and limitations. However, the allure of a magi-cal silver bullet can bring harms one step closer. Illusory diver-sions act to prop up and stabilize a system of extreme energy consumption and waste. Hype surrounding wind energy might even shield the fossil-fuel establishment—if clean and abundant energy is just over the horizon, then there is less motivation to clean up existing energy production or use energy more wisely. It doesn't help when the government maintains two ledgers of incompatible expectations. One set, based on fieldwork and his-torical trends, is used internally by people in the know. The sec-ond set, crafted from industry speculation and "unconstrained" by history, is disseminated via press releases, websites, and even by the president himself to an unwitting public.

It may be time for mainstream environmental organizations to take note of this incongruence, put away the clean energy pom-poms, and get back to work speaking up for global eco-systems, which are hurt, not helped, by additional energy pro-duction. Because as we shall see, the United States doesn't have an energy crisis. It has a consumption crisis. Flashy diversions created through the disingenuous grandstanding of alternative-energy mechanisms act to obscure this simple reality.

3. Biofuels and the Politics of Big Corn

Years ago, fairy tales all began with "Once upon a time . . ." Now we know they all begin with, "If I am elected." —Carolyn Warner

Iowa. That's the answer to a question that growing numbers of scientists, aid workers, reporters, and environmentalists are asking about ethanol and other biofuels. But before we can address the *question*, it would be helpful to understand what biofuels are and how they are affecting our energy infrastructure.

Biofuels in Sixty Seconds or Less

Like photovoltaics and wind turbines, biofuels are another way to harness power from the sun, but through photosynthesis. Unlike wind turbines and solar photovoltaics, biofuels are easily stored and dispatched as needed, much like oil, coal, and natural gas, making their energy far more valuable.

Before the industrial revolution, biomass materials (i.e., living and recently dead plant material, such as firewood, and biological material, such as dung) were humanity's primary sources of energy.[1] The world's first mass-produced flex-fuel vehicle, Ford's Model T, ran on ethanol. And even up through World War II, the United States Army distilled ethanol to supply fuel for combat vehicles. Nevertheless, after the war an abundance of low-cost petroleum washed America's biofuel industries down the drain.[2] For a time.

We'll dredge up the politics behind biofuel's reemergence in a moment. But first, let's consider the chief biofuels available today:

- *Solid biomass* such as wood, sawdust, agricultural waste, manure, and other products are burned directly, formed into pellets, and converted into charcoal.
- *Biogases* such as methane are produced from organic materials in anaerobic digesters or captured as they naturally emit from animal, agricultural, and landfill waste.
- *Bioalcohol*, most commonly ethanol, is distilled from starchy plants such as corn, sugar beets, and sugar cane.
- *Biodiesel* is chemically manufactured from oil-rich plant and animal feedstocks such as animal fats, rapeseed oil, palm oil, and algae.

Though the various biofuel techniques vary in style and complexity, the basic idea is the same: refiners convert plant and animal materials into usable energy products. In the United States today, biomass products serve about 5 percent of primary energy demand.[3]

Biofuel critics point out that the industry produces airborne heavy metals, copious amounts of wastewater, and a variety of other externalized environmental costs. As evidence, they point to Brazil, where ecologists declared many rivers and waterways biologically dead as early as the 1980s due to biofuel effluents (ethanol represents roughly a third of Brazil's automotive fuel).[4]

Perhaps the most cited drawback, however, is the risk that biofuels can spark land competition between food and fuel, inducing an upward pressure on global food prices. As biofuels become more valuable, farmers may opt to grow *fuel* crops instead of *food* crops on their existing fields or even level forests in order to expand croplands. High food prices do not significantly affect rich consumers because they spend just a small portion of their income on food. Not so for the world's poor. For years, researchers warned that expanding biofuels production would jeopardize food security worldwide. Eventually, they proved to be right.

Turning Food into Fuel

In 2008 riots ensued throughout the world in response to a dramatic increase in corn prices. The White House blamed the increase on rising food demand from fast-growing China and India.[5] Others disagreed. World Bank president Robert Zoellick acknowledged that by early 2008 it was evident that biofuel demand had become a "significant contributor" to grain price escalations, which put thirty-three countries at risk for social upheaval.[6] Washington was dismayed, maintaining that biofuel demand was responsible for less than 3 percent of the price increase—bad news for Zoellick, as the United States was the World Bank's major donor. Zoellick immediately backpedaled by sequestering a confidential report that the World Bank had painstakingly prepared to research the price shock. But the report did not remain secret for long. An informant leaked it to the *Guardian* that summer.[7] The report's authors concluded that biofuel demand was actually responsible for a hefty 75 percent of the food price jump.

To some, converting arable fields over to fuel crops was especially troubling given that much of the resulting biofuel would eventually burn away in inefficient vehicles driving through inefficient transport systems. The head of the International Food Policy Research Institute, Joachim von Braun, announced that world agriculture had "entered a new, unsustainable and politically risky period."[8] Around the same time, researchers at the Carnegie Institution and Lawrence Livermore National Laboratory published a paper claiming that even if nations were to divert the entire global harvest of corn to ethanol production it would satisfy just 6 percent of global gasoline and diesel demand. They observed that "even in the best-case scenario, making ethanol from corn grain is not an effective route for lowering the carbon intensity of the energy system . . . ethanol from corn is basically a way to make cars run on coal and natural gas."[9] Why coal and natural gas? We shall come to that soon.

In 2009 the National Academy of Sciences released a study detailing how the combined health costs, pollution, and climate change impacts from producing and burning corn ethanol were worse than simply burning gasoline, perhaps almost twice as bad. A prominent professor from Iowa State University's Agriculture and Biosystems Engineering Department published a biting attack on ethanol claiming that "while feedstock can be grown annually, ethanol is not renewable. Ethanol production is entirely dependent on nonrenewable (petro-) energy in order to get any energy out. The term 'renewable' is grossly overused by those promoting ethanol and other biofuels, indeed the promotions sound like a call for a perpetual energy machine."[10]

The criticisms didn't stop.

Academics and government agencies released a flurry of scientific research investigating ethanol. By 2011, when food prices spiked again and Congress let an ethanol tax credit expire, it was difficult to find any informed individual that didn't have some sort of opinion on the fuel. Critics continued to deride it for polluting water, consuming fossil fuels, spewing greenhouse gasses, endangering biodiversity, spreading deforestation, and of course destabilizing food supplies. If concerned citizens were disagreeing on the reason, they weren't disagreeing on the consensus: corn ethanol was a flop.

So the question arises—Why did Americans ever think it was a good idea to turn food into fuel in the first place? The answer is, of course, Iowa.

Big Corn

In 2008 the United States found itself in an election year, as it so frequently does, and as a matter of habit, turned to Iowa, the geographic heart of the nation, to sound off on that year's primary candidates. Perhaps the greatest fear of presidential and congressional candidates alike is being labeled "antifarmer" in the heat of the Iowa spotlight. Politically, this fear is justified. Americans are mesmerized by the romantic ideal of farmers who wear bib overalls, drive red tractors, and cultivate their own destinies—even though the real control of farming today lies firmly in the manicured hands of businesspeople who wear suits and drive Porsches. Nevertheless, this pastoral imagery guarantees there is always plenty of election angst in the air to scare up a veto-proof majority of Congress that will pass agri-anything. In that election cycle, it was a farm bill that handed out subsidies to big agribusiness and wealthy individuals—a list that included people such as David Letterman and David Rockefeller for their "farming" activities.[11] By the same token, political candidates were leery of voicing anything but praise for the then well-recognized dirty, wasteful, and risky practice of distilling corn ethanol.

Amidst the drive to realize economies of scale, most farms in wealthier nations no longer resemble the kind Old McDonald had, with an assortment of animals, vegetables, grains, and fruits. Farms once worked as miniature ecosystems where animal wastes fertilized plants, plants produced food, and animals ate the leftovers. By some measures, this system was inefficient—it no doubt required significant human labor. In contrast, modern farms utilize highly mechanized systems to produce just one or two high-yield products. The contemporary farming system realizes far greater harvests, feeding more people with less land, but it is not without its own set of costs and risks.

After World War II, farmers invested in plant breeding programs and agricultural chemicals in order to increase yields.

Over subsequent decades, they grew increasingly dependent on technological advances, intensifying their farming practices with outside capital along the way. Smaller farms consolidated into larger and larger farms and by the beginning of this century, the bulk of farming income in the United States came from the top few percent of the nation's farming firms. These superfarms wield a historically unprecedented degree of influence on agriculture, from purchasing, sales, and distribution to the investment and control of resources. While these select few corporations eagerly showcase the promise of their technologies to feed the world's poor, critics maintain that the bulk of these research efforts accrue toward increasing short-term profits—not toward addressing hunger-related or longer-term concerns of low soil fertility, soil salinity, and soil alkalinity. Critics also condemn agribusiness for downplaying the risks that their models of agriculture produce—risks stemming from superpests, declining crop diversity, and national overreliance on a few large transnational firms for food production. Furthermore, they maintain that centralized agriculture is responsible for the proliferation of dead zones, has bred distribution-related externalities, and has led to the demise of traditional rural communities.[12]

In the early 1970s, one of these large corporations, Archer Daniels Midland, had a solution that was looking for a problem. It wanted to market the byproducts of its high-fructose corn syrup, a product that was growing in popularity and would eventually come to dominate the sweetener market. The firm's politically savvy and well-connected president, Dwayne Andreas, knew that one byproduct, ethanol, could power automobiles. He launched an intensive lobbying and "educational" program to promote corn ethanol as a fuel source, an effort auspiciously timed with the 1973 oil embargo. Archer Daniels Midland's lobbying team convinced key senator Bob Dole (R-Kansas), as well as President Jimmy Carter, that corn ethanol production could circumvent the need for oil imports. With additional prodding from the corn and farm lobbies, Congress eventually passed the Energy Tax Act in 1978, which offered tax breaks for gasoline products blended with 10 percent ethanol. In 1980 Congress wove additional ethanol incentives into legislation—then again in 1982— and again in 1984.[13] For Archer Daniels Midland it was a windfall, though every dollar of ethanol profit was costing taxpayers thirty dollars, according to a critical report from the conservative Cato Institute.[14]

. Ethanol proponents received another thrust. Across the country, states increasingly chose the alcohol to replace the toxic gasoline oxygenate MTBE.[15] The quick switch launched ethanol prices higher, fueling both facility upgrades and new plant construction. In 2004 the American Jobs Creation Act provided a $0.51-per-gallon ethanol subsidy to oil companies that blended ethanol with their gasoline and instituted a protective tariff of $0.54 per gallon to ward off Brazilian ethanol imports. Ultimately, Congress required gasoline blenders to incorporate at least 4 billion gallons per year of ethanol with gasoline in 2006, 6.1 billion in 2009, 15 billion by 2015, and 36 billion by 2022.[16] Politicians assured agribusiness firms that they would enjoy predictable ethanol demand well into the future. But for Big Corn, being handed a guaranteed ethanol empire wasn't enough. They wanted taxpayers to pay for it.

It might have seemed a daunting task to both harness public support and motivate Congress to cede billions of public funds for one of the largest subsidies of big business (and by proxy, car culture) ever attempted. But Big Corn handled it deftly. Ethanol producers dispatched teams of lobbyists to federal, state, and local government chambers to ensure that legislators would subsidize the industry at every stage of development. They packaged their handout requests under various guises of helping farmers, increasing energy independence, protecting the environment, and keeping energy jobs at home, even though there was little real evidence to show that industry subsidies would serve any of these concerns. The rebranding was a success, prompting numerous legislative actions:

- *Loan guarantees*: The U.S. Department of Agriculture guaranteed biofuel loans and spent eighty million dollars on a bioenergy development program.
- *State tax breaks*: Individual states instituted retailer incentives, tax incentives, discounts for ethanol vehicles, and fuel tax reductions for ethanol, totaling several hundred million dollars per year.
- *Federal tax breaks*: The Internal Revenue Service reclassified biofuel facilities and their waste products into more favorable asset classes, allowing for billions of dollars in savings for the industry.
- *Research funds*: The U.S. Department of Energy released hundreds of millions of dollars to fund research and development and demonstration plants.
- *Labor subsidies*: State tax laws as well as the federal Domestic Activities Deduction introduced income tax reductions for workers in the biofuel industry, totaling about forty to sixty million dollars per year.
- *Farm subsidies*: U.S. farm policies long provided direct subsidies for numerous crops; billions of dollars of these funds accrued to crops for biofuel production.
- *Water subsidies*: County and state subsidies for water greatly benefited ethanol producers since every gallon of ethanol required hundreds of gallons of water to grow crops and process the fuel.[17]

And the total bill? The subsidies between 2006 and 2012 equated to about $1.55 per gallon-gasoline-equivalent of ethanol. For comparison, that's over one hundred times the subsidies allotted for a gallon of petroleum gasoline. Nevertheless, subsidies were only part of Big Corn's benefit package. There was more. Possibly the most egregious capitulation came in 2007 when the Environmental Protection Agency reclassified ethanol fuel plants, allowing them to significantly increase their federally controlled emissions. Government regulators also released producers from accounting for their fugitive emissions (pollutants not from the plant stack itself) and no longer required them to adopt the best available control technologies. Under these lax standards, biofuel refineries shifted away from using natural gas to power their energy-intensive distillation operations and instead began using cheaper, dirtier coal.[18] Additionally, regulators lowered corporate average fuel economy (CAFE) requirements for car companies that produced flex-fuel vehicles, which run on both ethanol and gasoline, even though customers opted to fill their flex-fuel tanks with standard gasoline over 99 percent of the time. This gaping loophole reduced the efficiency of the U.S. vehicle fleet across the board, increasing oil imports by an estimated seventy-eight thousand barrels *per day*.[19]

Energizing Iowa

Dr. Dennis Keeney has a front-row seat to the nation's electoral primary action, being a professor at Iowa State University. He maintains that Iowa's early primary position places the state's electorate in an especially influential role. "Any politician, be it dogcatcher or presidential candidate, speaking against ethanol in Corn Belt states has been doomed to denigrating letters, jeers from peers, and political obscurity," remarks Keeney, who asserts, "had ethanol expansion been subject to environmental assessment guidelines and or life cycle analyses, the ethanol support policies, in my opinion, would never have been adopted."[20] But if public perception of ethanol was high, politicians knew their poll numbers would be too, as long as they supported the fuel. To speak reason about corn ethanol would be to spit in the face of the Iowa economy, certain to prompt unwanted head shaking by voters come primary season.

In the run-up to the 2008 election, Archer Daniels Midland,

along with grain handlers, processors, and other corporate farming interests, formed cover for their operation and the politicians that supported it by forming an "educational" group called the Renewable Fuels Association. This corporate-funded research and lobbying group effectively silenced numerous economic, scientific, environmental, and social critiques of corn ethanol as they arose.[21]

In fact, up until the food riots of 2008 there was little well-developed resistance to the idea of a corn ethanol economy. It seemed everyone had something to gain. To start, security hawks saw corn ethanol as a step toward energy independence. And at that time, most mainstream environmental groups were still vibrating with excitement that Big Oil might be overpowered by a countryside brimming with yellow kernels of clean fuel. Automakers were similarly pleased; converting cars to run on ethanol was much easier than building electric or fuel-cell vehicles (the best greenwashing alternatives) and cheaper too—at only about one hundred dollars extra per vehicle.[22] Meanwhile, fossil-fuel firms likely found little reason to fret; every step of the ethanol scheme required their products—natural gas forms the basis for requisite fertilizers, oil fuels the tractors and shipping infrastructure for biofuels, and coal heats the high-temperature distillation and cracking processes.

It takes power to make power.

It's well known that crop yields grew dramatically during the green revolution—indeed from 1910 to 1983 corn production per acre in the United States grew 346 percent—but the corresponding caveat is rarely mentioned. The raw energy employed to achieve those gains grew at more than twice the rate—810 percent during the same period. One of the reasons it takes so much energy to create biofuels in America is that roads and parking lots entomb the richest soil. Early settlers built cities along rivers and near deltas, precisely where eons of annual flooding had deposited layers of nutrient laden silt. Suburban expansion subsequently pushed farms onto less fertile land, and farmers shipped the bounty back to the cities via a fossil-fuel-based transportation system. The system has changed little since.

Faced with less fertile soil, farmers started using synthetic fertilizers derived from petrochemicals to increase yields. Today, petrochemical fertilizer use is widespread. However, these nitrogen-rich fertilizers are not particularly efficient; only about 10–15 percent of the nitrogen makes it into the food we eat.[23] The rest stays in the ground, seeps into water supplies, and makes its way into rivers and eventually oceans to create oxygen-depleted dead zones, like the one that extends from Louisiana to Texas in the Gulf of Mexico.[24] This runoff flows through a very large loophole in the Clean Water Act. Where spring floods used to bring life, they now increasingly carry deadly concentrations of nitrogen, which artificially stimulate algal plumes, cutting off the oxygen to entire reefs of animal life. Hundreds of these dead zones afflict coastal regions around the globe.

It's worth noting that following the BP oil spill in the Gulf of Mexico and during the run-up to the 2010 and 2012 election cycles, the ethanol industry pointed out that "no beaches have been closed due to ethanol spills."[25] Another somewhat true statement. Even though dead zones may not generate the same attention as a cataclysmic gulf oil spill, they are increasing in number and their detrimental impacts are intensifying.[26]

In the end, experts debate whether the energy obtained from corn ethanol is enough to justify the energy inputs to plant, plow, fertilize, chemically treat, harvest, and distill the corn into useable fuel. At best, it appears there is only a small return on fossil-fuel investment. Meanwhile, experts largely agree that Brazilian ethanol delivers a whopping eight times its energy inputs because it is based on sugarcane, not corn. Why don't we do the same? The answer is simple—sugarcane doesn't grow in Iowa.

Illustration 4: Mississippi River dead zone Agricultural run-off is propagating a dead zone at the mouth of the Mississippi River. The dead zone now covers over five thousand square miles. (Image courtesy of NASA/Goddard Space Flight Center Scientific Visualization Studio)

Measuring the Scale of the Resource

But what if sugarcane did grow in Iowa? Could we grow our way out of the impending energy crunch? Industry forecasters measure the potential harvests of biofuel feedstock such as corn, sugarcane, and rapeseed using widely accepted yield tables. Yield tables simplify complex relationships between plant starch, sugar, and oil content into convenient estimates of biofuel output per acre. News articles, scientific papers, and policy documents give a voice to these wildly popular simplifications. In fact, through their repeated use, these numbers have crystallized to achieve an air of credibility that was never intended and under other circumstances may never have been achieved.[27] Well dressed in formal gridlines, they evolved as a mismatched patchwork of numbers plucked from various contexts at various times in various places. Some average regional growing data, some come from experimental farms (often with unusually high yields), and others reflect yields from random individual farms. These entries often lack controls for climate, location, length of growing season, soil type, availability of fertilizer, agricultural management, technological influences, and other factors that dramatically influence crop yields. In practice, researchers might unwittingly draw upon yield data from a specific farm in France in a given year and extrapolate the numbers to speak for expectations on the other side of the globe in a village with not just a different physical climate but also a different economic, political, and social climate. Researchers might call upon that French farm to estimate global harvests over a period spanning several decades.

In some cases, yield tables provide practical crop estimations. In other cases, biofuel proponents can employ them as reputable cover for coarse overestimations. A team of researchers from the University of Wisconsin, the University of Minnesota, and Arizona State University claim that extending such narrow figures to estimate global biofuel production is "problematic at best." Since the statistics usually come from farms within the prime growing regions for each crop, energy productivists routinely overestimate yields by 100 percent or more.[28] Related investigations back up this team's research.

In all, increasing crop yields globally will require new genetically altered plants, greater agricultural productivity, significant land use alterations, and more water—challenges that will be especially pronounced for poorer regions.[29] According to numerous studies, including a prominent report from the National Academy of Sciences, land-use alterations can lead to local and global warming risks, which may in turn decrease crop yields.[30] The Stanford Global Climate and Energy Project estimates that climate change will have varying effects on agriculture by region, but on average will decrease traditional crop yields as global temperatures rise.[31]

Another wildcard is water. Even though scientists have developed genetically engineered crops that are drought resistant, they have not been able to modify the fundamentals of transpiration. Therefore, they have been unable to design crops capable of producing significantly higher yields with less water— a limitation that will tighten further as water too grows dearer.

Carbon Dioxide and Climate Forcing

Biofuel proponents have long hyped their fuels for being CO_2 neutral. In their idealized case, biofuel crops absorb carbon dioxide from the surrounding atmosphere as they mature and release an equivalent amount when burned. Most researchers now agree that the biofuel carbon cycle is not so straightforward. Critiques from the scientific community coalesce around four central points.

First, soils and crops can precipitously increase net CO_2 where farmers employ destructive cropping methods or deforestation. In Indonesia, soil decomposition accelerated as developers drained wetlands to plant palm oil crops. As a result, every ton of palm oil production grosses an estimated thirty-three tons of CO_2.[32] Since burning a ton of palm oil in place of conventional fuel only has the potential to save three tons of CO_2, the process nets an excess thirty tons of the greenhouse gas. This realization led the Dutch government to publicly apologize for promoting the fuel; Germany, France, and eventually the European Union followed.

Biofuel crops may overflow into rainforests, as the authors of a recent article in *Science* point out:

> Regardless of how effective sugarcane is for producing ethanol, its benefits quickly diminish if carbon-rich tropical forests are being razed to make the sugarcane fields, thereby causing vast greenhouse-gas emission increases. Such comparisons become even more lopsided if the full environmental benefits of tropical

forests—for example, for biodiversity conservation, hydrological functioning, and soil protection—are included.[33]

Altogether, rainforests are magnificent resources for humanity; they stimulate rainfall, provide vital services for local inhabitants, and act as large sponges for CO_2. Biofuel proponents are quick to point out that sugarcane crops are not planted in Brazilian rainforests, but on pastureland to the south. They're correct for the most part. But since the demand for meat and other food products has not dropped (demand for Brazilian cattle is actually increasing), land scarcity rules. Higher land values push a variety of new developments into existing rainforests. Many admire Brazil for having figured out a way to make their cars run on domestic ethanol, but by proxy, they may have simply found a way to make cars run on rainforests.

The second biofuel concern involves the reflectivity of the earth's surface. In the earth's higher latitudes, dark evergreen regions absorb more heat from the sun than lighter vegetation such as grass and food crops.[34] In the tropics, this phenomenon reverses as evapotranspiration above forests generates reflective cloud cover. Rainforest development endangers this shield. While biofuel feedstocks may provide a short-term financial boon for many of the world's poor farmers, the resulting land competition will ultimately degrade their most valuable community asset.

It makes more sense to grow biofuel feedstock on *abandoned* cropland. As perennial biofuel grasses absorb airborne CO_2 and sink carbon into their roots, soil carbon content can modestly increase.[35] However, these tracks of empty land lie mostly in cooler climates and suffer from poor soil quality. Yields won't be impressive. Even if biofuel producers exploited every abandoned field worldwide, the resulting fuel would only represent about 5 percent of current global energy consumption.[36]

Third, researchers criticize the biofuel industry's reliance on fossil fuels, from the fertilizers derived from natural gas and petroleum to the fuels employed to plant, treat, plow, harvest, ferment, distil, and transport the fuels. It's difficult to calculate an entire life-cycle analysis for rapeseed biodiesel, corn-, and sugar-ethanol products since there are so many assumptions built into these models. Researchers who have undertaken the challenge come to different conclusions. Some argue these bio-fuels use more fuel and create more carbon dioxide than if we had simply burned fossil fuels directly. Others argue there is a benefit, even if it is not overwhelming.

Fourth, biofuel crop residues release methane, a greenhouse gas with twenty-three times the warming potential of CO_2.[37] Additionally, creating sugar, corn, and rapeseed biofuels yields considerable quantities of nitrous oxide, a byproduct of the nitrogen-rich fertilizers farmers use to grow the plants. Nitrous oxide's global warming potential is 296 times that of carbon dioxide and additionally destroys stratospheric ozone.[38]

Ethanol might seem more attractive if it didn't prompt food competition, net greenhouse gases, or require so much fossil fuel, water, and arable land. That's precisely the promise of cellulosic ethanol.

Woodstock: The Promise of Cellulosic Ethanol

Instead of using foodstuffs such as corn or sugar, cellulosic ethanol producers harvest trees and grass that can grow within a variety of climates and require less fertilizer and water. Cellulosic ethanol is expensive, but proponents say it could ease the food-fuel competition. Cellulosic feedstock is rich with carbohydrates, the precursors of ethanol, but these sturdy plants lock away these calories inside fibrous stems and trunks. Extracting them is a tricky process.

Ethanol forms when sugars ferment, which is why Brazilian producers can process sugarcane into fuel almost effortlessly. Corn requires one additional step. Producers must mix the corn meal with enzymes to create the sugars. Wrestling ethanol from the cellulose and hemicellulose in grasses and trees is even more involved. Refiners must liberate sugars via a cocktail of expensive enzymes. Numerous firms are working to reduce the cost of these enzymes and some have even taken to bioprospecting to locate new ones, such as the digestive enzymes recently discovered in the stomachs of wood-munching termites.[39] Yet bioprospecting brings its own set of ethical dilemmas.

Why bother? Because if perfected, cellulosic ethanol could potentially yield up to 16 times the energy needed to create the fuel—that compares favorably with corn ethanol, which arguably yields just 1.3 times its energy inputs, sugarcane ethanol, which yields about 8 times its energy inputs, and even regular old gasoline, which yields about 10 times its energy inputs. Furthermore, proponents claim that second-generation biofuels could someday cost as little as three dollars per gallon-gasoline-equivalent. But today, despite weekly breakthroughs in the field, cellulosic techniques remain prohibitively expensive and unproven on a commercial scale. Even the productivist-leaning *Wall Street Journal* calls for some degree of sobriety in formulating expectations for this moonshine of the energy world, stating that cellulosic ethanol "will require a big technological breakthrough to have any impact on the fuel supply. That leaves corn- and sugar-based ethanol, which have been around long enough to understand their significant limitations. What we have here is a classic political stampede rooted more in hope and self-interest than science or logic."[40]

The Forgotten Biofuels

Even as legislators flood cellulosic ethanol and other biofuel initiatives with funding, some biofuel opportunities go overlooked, mostly because they are boring in comparison. For instance, wastewater treatment facilities release methane, the main component of natural gas, but more than 90 percent of America's six thousand wastewater treatment plants don't capture it. As mentioned earlier, methane is a major greenhouse gas liability since its venom is more potent than that of carbon dioxide. The sludge output of the average American yields enough power to light a standard compact florescent light bulb without end. So skimming the methane from an entire city's wastewater would not only prevent harmful emissions but also would produce enough power to run the entire wastewater operation, perhaps with energy to spare.[41] Although not a large-scale solution, captured biogas is a reminder of the modest opportunities to draw upon biofuels without advanced technology.

Another biofuel product that is now starting to gain more attention is a convenient replacement for firewood. Burning firewood directly is a relatively dirty practice, emitting dangerous particulates, hydrocarbons, and dioxins.[42] In poor countries, the soot from firewood, waste, and dung kills about 1.6 million people per year. It's also a local climate changer; soot darkens air and darker air absorbs more solar radiation. But there's another way to extract energy from wood besides burning it—one that was widely employed before the Industrial Revolution but has since fallen by the wayside—charcoal (recently rebranded as *biochar*). When processors heat wood above 300°C with limited oxygen, in a process called pyrolysis, it spontaneously breaks into three useful fuels: biochar, heavy oil, and flammable gas. In addition to its use as a fuel, farmers can till their soil with biochar in order to reduce methane and nitrous oxide greenhouse-gas emissions.[43] Archaeologists uncovered ancient South American settlements in which buried charcoal has been sequestered for thousands of years, lending interest to the concept of using biochar as long-term storage for excess carbon.

In all, there may be many benefits to implementing biochar techniques in place of burning wood and waste for fuel directly. But this doesn't make biochar a global solution. Cornell researcher Kelli Roberts points out that large-scale biochar production, as envisioned by some eager biofuel productivists, could yield unintended consequences.[44] As with other biofuel methods, if producers clear virgin land to grow biochar inputs such as trees and switch grass, the process could ultimately do more harm than good. Alternately, if producers grow biochar crops on existing farmland, farmers may be forced onto new land, yielding the same negative effects on virgin land plus the added risk of local food price instability. And then there is the hitch with any method for increasing available energy supply—it inevitability leads to growth, expansion, and increasing energy consumption—a reminder that smart upgrades in energy practices for local communities may not have the same positive effects if implemented on a larger scale.

Dreary Expectations

Researchers from the Carnegie Institution and Lawrence Livermore National Laboratory neatly sum up the limitations of biofuel technologies: "The global potential for biomass energy production is large in absolute terms, but it is not enough to replace more than a few percent of current fossil-fuel usage. Increasing biomass energy production beyond this level would probably reduce food security and exacerbate forcing of climate change."[45] The U.S. Department of Energy's biofuel forecast is similarly tentative. It forecasts that biofuel use will only modestly expand, from 5 percent of primary energy supply today to about 9 percent in 2030. The agency lowered its previous expectations for the fuel, citing technical concerns about cellulosic ethanol development.[46] Even the International Energy Agency's "450 Scenario," which employs highly optimistic assumptions that the agency itself admits will be "very challenging" to realize, forecasts that biofuels will fulfill at most just 16 percent of primary energy demand by 2030.[47]

Not long ago, these unadventurous expectations for biofuels would have been heretical. America was in a fervor over rising oil prices, and pundits gleefully framed ethanol as the answer. In 2006 the National Corn Growers Association complained they were sitting on a surplus of corn, the Worldwatch Institute proclaimed that biofuels could provide up to 75 percent of transportation fuel in the United States, and Congress was trucking bales of public funds to Big Corn.[48] The subsequent collapse of ethanol's popularity may very well have been a dress rehearsal for wind and solar industries if the public comes to better understand the limitations of these schemes as well.

We now have every reason to suspect that large-scale biofuel production will require vast water resources, endanger areas reserved for conservation, intensify deforestation, and decrease food security. The net greenhouse-gas impact could be positive or negative depending on the type of feedstock plant materials, the biofuel production process, and the difference in reflected solar radiation between biofuel crops and preexisting vegetation. Ultimately, we might presume that biofuels will provide modest energy resources worldwide, but the most promising biofuel strategies are unproven on a commercial scale, may not be economical for some time, and will certainly entail side effects and limitations not yet well understood. Worthless? No. But certainly uninspiring.

Perhaps that's why many people in the money have shifted their bets to another energy production technique, one that they are slowly resurrecting from its grave.

4. The Nuclear-Military-Industrial Risk Complex

Boy, we're sure going to have some wrecks now!
—Walt Disney, upon constructing a model train
to encircle his house

On March 16, 1979, Hollywood released a run-of-the-mill film that might have been rather unremarkable had the fictional plot not played out in real life while the movie was still in theaters. *The China Syndrome*, starring Jane Fonda, Jack Lemmon, and Michael Douglas, features a reporter who witnesses a nuclear power plant incident that power company executives subsequently attempt to cover up. Many days pass before the full extent of the meltdown surfaces. Just twelve days after *The China Syndrome* premiered, operators at the Unit 2 nuclear reactor at Three Mile Island, outside Harrisburg, Pennsylvania, received abnormally high temperature readings from the containment building's sensors. They ignored them. Many hours passed before the operators realized that the facility they were standing in had entered into partial core meltdown. Power company executives attempted to trivialize the incident and many days passed before the full extent of the meltdown surfaced.

The China Syndrome went viral. When star Michael Douglas appeared on NBC's *The Tonight Show*, host Johnny Carson
quipped, "Boy, you sure have one hell of a publicity agent!" The
staged nuclear leak filmed in the back lots of Hollywood and
the real nuclear leak on Three Mile Island became conjoined,
feeding into one another, each event becoming more vividly
salient in the eyes of the public than if they had occurred independently. The intense media and political fallout from the leak
at Three Mile Island, perhaps more than the leak itself, marked
the abrupt end of the short history of nuclear power development in the United States.

Nuclear industry officials regularly accuse their critics of un-
fairly brandishing the showmanship of disaster as if it were characteristic of the entire industry while downplaying the solid safety
record of most nuclear facilities. Indeed, meltdowns like the
ones at Three Mile Island, Chernobyl, and Fukushima Daiichi
don't occur as frequently as oil spills. But then, the risks people
associate with nuclear leaks are inordinately more frightening.

As with oil spills, journalists, politicians, and industry officials frame meltdowns as *accidents*, almost without exception,
though, we could alternately choose to frame nuclear power activities as highly unstable undertakings that are bound to expel
radioactive secretions into the surrounding communities and
landscapes over time.

One of the largest single releases of atmospheric radiation
into American communities, about seven hundred times that
of Three Mile Island, cannot even plausibly be framed as an
accident.[1] The U.S. government deliberately planned and released this emission under the code name "Green Run" in a once-
secret government compound so infrequently acknowledged that
even President Obama claimed to have been unaware of the site
during his time in the Senate.[2] The facility would later rise to become the single largest recipient of his federal stimulus funds.

Green Run

In the early 1940s, a U.S. government convoy rolled into a small community in Washington State, inexplicably condemned private homes, shut down the high school, and hastily laid out foundations for over five hundred buildings on an area roughly half the size of Rhode Island.[3] For a time, nearby residents had no idea what happened behind the gates of the enormous secret facility, which was named the Hanford Site. But on August 6, 1945, when U.S. forces dropped an atomic bomb on Hiroshima, Japan, its purpose became abundantly clear. The United States built Hanford to enrich plutonium, in a hurry.

After the war, Hanford's purpose shifted (the first of many shifts). In an effort to judge how much plutonium the Soviet Union was processing during the cold war's infancy, the Pentagon decided to take measurements from the dispersion of a known quantity of radioactive iodine–131, a byproduct of plutonium production, to be released at Hanford. During the night of December 2, 1949, the U.S. Air Force deliberately executed a sudden and clandestine discharge of radioactive iodine intended to disperse and contaminate the fields, communities, and waterways surrounding Hanford.[4] The U.S. government kept the radioactive dispersion secret for nearly forty years until the Freedom of Information Act forced the Department of Energy to release the classified documents in 1986.[5] Selected intelligence purposes remained secret until 1993.

After scientists expelled the radioactive cloud from Hanford, radiation levels in surrounding communities jumped to 430 times the then permissible limits. Hanford's scientific team measured the highest levels of radioactivity in nearby plant life, 28 microcuries/kg, or 2,800 times the 1949 permissible limit.[6] For comparison, the Washington State Health Department now identifies any food product over 0.013 microcuries/kg as radioactive and unfit for consumption. Even given the high levels of free

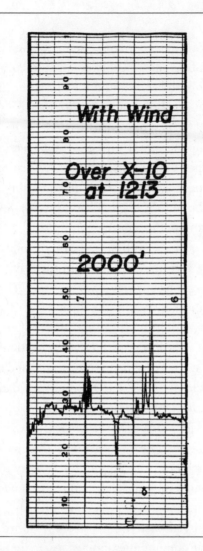

With Wind

Over X-10 at 1213

2000'

Shown here is
the record of the return
run which was made at the
level of the smoke and
haze top and which was
believed to be at the
inversion top. The
maximum return over the
stack is appropriately
lower than was received
on runs at lower levels.
The symmetry and strength
of this return is as
would normally be expect-
ed if the very top of
the gaseous cloud had been
traversed. The return
remains above background
for approximately 15
miles downwind.

Figure 6: Secret U.S. government document ORNL–341 This
1949 chart from the declassified document ORNL–341 details
fly-by radiation measurements following a planned govern-
ment-sponsored dispersion of radioactive materials into the
local landscape, waterways, and human settlements of southeast Washington State. This radiation release was roughly
seven hundred times that of the 1979 Three Mile Island meltdown and was kept secret for nearly forty years. (Image courtesy of Oak Ridge National Laboratory)

radiation in neighborhoods after the experiment, the scientists advocated for an even larger radioactive cloud in their official report, secret document ORNL–341. The passionless abstract of the report reads, "Very little information of a conclusive nature was gained concerning the diffusion. . . . Using a stronger source, is recommended."[7]

Today, local residents still have unanswered questions. Some details remain classified. Nevertheless, the nineteen thousand pages of declassified government documents detail a long history of radioactive emissions from Hanford—emissions that contaminated the air, soil, groundwater, and the Columbia River. Sadly, the experiences of families from Hanford and others exposed to radiation from the Green Run experiments are far from unique. In 1995, the Department of Energy released documents showing that the U.S. government has sponsored at least several hundred secret releases of radiation throughout the United States.[8]

In addition to intentional radioactive dispersions and unintentional radioactive leaks, nuclear processing and power activities also churn out large quantities of varied radioactive waste, which carry an assortment of protracted risks and costs of their own. One notable example comes from within the Hanford Site itself—a ponderous reservoir labeled "Tank SY–101."

Tank SY–101

If the construction, filling, and management of Tank SY–101 had been initiated in our day, it would surely become a leading news story throughout the world, though since government employees built it several decades ago, tucked away inside Hanford, its story is infrequently told. Over Hanford's forty years of operation, workers refined sixty-four metric tons of plutonium, enough to fill two-thirds of the nation's arsenal of roughly sixty thousand nuclear warheads.[9] Between the plutonium enrichment, energy generation, and other activities at Hanford, the facility produced a magnificent sum of radioactive waste.

Efforts to clean up the site began twenty years ago, but the cleanup is not yet halfway complete, and project funding is endangered. Originally expected to cost $50 billion, overruns have forced estimates much higher. For instance, a proposed waste treatment plant's budget skyrocketed from $4.3 billion in 2000 to over $12 billion by 2008. The Department of Energy expects to complete the plant in 2019 but it has already postponed construction three times.[10]

According the Department of Energy, Hanford contains:

- 2,100 metric tons of spent nuclear fuel
- 11 metric tons of plutonium in various forms
- about 750,000 cubic meters of buried or stored solid waste in 175 waste trenches
- about one trillion liters of groundwater contaminated above EPA drinking water standards, spread out over 80 square miles (contaminants include metals, chemicals, and radionuclides)
- 1,936 stainless-steel capsules of radioactive cesium and strontium, containing roughly 125 million curies of material in water-filled pools
- more than 1,700 identified waste dumps and 500 contaminated facilities
- more than 53 million gallons of liquid radioactive waste in 170 aging, underground single-shell tanks[11]

The faded hodgepodge of a sign marking the entry to the Hanford Site quietly deteriorates, along with a nearby assortment of massive storage tanks, which engineers originally designed as early as the 1940s to last no longer than a few decades.[12] According to the Department of Energy, sixty-seven of the tanks have sprung leaks—releasing a combined one million gallons of radioactive waste into the local soil and groundwater, which is seeping into the adjacent Columbia River.[13] Of the numerous tanks at Hanford, none has created more problems than Tank SY–101.

Like many of the multistory tanks at Hanford, Tank SY–101

Illustration 5: Entering Hanford This fading sign ingloriously marks the entrance to the Hanford Site, a historically multiuse nuclear complex, now a makeshift nuclear waste reprocessing site of unprecedented proportions. (Photograph courtesy of Tobin Fricke)

was filled with a largely unknown brew of mostly liquid radioactive waste. Since the tank's contents did, and still do, constantly enter into various sorts of unpredictable internal reactions, the concoction's chemical and reactive nature is no longer properly understood. For many years the slurry was not calm—a constantly evolving, hissing, sputtering brew topped with a crust sometimes prone to violent undulations and eruptions of radioactive and potentially explosive gases and fumes.[14] In 1991 operators caught the tank's contents on film, documenting what looked like an active lava flow. Lurching from side to side, splashing against the walls, and spitting gases exhausted by internal nuclear reactions, the entire mixture seemed alive. It occasionally agitated the tank, sometimes with enough force to bend metal components of its enclosure.[15]

Illustration 6: A four-story-high radioactive soufflé This 1989
photograph offers a rare glimpse into the interior of nuclear
waste containment Tank SY–101 at the Hanford Site. A lava-
like crust encapsulates the top of a highly reactive, little understood nuclear waste slurry four stories deep. The mixture
sputters angrily from around the edges of the tank in the up-
per part of the photograph. (Photograph courtesy of the U.S.
Department of Energy)

Dedicated Hanford technicians repeatedly attempted to calm
the angry concoction for years but the chemical chimera would
smother sampling tubes, morph unpredictably, and evade numerous efforts to bring it into submission. With the threat of an un-
controlled dirty nuclear explosion in Washington State, Hanford
workers developed special operating procedures to deal with the
tank. In 1993, the Department of Energy ordered a large pump
to circulate the toxic slurry so that its toxic gases would effer-
vesce more consistently, like bubbles gently rising from a glass
of radioactive carbonated soda. It worked—for a time—but then
the crusty top, being less agitated, began to thicken, harden, and
trap the gaseous effluents beneath. Subsequently the crust rose
up in what investigative reporter Matthew Wald characterized
as "a giant radioactive soufflé," growing to a height of thirty-

six feet and approaching the more fragile single-walled top of the tank.[16] In a risky move, operators lanced the top crust of the giant gas-filled cyst with air jets, which effectively thwarted the expanding mass's expansion. Some time afterward, the team was able to successfully pump out half of the container's waste into another holding tank and cut each remaining quantity with an equal amount of water, effectively doubling the volume of radioactive waste but diluting it enough to calm the potion's rambunctious nature.[17]

Even more rarely told than the story of Tank SY–101 is the story of how it changed Hanford. Over the years, the tank's unprecedented nuclear risks became a central concern of the Department of Energy, one that presented Hanford operators with uncertain and extreme challenges. The Department of Energy coordinated multiyear research projects around the tank, and academics published papers on its behavior. Over time, modes of work within Hanford changed to accommodate the special treatment and operations required to look after the gooey beast. These forces pulled apart existing networks of workers, operating procedures, goals, and facilities at Hanford and reassembled them in a novel way. For some, the activities and politics surrounding Tank SY–101 signaled the end of Hanford as a closed, secretive, armaments production facility and marked its reincarnation as an obligatory checkpoint for environmental research and understandings into the containment and handling of nuclear waste.[18] With the reorganization of Hanford, however, came a reorganization of risk perceptions and assessments. Imbedded researcher Shannon Cram argues that Hanford's practices for reducing uncertainty actually constitute a fictional reality "in which workers are to blame for nuclear accidents."[19]

Hanford is understood in different ways by different people. A government conspiracy project. An innovative waste-reprocessing facility. A dump. A place to work and make a living. A tombstone for the nuclear dream. The various meanings, ambiguities, and uncertainties tangled in Hanford's barbed-wire enclosures are a messy rendering of global nuclear anxieties writ small. However menacing these portents may be, they won't be enough to frighten away the gravediggers standing over nuclear power's tomb, shovels in hand.

The Resurrection

Soon after World War II, the United States initiated a self-described "peaceful" atomic energy program, during the presidency of Dwight D. Eisenhower, in an effort to assure a wary world that it was interested in more than just military deployments of nuclear science. Congress quickly ramped up nuclear energy funding. It also applied legislative lubricants, such as the 1957 Price-Anderson Act, which limited the nuclear insurance industry's liability to just $540 million per nuclear accident. These helped nuclear energy slide into America's power grid and into the psyche of its citizens. At the time, very few critics stood up to nuclear power. When they did pop up, holding out technical and economic risks in their hands, nuclear proponents handily smacked them back down. Former Environmental Protection Agency special assistant Professor Walter Rosenbaum recalls that "when problems could not be ignored, they usually were hidden from public view; when critics arose, they were discredited by Washington's aggressive defense of the industry."[20] By 1975 the United States had 56 commercial nuclear reactors online, 69 under construction, and 111 more planned. The nation hummed with a kind of nuclear intoxication that burrowed deep into remote nooks of popular culture, from industrial design to literature. Even a colorfully packaged children's toy, called the Atomic Energy Lab, came complete with radioactive materials for young nuclear scientists to test and measure.

And the rest is history.

After a boom-boom here and a boom-boom there, fears grew that there might everywhere be a boom-boom. So the industry came to an abrupt halt, right there in the middle of the nuclear highway. And there it sat with its parking brakes on.

That is, until the summer of the year 2004.

It was that summer when the Department of Energy murmured support for the nuclear industry's pleas to extend the lives of numerous aging nuclear power facilities as well as its plans to add fifty new power plants to the nation's electrical grid.[21] The next year, Congress coughed up enough money to push the plan into action. In fact, the *Washington Post* reported that the nuclear industry was the 2005 Energy Policy Act's surprise "biggest winner."[22] The act released more incentives to the nuclear industry than to wind, biomass, solar, geothermal, hydroelectric, conservation, and efficiency initiatives *combined*.[23] In 2008 AmerGen Energy Company submitted an application to the U.S. Nuclear Regulatory Agency requesting a license to allow Three Mile Island Unit 1 nuclear power facility to operate until April 2034. On October 22, 2009, during the height of a national swine-flu panic, the application was quietly approved.

The Legend of the Peacetime Atom

Whether or not this renewed interest in nuclear power is good or bad depends upon whom you ask. For some, nuclear power marks an opportunity for low-carbon and independent energy generation while for others it represents a prescription for nuclear proliferation and fallout risks. Environmentalists in Germany, for instance, overwhelmingly rail against nuclear power, but environmentalists in Britain tend to support it. In Japan, nuclear power risks remained conceptually separated from the fallout horrors of World War II until the 2011 meltdowns at Fukushima collided them together.

In 2008, the Nuclear Suppliers Group, a cartel of forty-five nations that limits the trade of nuclear materials and technology, agreed to bend their rules in order to allow India access to uranium imports.[24] When the waiver was first introduced, political sparring ensued between those who claimed such a move would lead to nuclear armament proliferation in the region and others who claimed the additional uranium marked a peaceful development of electricity that would benefit millions of Indians. So who's right? Is nuclear power a way to produce electricity or a path toward building deadly weapons?

In reality, it is both.

The often-cited division between civilian nuclear power and military nuclear weaponry is problematic for several reasons. First, countries often end up desiring a bit of both—a little civilian electricity and a little nuclear weaponry. Political desires rarely congeal into exclusively one form or the other. Second, peacetime and wartime nuclear technologies are intermingled. For example, the power plant fuel rods, once spent, contain high concentrations of plutonium, which is useful for building bombs. Third, nation-states are in constant flux—politically, economically, and culturally—the motivations of a country today cannot be assumed to hold in the future. In practice, an exclusively peacetime uranium atom is as inconceivable as a coin with just one side. We'll review each point in order.

First, in his book *The Light-Green Society*, historian Michael Bess illustrates how a nuclear-armed France was not born from a single directive, like America's Manhattan Project, but instead rose up from a series of smaller, more subtle nudges:

> Perhaps the most striking fact about France's emergence as a nuclear power is that no single meeting or series of meetings, no single group of individuals working together, no single confluence of key events, can be identified as *the* point at which the nation's leadership decided to endow France with nuclear weapons. Rather, what we find is a sequence of incremental "mini-decisions," some technical in nature, some budgetary, some administrative—a long series of tacit compromises and gradual techno-

logical advances accumulating, sliding into one another, over a decade and a half. French politicians and scientists in the Fourth Republic appeared extremely reluctant to come out and openly say, "Let's build a Bomb!" Instead, they always opted for keeping the door open, for continuing lines of research and technical development that would leave available the option of building atomic weapons in the future—without committing anyone to an explicit military policy in the present. And so, one by one, the pieces of the puzzle quietly came together: a steady expansion of funding, allowing the newly created CEA, or Commissariat à l'Energie Atomique, to double its facilities every two years; a decision among scientists in 1951 to build a new generation of high-powered reactors, capable of producing weapons-grade plutonium; a top-secret cabinet meeting in 1954 under the prime ministership of Pierre Mendès-France, giving the green light for exploratory studies of a nuclear bomb; an international environment of increasing East-West tensions, coupled with the chilling prospect of West German rearmament within NATO; and growing pressures from the French military, which became ever more keenly interested in the physicists' string of successes. All these developments gradually accumulated, with a weirdly impersonal but also seemingly irresistible momentum.[25]

The choice of France to build up a nuclear arsenal might better have been dubbed the "gigantic decision that no one made," for by the time the de Gaulle government had announced it would take up nuclear arms in 1958, the thought had already tunneled its way into the heart of French institutions over the course of a decade, implicitly aligning a ready-made framework of chutes and ladders for weaponry development. In fact, less than two years passed between de Gaulle's announcement and the detonation of the first French plutonium bomb at a Sahara test site.[26]

It didn't have to turn out this way. The allure of nuclear weaponry attracted governments to uranium, rather than the arguably much safer element thorium, in building their reactors. Thorium is now gaining attention as an alternative to uranium. It's less suited for advanced nuclear weaponry and therefore less likely to induce proliferation risks (though it's perfectly suitable for low-tech dirty bombs). Thorium is also more naturally abundant than uranium and its waste products are easier to store because of their much shorter half-life. Nevertheless, since government interest in the fuel waned during the early years of the cold war and did not recover until recently, the thorium cycle will require significant research, development, and testing in order to overcome the technical hurdles that are currently preventing it from becoming a competitive alternative.

The second reason we can't neatly separate nuclear technologies into piles of "peacetime energy" and "wartime weaponry" is that they share many of the same technological foundations. Technological optimists fervently assert that scientific advancements will deliver safer forms of civilian nuclear power. There is little argument that they're correct. The caveat, however, is that to whatever degree they are correct, related technological advances for nuclear weaponry will advance as well. Consider, as an analogy, downhill ski resorts designed for recreational snow skiing. During the 1980s and 1990s, growing numbers of snowboarders realized that ski resorts provided easy access to some of the world's sickest shank-high powder. Snowboarders promptly moved in, and in some cases smoked out the skiers altogether. Today, resorts throughout the world have rules, obligations, and practices that dictate who can use what hills, but the underlying technologies, the snowmaking equipment, plows, and gondolas, are the same.

Nuclear facilities are a tad more intricate, but the analogy holds: a ski hill could be repurposed for snowboarding just as a nuclear energy enrichment facility could be repurposed to enrich devices of destruction. However, unlike skiers and snowboarders, military contractors and arms dealers have a seemingly limitless supply of financial resources and political backing. It's treacherously naïve to assume that such a force will lie dormant during a global expansion of nuclear enrichment activities. Indeed, the armaments industry has even devised and tested bombs using byproducts from the purportedly peaceful thorium cycle.[27] Just in case.

Due to all of this interchangeability, the United Nations formed an oversight organization called the International Atomic Energy Agency (IAEA) to prevent theft or diversion of fissile materials into the wrong hands. A two-year study by the Nonproliferation Policy Education Center, however, paints the picture of an IAEA that is overloaded with responsibilities, working with outdated assumptions, and essentially being asked to do the impossible. Over recent decades, the United Nations doubled the IAEA's budget, yet it expects the agency to track six times as much highly enriched uranium and plutonium. Over a recent six-year period, IAEA inspectors, who visit monitoring locations every ninety days to manually download monitoring video, discovered twelve blackouts that lasted longer than thirty hours, plenty of time to covertly divert nuclear material to people with ominous intentions.[28] Even in the best-run nuclear facilities, the IAEA reports substantial sums of "material unaccounted for." When the IAEA could not account for sixty-nine kilograms of plutonium at a Japanese fuel fabrication plant in Tokai-mura, enough to build eighteen warheads, it ordered a $100-million plant disassembly in order to locate the material; ten kilograms were never found.[29] In a report to Congress, the U.S. Government Accountability Office detailed several corresponding accounts of lost fissile material. Apparently, entire spent nuclear fuel rods occasionally go missing.[30]

If nuclear fuel enrichment operations expand, so will plutonium detours. So far, we have not discovered appropriate political, technical, ethical, and economic strategies to restrict fissile material to exclusively peacetime activities over the long term. Indeed, such strategies may not be plausible. Even if nations agreed upon preventative regulations and inspections for nuclear fuel materials today, subsequent ruling regimes may not recognize them. The political realities of an era greatly influence how leaders perceive and implement nuclear enrichment technologies. Peacetime nukes beget wartime nukes beget peacetime nukes in an unrelenting historical seesaw between prosperity and destruction. These risks call for contemplation to which we are generally ill suited. As the Department of Energy has strikingly acknowledged, there is no precedent to assume that the United States will remain a contiguous nation-state throughout the period required for nuclear waste to decay below safe levels. It is probably far more likely, some would argue certain, that America's nuclear waste will someday fall under the direction of some other form of government.

This is complicated by the fact that containment facilities are not as simple as we might imagine. They aren't as they appear in the beginning of mutant films—vast warehouses of vacant gray hallways idling silently except for the occasional clip-clop of a security guard's shoes. These facilities require immense staffs to monitor sensing and tracking information on nuclear waste pools, perform regular analysis and research on unstable and evolving chemical waste slurries, maintain and repair vessels, and countless other tasks. The parking lots of these facilities are full. If they weren't, we'd all be in a lot of trouble, quickly. If those people don't show up to work some day, perhaps due to a disease pandemic, economic depression, political turmoil, or a natural disaster, the infrastructures established to maintain fissile material in a stable state could deteriorate or even tumble into chaos. With nuclear energy, we risk not only our own well-being but also the contamination of many, if not all humans, who occupy this planet after us.

Carbon, Again

These risks may be worth it, some say, since nuclear power generation produces less carbon dioxide than the alternatives and therefore promises to mitigate the potentially far greater risks of catastrophic climate change. For solar, wind, and biofuel power generation, the projected costs to mitigate a ton of CO_2 are very high. Does nuclear fare any better?

Not really.

Assuming the most favorable scenario for nuclear power, where nuclear power generation directly offsets coal-fired base-load power, avoiding a metric ton of CO_2 costs about $120 ($80 of which is paid by taxpayers). This figure does not include the costs of spent-fuel containment and the risks of proliferation and radiation exposure, burdens that are especially difficult to quantify. Again, this is far more expensive than boosting equipment efficiency, streamlining control system management, improving cropping techniques, and many other competing proposals to mitigate climate change. Why spend 120 bucks on nuclear to avoid a single ton of CO_2 when we could spend the same money elsewhere to mitigate five tons, or even ten, without the risks? Nuclear energy might be a plausible CO_2 mitigation strategy after we have exhausted these other options, but we have a long way to go before that occurs. This won't, however, stand in the way of the nuclear industry's expensive expansion plans. And by the way, the bill is on us.

The Total Bill

Every single nuclear plant was built with our help. Additionally, the nuclear industry incurs substantial capital write-offs, through bankruptcies and stranded costs, which leave the burden of their debt on others—a hidden and formidable set of overlooked costs.[31] To make matters worse, economies of scale don't seem to apply to the nuclear industry. Just the opposite in fact.

Historically, as the nation added more nuclear energy capacity to its arsenal, the incremental costs of adding additional capacity didn't go down, as might be expected, but rather went up.[32]

If the costs to taxpayers are so high, the risks so extreme, and the benefits so unexceptional, why do nations continue to subsidize the nuclear industry? It's partly because so many of the subsidies are hidden. Subsidy watchdog Doug Koplow points out, "Although the industry frequently points to its low operating costs as evidence of its market competitiveness, this economic structure is an artifact of large subsidies to capital, historical write-offs of capital, and ongoing subsidies to operating costs."[33] The nuclear industry often loops taxpayers or local residents into accepting a variety of the financial obligations and risks arising from the planning, construction, and decommissioning of nuclear facilities, such as

- accepting the risk of debt default;
- paying for cost overruns due to regulatory requirements or construction delays;
- dropping the requirement of insurance for potential damage to surrounding neighborhoods; and
- taking on the burden of managing and storing high-level radioactive waste.

Since these handouts are less tangible and comprehensible to the public than cash payments, the nuclear industry and its investors have found it relatively easy to establish and renew them. One in particular is especially problematic.

The Decommissioning Subsidy

Travel two hundred miles off the northeast coast of Norway into the Arctic Ocean toward the shores of Novaya Zemlya Island and you'll see seals, walrus, and aquatic birds as well as numerous species of fish, such as herring, cod, and pollack, much as you'd expect. But some of them will be swimming around an article less anticipated—a curious fabricated object rising above the dark sea floor like an ancient monument, identifiable only by the number, "421." Inside the corroded steel carapace lies a nuclear reactor. Why, we might wonder, has someone installed a nuclear reactor under the sea so far from civilization? It wasn't built there. It was dumped there—along with at least fifteen other unwanted nuclear cores previously involved in reactor calamities. These cores lie off the coasts of Norway, Russia, China, and Japan.[34] Many of the reactors still contain their fuel rods. Resurfacing them and processing them in a more accepted manner would be risky and expensive. But even disposing of the world's existing nuclear reactors that haven't been tossed in the ocean won't be a straightforward proposition. It costs hundreds of millions of dollars to carefully assemble a nuclear power plant, and it costs hundreds of millions to carefully disassemble one as well. The largest problem being, of course, what to do with the radioactive waste?

The Department of Energy started to construct a repository in Yucca Mountain, Nevada, to store the nation's spent reactor fuel. It was to accept spent fuel starting in 1998, but management problems, funding issues, and fierce resistance by the state of Nevada pushed the expected completion date back to 2020.[35] President Obama called off the construction indefinitely, slashing funding in 2009 and finally withdrawing all support in 2011. If completed, the Yucca Mountain crypt will cost about $100 billion.[36] Even then, it's designed to house just sixty-three thousand tons of spent fuel. More than that is already scattered around the country today.[37]

In the meantime, utility companies have been storing waste in open fields surrounding their plants. A large nuclear power reactor typically discharges twenty to thirty tons of twelve- to fifteen-foot-long spent fuel rods every year, totaling about 2,150 tons for the entire U.S. commercial nuclear industry annually.[38] Taxpayers will end up paying billions to temporarily store this waste.[39]

Another option is to "recycle" the spent fuel into new fuel. However, reprocessing is expensive and leaves behind separated plutonium. Since plutonium is ideal for making bombs, many countries, including the United States, consider reprocessing a proliferation risk. Meanwhile, the United Kingdom, France, Russia, Japan, India, Switzerland, and Belgium reform their spent rods. They have separated a combined 250 metric tons of plutonium to date, more than enough to fuel a second cold war. Alternately, fast-neutron "burner" reactors can run directly on the spent fuel. This presumably sidesteps the plutonium issue, though such plants may not be commercially feasible to build.

A Hard Sell

The Colorado River flows through one of the largest natural concentrations of radioactive surface rock on the planet, containing about a billion tons of uranium in all. The levels of radiation are twenty times the proposed limit for Yucca Mountain and unlike the glass-encapsulated balls used to store radioactive waste, Colorado's uranium is free ranging and water soluble. Berkeley physicist Richard Muller claims, "If the Yucca Mountain facility were at full capacity and all the waste leaked out of its glass containment immediately and managed to reach groundwater, the danger would still be twenty times less than that currently posed by natural uranium leaching into the Colorado River."[40] Does this mean Coloradans are exposed to more radiation than the rest of us? Yes—along with those in Los Angeles who regularly bathe and drink water piped in from the Colorado River. Yet the residents of Colorado and California, together with those of the nearby states—South Dakota, Utah, and New Mexico—experience the lowest cancer incident rates anywhere in the contiguous United States according to the National Cancer Institute—which all goes to show how tricky it is to assess complex radiation risks.[41]

According to early documentation of the Chernobyl nuclear reactor meltdown in 1986, the catastrophe exposed thirty thousand people living near the reactor to about 45 rem of radiation each, about the same radiation level experienced by the survivors of the Hiroshima bomb.[42] According to a statistical scale developed by the National Academy of Sciences, 45 rem should have raised cancer deaths of residents near Chernobyl from the naturally occurring average of 20 percent to about 21.8 percent—or roughly five hundred excess fatalities. Nevertheless, deaths are only one of many measures we might choose to evaluate harm, and even then, what counts as a radiation fatality in the first place is not so clear and has changed over time. In 2005 the United Nations put the death toll at four thousand. And in 2010 newly released documents indicated that millions more were affected by the fallout and cleanup than originally thought, which in turn led to tens of thousands of deaths as well as hundreds of thousands of sick children born long after the initial meltdown.[43] To make matters more complex, the concrete sarcophagus entombing the reactor is now beginning to crack— a reminder that it is far too early to complete a history of Chernobyl and its aftermath. We will have to wait equally long to assess the fallout at Fukushima Daiichi, which is now long after the tsunami still posing new challenges to our conceptions of acceptable radioactive risk.

Humans won't be able to calculate nuclear risks as long as humans have nukes. Perhaps it is this very uncertainty that evokes particularly salient forms of nuclear unease. The emotive impulse that wells up in response to free radiation is a more visceral phenomenon than one bound to the shackles of calculation. Fossil-fuel executives should consider themselves lucky

Illustration 7: In the wake of Chernobyl Radioactive bumper cars lie silent in the abandoned city of Priypat near the Chernobyl reactor. (Photograph courtesy of Justin Stahlman)

that the arguably more dangerous fallout from fossil-fuel use, which kills tens of thousands of people year after year, has not elicited a corresponding fear in the minds of the citizenry. As a society, we begrudgingly tolerate the fossil fuel–related risks of poisoning, explosions, asthma, habitat destruction, and spills, which regularly spawn tangible harms. Yet when it comes to nuclear power we slide our heads back on our necks and purse our lips with added skepticism. Whether the degree of our collective skepticism toward nuclear power is appropriate, or even justified, doesn't really seem to matter. The public doesn't need experts to tell them when to be terrified.

As simple as fear, and as complex as fear, public angst has been the nagging bête noire of the nuclear industry. The relatively small leak at Three Mile Island provided ample motivation for the American public to yank back the reigns of nuclear power development for decades. The Fukushima meltdowns prompted nuclear cancellations across the globe. Could it happen again? Is it possible that taxpayers and investors could spend billions of dollars constructing a new generation of nuclear reactors just to have a hysterical public again shut the whole operation down following the next (inevitable) mishap? Absolutely. As taxpayers subsidizing the nuclear industry, we must worry not only about the risk of a hypothetical nuclear event with *tangible* consequences but also about an event with *imagined* consequences, especially if it should strike during a slow news week. Whether governments, taxpayers, politicians, and investors are willing to increasingly place these wagers will, more than technical feasibility, become the central nuclear issue in coming decades. Then again, some day we may find our choices on the matter to have dwindled. The more nuclear power we build today, the less choice we'll have about it tomorrow.

Either way, should environmentalists make it their job to promote nuclear power? Proponents argue that nuclear energy produces less carbon dioxide than coal or natural gas. But this might not matter in the contemporary American context. There is little precedent to assume that nuclear energy will necessarily displace appreciable numbers of coal plants. In fact, historically just the opposite has occurred. As subsidized nuclear power increased, electricity supply correspondingly increased, retail prices eased, and greater numbers of energy customers demanded more cheap energy—a demand that Americans ultimately met by building additional coal-fired power plants, not fewer.[44]

It didn't turn out that way everywhere. Take France or California, for instance. Residents of these regions enjoyed a different set of economic and legal rules that thwarted this perverse feedback loop.[45] We'll come back to them later.

Without first addressing the underlying social, economic, and political nature of our energy consumption, can we assume that nuclear power, or any alternative production mechanism, will automatically displace fossil-fuel use? Should the environmental movement address these underlying conditions before cheering on nuclear or alternative energy schemes? Should we perhaps view alternative energy as the dessert that follows a balanced meal? If so, we have plenty of succulently buttered vegetables to eat before moving on to the brilliantly frosted assortment of saccharine alternatives perched on the silver-plated platter before us.

5. The Hydrogen Zombie

Ignorance is the undead's strongest ally, knowledge their deadliest enemy. —Max Brooks, *The Zombie Survival Guide*

By the close of the first decade of the twenty-first century, the hydrogen dream might have seemed dead to any casual observer that happened to pass its rotting corpse on the side of the street. The financial foundations upon which the hydrogen economy stood had been reduced to a shadow. Numerous governments had slashed, yanked, and all but completely eliminated hydrogen funding. Corporations that hastily filled their pockets bringing hydrogen fuel cells to market eventually witnessed their balance sheets tumbling in flames just as quickly. Finally, after the crash and burn of the hydrogen economy, credit crises and financial upheavals swept away the smoldering ashes left behind. But soon after the fatality, something curious started to occur. Citizens beheld the *New York Times* dedicating a full-spread feature to the hydrogen economy and witnessed CBS News claiming that General Motors' new hydrogen fuel-cell car was "a terrific drive with almost no environmental impact."[1]

Long after the practical infrastructure for the hydrogen economy died, the hollow shell of the dream pressed on—a technological zombie.

Characterizing hydrogen as a zombie technology might seem a bit harsh for those enchanted by the idea of a hydrogen economy, but in fact, it has been called much worse by others—a pipedream, a hoax, or even a conspiracy. Nevertheless, these concepts are too blunt to carve out an intricate appreciation for the rise and fall of the hydrogen dream. A more nuanced rendering offers a peek into how diverse groups can coalesce around a technological ideal to offer it not only a life it would never have achieved otherwise but an enigmatic afterlife as well.[2]

The Hydrogen Economy in Sixty Seconds or Less

The idea of a hydrogen economy is based on two central components, hydrogen (the gas) and fuel cells (the contraptions that combine hydrogen and oxygen to create electricity). At the outset, it is important to correct the common misconception that hydrogen is an energy resource. Hydrogen is simply a carrier mechanism, like electricity, which energy firms must produce. Unlike sunlight, tides, wind, and fossil fuels, hydrogen gas does not exist freely on earth in any significant quantity. Processors must forcibly separate hydrogen from other molecules and then tightly contain the gas before distributing it for use. They most commonly derive hydrogen from natural gas (through steam hydrocarbon reforming) or less frequently from water (through electrolysis). Both processes are energy intensive; it always takes more energy to create hydrogen than can be retrieved from it later on. Hydrogen firms presumably won't be able to change this restriction without first changing the laws of thermodynamics and conservation of energy.

Historians generally credit Sir William Grove for devising the first fuel cell in 1839, although it was another fifty years before chemists Ludwig Mond and Charles Langer made them practical. The internal combustion engine revolution overshadowed early fuel cell research but proponents slowly coaxed the technology along. Eventually NASA and General Electric unveiled the first modern platinum fuel cell for the Gemini Space Project. In the 1970s the U.S. government, scrambling to respond to oil embargos, began working more closely with industry to advance fuel cell research. In the 1980s car manufacturers joined in.

The Early Years

By the early years of the twenty-first century, almost every automotive company had initiated a fuel cell program. At the 2006 Los Angeles Autoshow, then California Governor Arnold Schwarzenegger stood on a stage to christen fuel cell vehicles as "the cars of the future." Shortly after, BMW CEO Dr. Michael Ganal took to the podium and declared: "The day will come when we will generate hydrogen out of regenerative energies, and the day will come when we will power our cars by hydrogen. This means no exploitation of natural resources anymore; this means no pollution anymore. We know there is a far way to go, and the new BMW Hydrogen 7 is a big step towards the future."[3]

BMW's Hydrogen 7, along with its numerous American counterparts, such as the Chevrolet Equinox and Jeep Treo, might never have been built if not for one of George W. Bush's earliest projects, the National Energy Policy Development Group, headed by Dick Cheney and charged with identifying future energy markets. The group immediately locked in on hydrogen. It identified the elemental gas as the "future," dubiously referring to hydrogen as an "energy source" that produced but one "byproduct," water.[4] They mentioned few details about how hydrogen might be produced, beyond the claim that it could be created with renewable resources. Most shockingly, the report explicitly considered nuclear power and fossil fuels to be "renewable energy sources."[5]

This remarkably generous definition proved quite useful—especially when it came to enrolling supporters. The commission invited CEOs of British Petroleum, DaimlerChrysler, Ford, Exxon, Entergy Nuclear, the National Energy Technology Laboratory, Texaco, Quantum Technologies, and the World Resources Institute to help draft the boilerplate language describing hydrogen that would be adhered to by all constituents and subsequently copied and pasted into ad campaigns, PR initiatives, and annual reports, more or less word-for-word. In 2002 the Department of Energy (DOE) formalized the marching orders with two reports. One of the reports concluded that the government should treat dissenting views of hydrogen as "perceptions based on misinformation," which should be "corrected."[6] A subsequent report claimed, "The government role should be to utilize public resources to assist industry in implementing this massive transition and in educating the public about fuel cell vehicles' safety, reliability, cost and performance."[7] The DOE determined that the public reeducation campaigns were to start as early as grade school. The European Commission adopted corresponding language and education campaigns following the same script, which energy multinationals presumably transferred overseas from Washington.

It may not be immediately evident why traditional energy giants were so keen on hydrogen. But it may start to make sense when we consider the enormous quantity of energy required to pry hydrogen atoms from their molecular resting places. Hydrogen reformation can easily consume more fossil fuel than simply deploying natural gas and coal in their traditional dirty manners. Coordinators of California's "Hydrogen Highway" even admitted that their vehicles led to more particulate matter and greenhouse-gas emissions than gasoline-powered vehicles on a well-to-wheel basis.[8] Still, promoting hydrogen as a clean fuel that energy companies could create using "renewable energy sources" promised to offer particularly valuable environmental cover for dirty fossil-fuel operations. If only mainstream environmental organizations could be brought on board . . . but how?

Throughout the world, numerous geothermal sources, wind farms, and industrial processes emit excess energy that producers can capture, convert, and store in the form of hydrogen, at some cost. These overflows were, and are still today, far too rare and inadequate to produce appreciable sums of hydrogen— certainly not enough to run an economy on the stuff. But the concept was nevertheless alluring, even if it was far-fetched. Mainstream environmental organizations took the bait. Soon they were walking hand-in-hand with fossil-fuel giants to celebrate the hydrogen dream together.[9] So long as the public associated hydrogen with windmills, it didn't really matter how much production occurred behind the scenes using natural gas or coal reformation. The trope was complete, almost.

Automotive companies danced with fuel cells in the public spotlight, world governments got to task reeducating the public, environmentalists cashed in their principles for power at a suspect rate of exchange, and fossil-fuel companies stood guard over the whole operation. But another industry waged its bets less conspicuously. In fact, insiders sometimes characterize it as the scowling director of the hydrogen ballet—a cane in one hand and a cigarette in the other, winkled inconspicuously into an offstage seat overlooking the scene. The nuclear power industry.

The nuclear industry's ambitions corresponded to those of the fossil-fuel industries, save for one small twist. Given the limitations of solar and wind power, the nuclear industry expected to position itself as the only "clean" solution for manufacturing hydrogen on a *large scale*. A new nuclear plant had not been built in the United States for decades. Hydrogen provided a convenient opportunity to reposition nuclear power as an environmentally progressive undertaking for the nation to pursue—one that would eventually end up costing taxpayers dearly.

Many years before BMW unveiled the Hydrogen 7 at the Los Angeles Autoshow, the nuclear industry lobbied Congress to direct a sizable chunk of the energy budget toward a Next Generation Nuclear Plant (NGNP). Like any nuclear power plant, the NGNP would generate electricity. It would also produce hydrogen. The nuclear industry had a persuasive proponent, Congressman Darrell Issa, a powerful energy productivist and Congress's wealthiest member, worth about $250 million according to the Center for Responsive Politics.[10] During a 2006 congressional hearing before the House Subcommittee on Energy and Resources, Representative Issa could not have been more explicit about the link between the nuclear industry and the hydrogen dream. He stated that the Next Generation Nuclear Plant project was "a key component in the Administration's plans to develop the 'hydrogen economy' because an associated purpose of the advanced demonstration plant is to produce hydrogen on a large scale."[11] If anyone had been uncertain of the nuclear industry's interest in the hydrogen wager, Issa put those concerns to rest. The nuclear industry was in.

In addition to the nuclear industry, car companies, California politicians, environmental groups, and fossil-fuel giants, there were plenty of others excited to get in on the action. Academic researchers, scientists, journalists, and of course the fuel cell manufactures themselves all had something to gain from the hydrogen promise. Had hydrogen interests not gravitated to form this emergent superstar, the powerful ideal behind this tiny molecule, the hydrogen dream itself, might never have been. It's worth mentioning that this particular alignment of interests looks a lot like the unorganized gravitations that stabilize solar, wind, and biofuel technologies. However, unlike the strong bonds holding together the larger dream of an alternative energy future, the ties binding the hydrogen dream eventually began to loosen.

The Undoing

A growing minority of energy analysts started discounting hydrogen as nothing but hype. Others characterized it as an outright hoax. Hydrogen critics argued that any meaningful hydrogen economy would have required not just a couple of large breakthroughs but rather numerous breakthroughs, each monumental in their own right. Their attacks came from multiple fronts, covering hydrogen production and transport as well as its eventual use in fuel cells.

Once hydrogen is created, critics maintained, the challenges to employ it as a fuel multiply. First, hydrogen must be contained, either as a supercooled liquid below -253°C or as compressed gas. Both processes are energy intensive. High-pressure pumps consume about 20 percent of the energy in the hydrogen for compression, while liquefaction wastes 40 percent of the embodied energy.[12] Maintaining hydrogen in its liquid form requires specially insulated vessels and massive refrigeration power. For instance, BMW's car of the future stored cryogenic liquid hydrogen in its tank. While parked, the supercooled liquid warmed inside the car's thirty-gallon tank. Internal sensors allowed the explosive gas to build to a maximum pressure of 5.5 bar, at which time (assuming everything worked correctly) the hydrogen would overflow through a pressure valve, combine with oxygen, and drip onto the ground as water. As long as the car was parked, the process would continue until the tank was empty. Comedian and automobile enthusiast Jay Leno drove the car around Burbank, California, during a celebrity test drive, later joking: "So a guy drives into a bar in a BMW Hydrogen 7 and the bartender says, 'What do you want?' The guy answers, 'I just came in to take a leak.'" Reporter Matthew Phenix quipped in a *WIRED* magazine report that the derisory Hydrogen 7 was "saving the world, one P.R. stunt at a time."[13] Not a particularly glowing endorsement from perhaps the nation's most techno-friendly news source.

In other concept vehicles, on-board supercoolers held hydrogen in its liquid form, but these units had to be powered 24-7. Another option involved blocks of solid metal hydrides—essentially giant heavy sponges capable of soaking up hydrogen—but even after years of development such schemes proved about as clumsy as one might imagine. The most promising concept vehicles stored their hydrogen as a compressed gas, but the tanks were ponderously heavy unless crafted from expensive carbon fiber, which is still prone to exploding in a crash.[14] Even though many researchers believed these tanks would be as safe as natural gas or gasoline storage, convincing the public would be the larger hurdle. Might drivers feel hesitant to zoom down the freeway atop pressurized tanks of a hot-tempered gas that has an atomic bomb named after it and brings to mind an exploding Zeppelin?[15]

Critics also attacked the proposed hydrogen transportation and filling-station infrastructure. There's the obvious chicken-and-egg problem: Why would car companies produce hydrogen vehicles without hydrogen filling stations and why spend billions for an infrastructure without an existing market of hydrogen-vehicle owners waiting to fill up? True, legislative tools could have spurred construction of a national network, as was begun in California, but it would have cost half a trillion dollars in the United States alone, according to a Department of Energy chief.[16]

Even then, distributing the gas to filling stations would have been grueling. A tanker truck can carry enough gasoline for eight hundred cars but can only hold enough hydrogen for eighty.[17] The requisite back-and-forth trips would have consumed enough diesel to offset 11 percent of the hydrogen's energy after just a 150-mile jaunt, according to one critic.[18] Distributors could have shortened the trips by building pipelines, but hydrogen requires novel pipeline technologies because it embrittles metal pipeline walls, couplings, and valves. Moreover, hydrogen molecules are excellent escape artists; they are tiny enough to squeeze their way through the narrow molecular-level gaps within solid materials. As a result, hydrogen can seep right through the walls of a solid pipe, a serious consideration given the large surface area of a distribution pipeline.[19]

Reforming natural gas into hydrogen right at filling stations could have bypassed some distribution concerns, but small-scale reformers are expensive and inefficient, especially if they must store the carbon dioxide exhaust. Hydrogen hecklers argued that the nation's drivers would be better off simply pumping natural gas directly into their vehicles rather than going through the trouble of reforming it into hydrogen, which would contain less energy, yield the same greenhouse gases, occupy three times the volume, and limit use to fuel-cell vehicles that were incredibly expensive.

However, hydrogen proponents proposed a cleaner way to secure hydrogen: electrolysis, a process wherein electricity separates water into hydrogen and oxygen. Environmentalists envisioned that wind turbines and solar panels would power electrolyzers. Meanwhile, fossil fuel and nuclear industry executives knew they didn't have to worry about solar or wind taking over anytime soon. In 1994 researchers erected a solar-powered hydrogen station, called Sunline, outside of Palm Springs, California, but it took ten hours of solar electrolysis to produce just one kilogram of hydrogen, the energy equivalent of about one gallon of gasoline. Hooked to the grid and drawing power from nearby wind turbines, the station could have theoretically produced up to sixteen kilograms of hydrogen per day, "assuming optimal season conditions," according to Sunline's own calculations.[20] Even if there was enough excess solar and wind power available in the American grid to electrolyze hydrogen, the costs of the massive sums of raw electrical power along with the requisite transformers, electrolyzers, explosion-proof compressor pumps, cryogenic refrigerators, and insurance would have been "economic insanity," claimed critic, Robert Zubrin. In a critical exposé, he remonstrated that for an ongoing investment of "$6,000 per day, plus insurance costs, you could make $200, provided you can find fifty customers every day willing to pay triple the going price for automobile fuel." He continued, "I don't know about you, but if I were running a 7-11, I'd find something else to sell."[21]

Nevertheless, another concern was eclipsing the difficulties of creating the hydrogen fuel itself. Critics narrowed in on the high cost and durability of fuel cells, the electro-chemical devices that could combine hydrogen and oxygen to create usable electricity for cars, buildings, laptops, and other devices. There was no doubt that when coupled with an electric motor in a vehicle, fuel cells were more efficient than internal combustion engines. However, critics pointed out that, as designed, the fuel cells only had an operational life of about thirty thousand miles and therefore would have to be replaced more frequently than a car's brake pads—and the fuel cells weren't cheap.[22]

Despite the billions spent to commercialize fuel cells, they remained unaffordable partly due to the high cost of platinum, a catalyst sparingly applied to fuel-cell membranes in layers just a few atoms thick. Even at these reduced concentrations, economists warned that large-scale fuel-cell production could spark platinum price bubbles, tilting the overall scheme into an uphill economic challenge unless manufacturers could identify a cheap platinum substitute.[23]

In short, critics argued that automotive fuel cells had proven to be extraordinarily expensive, finicky in bad weather, short-lived, and prone to molecular clogging, which dramatically reduced their efficiency. Joseph Romm, former director of the U.S. Department of Energy's Office of Energy Efficiency and Renewable Energy, observed, "If the actions of Saddam Hussein and Osama Bin Laden and record levels of oil imports couldn't induce lawmakers, automakers, and the general public to embrace existing vehicle energy-efficiency technologies that will actually pay for themselves in fuel savings, I cannot imagine what fearful events must happen before the nation will be motivated to embrace hydrogen fuel cell vehicles, which will cost much more to buy, cost much more to fuel, and require massive government subsidies to pay for the infrastructure."[24]

Hydrogen's future wavered.

The Fall

It began just like any other bubble. By the early years of the twenty-first century, the costs of commercially viable fuel cells had triumphantly dropped from the tens of millions of dollars per unit in the 1960s to below $100,000. Stocks of fuel-cell and hydrogen-related component manufacturers, such as Ballard Power Systems and Millennium Cell, sprang to all-time highs. And in 2006 *Popular Mechanics* magazine predicted that fuel cells could cost as little as $36,000 if mass produced.[25] That year, a company called Smart Fuel Cell hit the stock market commanding an impressive market capitalization trading at $150 per share. But the banana peel was already laid out.

Platinum prices were rising. In the early years of the century, spot prices for platinum doubled. Even as pundits proclaimed that the rare metal was overpriced, it doubled again by 2008. It wasn't just platinum prices that were making investors jittery; traders took note that even though fuel-cell firms were not burning oil, they were quickly burning though cash. "The bread-and-butter profits we need to see are years away. It's not even a niche market yet," observed John Webster, coauthor of a Pricewaterhouse-Coopers investigative study on fuel cells.[26] Esteemed industry analyst David Redstone pointed out that even though Ballard Power Systems had "a great public relations machine," and politicians were "interested in fuel cells," the industry as a whole had overpromised. "There is not a stream of commercial revenue. There are not products. Overpromising and underperforming leads to investor disappointment," claimed Redstone.[27] Investors fled.

A year after Smart Fuel Cell's issue, the stock had dropped from $150 to $50, and by 2008 it was trading below $15. Ballard Power Systems, which had been trading at over $100 per share, plummeted to $4. Investors slashed the market capitalization of Millennium Cell in half by 2002, then again by 2004, then again by 2006—the same shares that attracted investors at $25 in 2000 were having a difficult time finding support at the five-cent level in 2009. By 2011, keepsake investors could buy twenty-five shares for less than a penny.

The smart money left, and so did the politicians. Posthaste. Originally, Schwarzenegger had forecasted a "Hydrogen Highway" with some two hundred filling stations by 2010, but by 2009 the state had completed only a couple dozen and the project had stalled even before he left office in 2011. Governor Schwarzenegger dropped the hydrogen dream as quickly as he had picked it up. So did President G. W. Bush. After becoming a central feature to his energy plan in 2003, he didn't even utter the word "hydrogen" during his State of the Union address in 2007, or at any time thereafter. Finally, late one evening in 2008, Bush slipped into hydrogen's bedroom and slid a dagger into the movement's frail heart by quietly pulling funding for a FutureGen coal-to-hydrogen production facility, which proponents considered a key element in realizing their dream.[28] During his first months in office, president Obama marched on to the scene, grabbed the dagger, and twisted. He finished the kill by proposing to eliminate the remaining $100 million of funding from the federal government's hydrogen fuel-cell venture with carmakers.[29] At the wake, Obama's energy secretary, Steven Chu, stood up to say a few words. "We asked ourselves, 'Is it likely in the next 10 or 15, 20 years that we will convert to a hydrogen car economy?' The answer, we felt, was 'No.'"[30] Chu renamed the hydrogen fuel-cell group and recommended reorienting remaining fuel-cell research away from vehicles and toward a few low-prestige applications such as building power backup and battery replacement.

Betrayed, broke, stabbed in the back, and finished off—the hydrogen economy was most certainly dead. But it was still moving.

The Undead

After yielding the stage to the hydrogen skeptics, it's difficult to imagine the "hydrogen economy" as anything more than a smokescreen designed so that political and corporate elites might dazzle us as they shuffled energy subsidies behind their backs. While there may indeed be a good bit of shuffling going on in Washington, the lesson behind hydrogen, as with many other energy technologies, is far more nuanced than the existing assemblage of pyrotechnic literature on the subject might indicate.

The high-flying hopes and dreams for a hydrogen economy encountered severe turbulence around 2006—not a good sign for an industry premised on hopes and dreams. Early on, investors determined that automotive fuel cells were nothing more than glorified science-fair experiments, hardly a reasonable basis for alleviating smog, CO_2 emissions, conflicts, and costs associated with the nation's ever-raging dependence on fossil fuels. Steven Chu, the Nobel Prize–winning energy secretary, was similarly put out. Nevertheless, well-established hydrogen promoters continued to pawn off these alleged snake oil liniments on the public. It wasn't just environmental groups, carmakers, and mainstream energy companies, but also political representatives through many levels of state and federal government and, for a time, all the way to the Oval Office.

Critics claim that we spent billions of our hard-earned dollars and all we have to show for it are a few hydrogen vehicle "screwvenirs." Zubrin concluded, "The hydrogen economy makes no sense whatsoever. Its fundamental premise is at variance with the most basic laws of physics. The people who have foisted this hoax on the American political class are charlatans, and they have done the nation an immense disservice."[31] Many hydrogen detractors conclude that the hoax was intentional and that the truth was somehow kept secret. But the formal and informal coordination between regulators, politicians, scientists, environmentalists, and corporations did not present the possibility for an outright conspiracy of the sort plotted by suspense novelists. Only the most tightly controlled organizations can hold a slippery secret in their grasp without it being leaked. Had such a diversity of people been in on a hydrogen ruse, someone would have eventually squealed. Yet if nobody manufactured a hoax, then how was the effect of a hoax created?

This question becomes even more puzzling. Even though Wall Street had handily dismembered the hydrogen industry, the reversal in fortunes didn't faze the public, scientific, or media enthusiasm surrounding hydrogen in subsequent years. Numerous government and university research budgets and disbursements, which had been preplanned in years prior, were still flowing. The nuclear establishment, hardly prepared to loose face, nervously kept its hydrogen sights on autopilot. Car company PR and advertising departments still found it useful to tout their fuel-cell concept cars to a public that was apparently unable to recognize the promotions as outdated greenwashing. Journalists were evidently no savvier; with prohydrogen press releases still streaming into its news office, the *New York Times* published a prohydrogen feature in 2009, in which it embarrassingly cheered on an industry and its associated product lines that had essentially been bankrupted years previously. The *Times* was not alone.

Even though the hydrogen economy had died, it was still busy posing for photo shoots, presenting at environmental conferences, speaking for the automotive industry, booking international trips, and eating at fine restaurants. It had even orchestrated a coup d'état in Congress to partially reinstate its funding. The hydrogen economy was not dead, but undead.

How was this possible?

Some might claim the hydrogen economy was never really alive to begin with; it surely never existed in any tangible way. Few people had ever seen a hydrogen vehicle, let alone driven one. The hydrogen economy was nothing more, and nothing less, than a dream—a damn good one. It allowed people the luxury of imagining a world of abundant energy, a clean utopia where the only pollution would be water vapor, with enough credible science mixed in to make the whole affair seem plausible.

It isn't the first time that critical environmental inquiry has been displaced by such utopian lobotomies. Since the 1970s, environmentalism in America and Europe has gravitated toward the theme of "ecological modernization"—the idea that the treadmill of technological progress will solve all environmental troubles. Martin Hultman, a researcher at Sweden's Linköping University, compares the utopian visions surrounding the hydrogen economy with the ones once envisioned by proponents of nuclear power. "They are similar in that they both invoke the dream of controlling a virtual *perpetuum mobile*, propose an expert-lay knowledge gap, downplay any risks involved, and rely on a public relations campaign to ensure the public's collaboration with companies and politicians," asserts Hultman. "The idea that the *level* of energy use is unimportant and not connected to environmental problems is constructed by describing fuel cells as intrinsically clean in themselves and producing only water as exhaust."[32]

Others might say the hydrogen economy never died. After all, technologies are more than just physical artifacts—the gears, the batteries, the circuit boards. Technologies are a hybrid of intentions, interests, promises, and pretensions. Technologies are stories. If they weren't, they'd never catch on. The story of the hydrogen perpetual motion machine could not have been formed and fueled by just any single interest group, any single conspirator, or any single hoaxer as it were. As with solar, wind, and biofuel technologies, the hydrogen dream arose from a complex alignment of interests coalescing to synchronize a future narrative—one that featured selected benefits and diminished or overlooked associated side effects and limitations.

Elected officials, many of whom had worked in the energy sector and were tacitly imbued with its productivist cant, stood to gain both donors and constituents by supporting clean hydrogen.[33]

Gas, coal, and nuclear industrial elites knew there was money to be made and valuable cover to be gained by articulating clean hydrogen visions.

Researchers knew their work would be funded if it was framed as a national priority.

Environmentalists could feel good about their work, while gaining public and financial support by pledging allegiance to the clean fuel of the future.

Automotive manufacturers saw opportunities for subsidies, profits, and most of all a clean public relations cloak, offering protection from those who saw their industry as socially and environmentally destructive.

And the greater public, primed with the verve of ecological modernization, was willing, perhaps even eager, to be convinced that hydrogen was, in fact, the future of energy.

It is perhaps too early to write a history of the hydrogen economy (though the Smithsonian's National Museum of American History has begun just that). Lesser stories have created much more.[34] The cluster of technologies surrounding the hydrogen dream is sure to be resurrected by various interests. They will push to develop more economically viable hydrogen systems, especially for niche applications such as power backup, battery replacement, and the like. Nuclear hydrogen will likely reach a hand up from the grave at some point, but critics are certain to shackle it to their litany of limitations. For our purposes today, however, this technological zombie hardly presents a passable response to our energy problems. Its foul smell alone indicates it has no place in mannerly conversation on the subject.

In any case, there is a more chilling issue to address. It appears that there's more than one zombie in our midst.

6. Conjuring Clean Coal

Call it a lie, if you like, but a lie is a sort of myth
and a myth is a sort of truth. –Edmond Rostand,
Cyrano de Bergerac

The first major earthquake recorded in Australian history rocked residents of Newcastle on December 28, 1989. Ten years later, on the other side of the planet, an earthquake hit Saarland, Germany. Separated by time and space, these anomalous quakes might have seemed completely unrelated. They weren't.

Geophysicists have controversially identified a common trigger: coal mining. They point out that coal-mining operations can collapse land surfaces, divert waterways, and drain wetlands. Generations of mining can induce quakes that are now compromising previously seismically stable regions throughout the world. Newcastle's earthquake led to deaths, injuries, and $3.5 billion in damage, more than the value of all of the coal ever extracted from the region.[1] Nevertheless, earthquakes may rank among the lesser concerns of mining, processing, and burning this fuel.

Coal in Sixty Seconds or Less

Archeologists trace coal use back to China and Roman Britain during the Bronze Age, about three thousand to four thousand years ago. Today, nations burn coal to generate electricity, distill biofuels, heat buildings, smelt metals, and refine cement. Half of America's electricity comes from coal, along with 70 percent of electricity in India and 80 percent in China.[2] Coal is more widely available throughout the world than hydropower, oil, and gas. It's got a lower sticker price too. Understandably, coal attracts world leaders concerned about regional energy security. For these geographic, economic, and political reasons, it is problematic to assume that countries will willingly stop unearthing their coal reserves unless cheaper and equally secure alternatives arise. Even in rich Australia, a federal minister quipped, "The coal industry produces 80 percent of our energy and the reality is that Australia will continue to rely on fossil fuels for the bulk of its expanding power requirements, for as long as the reserves last."[3] In fact, the International Energy Agency (IEA) expects the coal sector's growth to outpace all other sectors, including nuclear, oil, natural gas, and renewables.

The problem, of course, is that despite its many benefits, coal is still dirty—in so many ways. Burning coal releases more greenhouse gases than any other fossil fuel per unit of resulting energy; it yields more than two times the CO_2 of natural gas. Coal features infamously in dialogs involving international ethics, workers' rights, and community impacts more broadly. Here's the critics' short list:

- *Air pollution*: The sky's vast quantity of visible stars reportedly shocked Beijing residents during a coal-burning ban in preparation for the Olympic Games in 2008. During the late 1800s and early 1900s, London, New York, and other industrialized cities were known for their characteristic coal smog that killed thousands of people. Today, air quality in these cities is much improved due largely to better coal-burning practices. However, combustion is still the primary source of heavy metals, sulfur dioxide, nitrogen oxides, particulates, and low-level ionizing radiation associated with coal use. Toxic emissions from mining and transportation are also significant.[4]

- *Water contamination*: Mining, transporting, storing, and burning coal also pollutes water aquifers, lakes, rivers, and oceans. Coal-washing facilities alone eject tens of millions of tons of waste into water supplies every year.[5]
- *Land degradation*: Above-ground coal mining destroys prairies, levels forests, and lops off mountain peaks.
- *Fly-ash waste*: Coal plants generally capture fly ash, a byproduct of coal combustion, which often ends up in unlined landfills, allowing toxins to leach out or blow away.[6]
- *Occupational risks*: Poisonous gases, tunnel collapses, flooding, and explosions kill thousands of coal miners every year. Tens of thousands are seriously injured and exposed to long-term respiratory hazards including radioactive fumes.
- *Community health risks*: One prominent study published in *Science* reviewed the widespread practice of mountaintop removal coal mining in the United States and found that local residents also suffer from unusually high rates of chronic pulmonary disorders, hypertension, lung cancer, chronic heart disease, and kidney disease.[7] The report's lead author, Margaret Palmer, of the University of Maryland, stated, "Scientists are not usually that comfortable coming out with policy recommendations, but this time the results were overwhelming . . . the only conclusion that one can reach is that mountaintop mining needs to be stopped."[8]

This list contains but a fraction of the concerns that critics voice regarding coal. We might think we should use a lot less of the stuff—unless, that is, we start to believe it too can be clean.

Ladies and Gentlemen, Look at My Right Hand

In recent years, the coal industry has applauded itself for introducing smokestack "scrubbers" into many of its plants, which spray water and chemicals directly into exhaust fumes to filter out contaminants. Less celebrated is the story of the resulting sludge. As a matter of general practice, it's simply dumped into nearby lakes, rivers, and streams—the same waterways that supply drinking water to the general public—right where much of it would have ended up if it hadn't been scrubbed out in the first place.[9] Fierce lobbyists shield this dumping from regulation.[10] The Clean Water Act limits some pollutants but not the most dangerous ones, such as arsenic and lead, which coal-fired power plants emit.[11] Even then, plants routinely violate provisions of the Clean Water Act, according to a *New York Times* investigative report.[12] Of the over three hundred coal-fired power plants that have violated regulations since 2004, 90 percent haven't faced a single fine or sanction for doing so. After a plant in Masontown, Pennsylvania, violated the act thirty-three times in the three years between 2006 and 2009, it was fined a total of just $26,000 during a period when the parent company reported $1.1 billion in profits. The Environmental Protection Agency (EPA) has attempted to institute stricter controls, but regulators must swim upstream in a current flowing with tens of millions of coal industry dollars.

Cleaner options exist, but these too come with risks of their own. Wastewater facilities can extract many of the toxins and solids from scrubber waste and place them into landfills. These synthetically lined tombs can cover over a hundred acres each. But liners are not foolproof; they can burst or leak over time. According to a recent EPA report, residents living near some leaky landfills face cancer risks that exceed federal health standards by a factor of two thousand.[13]

Of coal's almost countless effluents and side effects, the industry has notably reduced one: sulfur dioxide emissions. Some coal deposits naturally contain less sulfur, which converts to sulfur dioxide when burned, and modern coal facilities can remove most of the remaining sulfur dioxide from combustion fumes. Giving themselves a brisk pat on the back, coal promoters deployed public relations teams in the late 1990s to advertise their achievement in attempts to convince others that the reduction of this one pollutant could speak for the cleanliness of the entire industry. They correspondingly felt justified in christening their strategy "clean coal."

Anyone would agree that preventing sulfur dioxide from entering the atmosphere is a good thing. However, using one of the dirtiest and most destructive practices on earth as a benchmark to judge a fuel "clean," and then only reducing one of its many side effects, could certainly be interpreted as less than genuine. It's like claiming to have done the dishes after just washing one fork.

Coal advocates had a big job ahead of them but they were well equipped. To start, the industry courted journalists to assure that coal's self-appointed green credentials could achieve a level of credibility. They also began advertising. Unlike tobacco ads, which legislators have strictly limited due to public health concerns, advertisements for burning coal run freely. They'd probably run more frequently if we weren't already addicted to the stuff. Public relations firms design these big-blue-sky ads for clean coal to leave us with a sense of "Huh, I guess that stuff isn't so bad after all." They are pervasive, persistent, and often successful.

Coal companies also deliver loads of money to political campaigns and rally their employees to support candidates who support coal. For this, they are greatly rewarded. The already very profitable coal industry receives tens of billions of dollars every year in government subsidies throughout the world. Why subsidize such a rich and dirty industry? Professor Mark Diesendorf at the University of New South Wales contends that politicians and governments, eager to stay in office, see opportunities in appealing to well-organized interest groups whose activities spread costs over many less-organized and less-informed constituents.[14] During the 2008 U.S. presidential election, all of the major candidates supported coal use by evoking "clean coal" technology. Incidentally, not one of them suggested that the nation could instead, or conjointly, aim to cut coal consumption to European levels by wasting less electricity.

Coal promoters now extend the term "clean coal" to include carbon dioxide sequestration, or more accurately, the *mere promise* that carbon sequestration could be employed on a large scale. It turns out this is quite a large promise. It centers on capturing carbon dioxide from coal plant exhaust and storing it underground, where proponents claim it will dissolve and eventually, after millennia, react with other elements in surrounding rocks to form harmless mineral deposits. If this sounds kind of dreamy, that's because it still is. Proponents have shown that some underground carbon mitigation techniques can work on a small scale, but it's less clear that they could be ramped up to generate appreciable effects on climate change. Many scientists hotly contest the very plausibility of such schemes. Beyond the cost, which both proponents and detractors agree would be extremely high, there are several other hurdles to jump before carbon sequestration could become a practical and widespread option for dealing with excess CO_2. First, fossil-fuel plants must capture CO_2 either from flue gases or during the chemical process employed in "integrated gasification combined cycle" (IGCC) facilities. Capturing and compressing carbon dioxide from flue gases requires so much extra energy and equipment that doing so adds 60 percent to the cost of the resulting electricity.[15] IGCC carbon capture is easier, but these plants can only use certain types of coal; modifying one to run on lower-quality coal, such as Texas lignite, would add 37 percent to the cost of the plant and reduce efficiency by 24 percent.[16]

Second, after the industry captures and compresses the carbon dioxide, it must store the liquid or gaseous CO_2. Ubiquitous saline aquifers are one storage option, but these are prone to seismic instability and uncertainties of storage life.[17] Depleted oil and gas fields are obvious storage sites, but many of these deep underground crypts are structurally compromised after having been drained of their pressurized oil or inundated by multiple well piercings.[18] If the U.S. coal industry captured and liquefied just 60 percent of its annual CO_2 emissions, the effluent's volume would equal the volume of oil that Americans consume over the same period.[19] Geologists will be hard pressed to locate favorable storage sites on such a monumental scale. This may force the industry to risk even less secure formations.

This brings us to a third concern. Since economic factors will likely favor large stores over smaller ones, we must likewise consider the risk of large CO_2 releases. A 1986 tragedy in the African Republic of Cameroon, where CO_2 bubbled up from a volcanic crater and killed 1,800 people, portends what might ensue from such a release.[20] Since carbon dioxide is heavier than air, the gas quickly formed a thick blanket over the landscape, thirty miles in diameter. Rescue crews arrived to complete silence— no laughter of children, no bird songs; thousands of dead cattle did not attract the buzz of even a single surviving fly. If ever realized, geosequestration sites will be prone to slow leaks, abrupt escapes, sabotage, and attacks.[21]

Ecosystem health is a fourth concern of pumping carbon dioxide where it doesn't naturally occur. Mixed with water, CO_2 partially dissolves to form carbonic acid. Excess soil and waterway acidification can harm microorganisms and, in turn, the species that rely on them for survival (including us). The planet's oceans currently absorb about a million tons of carbon dioxide per hour, a third of the rate at which we produce the gas. Presently, this absorption is slowing climate change but is making naturally alkaline seawater more acidic. Ocean acidity endangers the shells and skeletons of sea life, in much the same way that carbonated drinks can soften and dissolve tooth enamel. Following current trends, parts of the oceans will become too acidic to host much of the life that lives there today.[22] *The Economist* considers the risk forbidding enough to warn, "No corals, no sea urchins and no who-knows-what-else would be bad news indeed for the sea. Those who blithely factor oceanic uptake into the equations of what people can get away with when it comes to greenhouse-gas pollution should, perhaps, have second thoughts."[23]

Scientists and legislators may ultimately decide that the risks and costs of carbon sequestration are worth it to reduce carbon dioxide buildup in the atmosphere. Even then, optimists suggest that carbon sequestration technologies won't be ready for mainstream deployment for at least another twenty years. A lot of coal will have been burned by then.

Assuming that nations could muster the political will, technologies, and funding to develop carbon capture and storage, how effective would it be? A study group in Australia, one of the largest coal-producing nations, set out to answer this question. Their findings are humbling. They determined that the cumulative CO_2 emissions reduction over the first thirty years of a sequestration program would be just 2.4 percent—not terribly impressive given the costs and risks that such an undertaking would involve.[24]

Why would the impact be so small? The decades-long wait for commercial viability is just part of the problem. The larger problem is regulatory. Coal firms aren't proposing to close down heavily polluting plants in order to build ones that capture carbon dioxide, but rather to *add* new plants to existing capacity.[25] In fact, the industry is using the very *promise* of carbon capture and sequestration to deflect calls to clean up their industry. For instance, Arch Coal chief Steven Leer touted carbon capture and sequestration over regulation in a recent *St. Louis Post-Dispatch* interview. When the journalist pointed out that

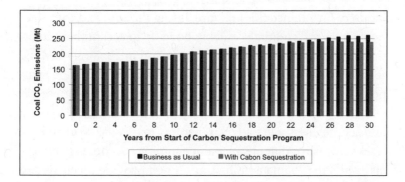

Figure 7: Clean coal's lackluster potential Even if aggressively deployed, carbon sequestration would have little impact on total emissions over the coming decades, according to an Australian study. The taller bars display CO_2 emissions (in metric tons) of a business-as-usual trajectory for coal-intensive Australia. The slightly shorter bars indicate CO_2 emissions assuming an aggressive carbon sequestration rollout. (Data from the Australia Institute)

these processes are twenty years away from becoming a reality, Leer conceded, "Probably." He swiftly translated this twenty-year lag into something much less foreboding by dismissively stating, "Twenty years in the energy world is right now. If you think about the infrastructure needed for carbon capture and sequestration, twenty years is a very short period of time."[26] Yes, a short period of time considering the obstacles, but nevertheless a very long time to wait for a modest improvement in CO_2 releases, especially, as we shall explore later, when there are far better options that we can deploy much sooner.

If the process of carbon capture and storage will require decades of research, billions of dollars, risky uncertainties, and meager paybacks, why has it become such a central focus of energy policy today? One needn't follow the path of money too far to discover the answer.

During the 2008 U.S. election cycle, the coal industry increased its budget for the National Mining Association, an industry lobbying group, by 20 percent, to $19.7 million.[27] The industry maintained this pressure during the 2010 and 2012 election cycles.[28] Candidates quickly lined up to accept their donations. Barack Obama's election campaign in 2008 released the following pronouncement: "Carbon capture and storage technologies hold enormous potential to reduce our greenhouse gas emissions as we power our economy with domestically produced and secure energy. As a U.S. senator, Obama has worked tirelessly to ensure that clean coal technology becomes commercialized. An Obama administration will provide incentives to accelerate private-sector investment in commercial-scale zero-carbon coal facilities."[29] And their specific recommendation? Out of the thousands of American coal plants, they proposed to convert just five to ten plants over to capturing carbon. It is hardly worth calculating the effects of such a plan even if it were to be wildly successful. It could be better understood as a pilot project. In 2010 the administration initially set a target completion date of 2016 but later shifted to announce a verdict on carbon sequestration that might be best described as wishy-washy: "While there are no insurmountable technological, legal, institutional, regulatory or other barriers that prevent ccs [carbon capture and sequestration] from playing a role in reducing [greenhouse gas] emissions, early ccs projects face economic challenges related to climate policy uncertainty, first-of-a-kind technology risks, and the current high cost of ccs relative to other technologies."[30] Not a glowing prediction but perhaps about the best the industry could have hoped for. Nevertheless, Obama's Interior Secretary Ken Salazar opened massive stretches of public land to coal mining corporations in 2011, and Obama continued to support coal during his 2012 reelection bid. Not everyone, however, was as friendly with the coal industry as the president.

Community groups across the country stood up to protest coal power plant construction. In Kansas, the state legislature blocked two coal-fired power plants from being built, an obstruction that residents overwhelmingly supported. In retaliation, the industry launched an initiative to reeducate the public in an effort to provide cover for coal-supportive politicians. Coal and utility companies spent $35 million to fund a public relations organization to support coal-based electrical generation, combat legislation designed to reduce climate-changing emissions, and portray their industries as more responsible and concerned about environmental problems. The organization, Americans for Balanced Energy Choices, ran an advertisement claiming coal is "70 percent cleaner" than in previous eras. That's somewhat true. But it failed to acknowledge that the coal industry didn't initiate the transformation. In reality, tighter federal regulations forced the clean-up—regulations that the industry vehemently *fought*.[31]

Cleaner and more efficient coal plants can and should be part of our future. While they are more expensive than older designs, they are far cheaper than carbon sequestration and can be deployed more quickly. Just don't expect the industry to take the initiative.

As for carbon sequestration? Some day it may indeed become a realistic option or perhaps even a reality. After all, most technologies we take for granted today started out as promises. But even if carbon sequestration does prove successful, will it make coal clean? What about resource limits, air pollution, water use, earthquakes, injuries, deaths, land degradation, and other negative side effects due to coal extraction and combustion practices? Keep in mind that the energy-intensive carbon sequestration process itself requires *additional* coal. The concept of clean coal directs our attention to just a couple of the many concerns about coal use. It is the promise itself that lends the term power. Clean coal ends up being a rhetorical cleaning more than a physical one.

The Rhetorical Cleaning of Coal

When I stand up to provide a critique of solar cells, wind turbines, and other alternative-energy technologies in front of student groups, foundation boards, philanthropists, and others, their keen senses occasionally signal that I might secretly work for the fossil-fuel industry. They are right to be wary, yet this chapter should help put those suspicions to rest. In reality, the alternative-energy project might be one of the best things ever to have fallen into the fossil-fuel industry's lap. By drawing upon the symbolic power of alternative-energy schemes, these industries can paint their operations as clean and civically engaged. But more significantly, the promise of alternative energy entices concerned citizens to overlook extreme consumption patterns and instead frame energy problems as a lack of clean energy *production*. And as we have witnessed, fossil-fuel giants are all too eager to fill the outstanding order.

In the end, the question isn't about whether or not clean coal exists. The more pressing question goes deeper: Why do so many Americans believe clean coal *could* exist?

The term "clean coal" was likely first uttered in a marketing boardroom. From there the coal industry's public relation teams refined and expanded the concept, transforming it into an option. Media pundits argued for or against its existence, lending the term legitimacy. Politicians ran on the convenient platform and were obliged to direct funding to technologists who pursued it. The concept of clean coal sat upon a carefully manicured set of definitions, promises, and possibilities. It became a slick technological ideal that could morph over time to evade hostile advances. It represented different things to different people, all while maintaining a common façade by name—a façade that hid injustices that might have otherwise been exposed and addressed.

If citizens could be drawn to believe in one part of the story, then their partial beliefs could stand in for the whole concept. They could believe in clean coal. Perhaps this could only have occurred in a society that prioritized quick fixes and was primed to apply technological solutions to any given problem. No doubt, this is the path that countless successful inventions have followed. It's just that humankind's most meaningful undertakings are not, at their core, marketing illusions.

7. Hydropower, Hybrids, and Other Hydras

The door to novelty is always slightly ajar: many pass it by with barely a glance, some peek inside but choose not to enter, others dash in and dash out again; while a few, drawn by curiosity, boredom, rebellion, or circumstance, venture in so deep or wander around in there so long that they can never find their way back out. —Tom Robbins, *Villa Incognito*

We've considered several alternative-energy novelties, a few imposters, and a rogue zombie, so it might seem there is little left to discuss. But there's more to be sure—perhaps too much to cover fully in one book. So I will take a few pages here to briefly review a selection of remaining topics before moving on. These remainders are either not being publicly held up as solutions, offer only restricted geographic potential, or fall outside the core scope of this book. Still, I have chosen to briefly touch on a few that can each lend something distinctive to the greater picture. Let's begin with an energy-production technology that has been largely forgotten.

Hydropower

As recently as 1950, hydropower fulfilled about a third of electrical demand in the United States, but growing energy consumption has eroded most of hydro's value. Even though hydropower output has expanded since the fifties, it now serves just 5 percent of total U.S. electricity demand. Hydropower is still a monumental source of energy in other parts of the world—in Norway, for instance, dams high in the mountains quench virtually the entire electrical grid. Like wind and solar systems, dams generate electricity using a freely available and renewable resource. But unlike wind and solar systems, dams provide scalable supply whenever it is needed. And once built, dams provide inexpensive electrical power for a very long time.

Worldwide, hydropower provides 15 percent of electrical power. Some enthusiastic proponents claim that hydro could grow three-fold if the planet's capacity were fully exploited.[1] However, tapping into that capacity would displace many thousands of people, disrupt fishing industries, place neighborhoods at greater risk of flooding, and lead to a long list of other problems. Canadian filmmaker James Cameron joined hands with indigenous residents in Brazil to protest the construction of one such problematic dam. He compared the struggle of twenty-five thousand indigenous people who were to be displaced by the Belo Monte Dam to the struggle of the imaginary indigenous population in his blockbuster film *Avatar*: "There is no plan for where they go—they just get shoved out of the way. They were promised hearings by law—the hearings didn't take place . . . and the process is not transparent to the public." Cameron maintains that, "in fact, the public are being lied to."

> Brazil is rapidly developing and its economy is growing very fast and they are running out of power . . . so the government tells urban Brazilians, "You're going to get power; we're building a dam," and so they shrug and say "Well, that's a good idea." Except, the power from the dam is not going to them—the dam is 1,500 miles away—the power is going to go to aluminum smelters. Aluminum smelters are incredibly energy intensive and they make very, very few jobs per megawatt so it's really a bum deal. Plus, the profits go offshore.[2]

Controversy over hydropower expansion is the norm throughout the world, not the exception. Uzbekistan is alarmed by Tajikistan's plan to build the highest dam in the world, which would take eighteen years to fill and leave little water for Uzbekistan's cotton-growing region. Proposed dam projects are fueling disputes between Pakistan and India, a border already strung tight with nuclear tensions. In this region the 1960 Indus Waters Treaty is in danger of collapsing as India develops new forms of hydropower that were unforeseen at the treaty's signing. In all, internationally shared rivers flow across the borders of 145 countries (the Congo, Nile, Rhine, and Niger are each split between nine to eleven countries). Global conflict risks alone are enough to bring into question the real potential for hydropower expansion. Still, there are other concerns. For instance, the Aswan High Dam, built across the Nile, generates enough electricity to power all of Egypt, but detractors blame it for polluting irrigation networks, invasions of water hyacinth, coastal erosion, and outbreaks of schistosomiasis, a remarkably gruesome parasitic disease. Silt can also reduce a dam's output over time and many dams are hampered by poor site planning, maintenance troubles, or design flaws. Such limitations have plagued India's hydropower industry. Even though hydropower capacity grew 4.4 percent per year between 1991 and 2005, hydroelectric generation actually declined.[3]

Given the drawbacks and limitations of dams, hydropower proponents are shifting their focus to microhydropower, wave power, and tidal power. These may make sense in many nooks of the world, but high cost and geographic constraints will likely restrict these energy sources to isolated markets.

With limited expansion potential, perhaps the best way to return dams to power is simply to waste less of the scarce energy we already obtain from them. For instance, instituting electrical pricing and efficiency strategies in the United States at levels already attained in Europe and Japan would effectively double the share of hydropower in the American grid from 5 percent to 10 percent without building a single additional dam. As we shall consider later, other strategies could elevate hydropower's share even higher.

Geothermal

Geothermal systems capture warmth from the earth's crust to produce heat or generate electrical power. Our planet's internal nuclear reactions generate a lot of heat, but it is spread thinly across the globe, save for a scattering of hot fissures. The resulting average energy density at the earth's surface is just 0.007 watts per square foot, or about fourteen thousand times less dense than energy from the sun.[4] Small-scale geothermal systems draw upon this small amount of heat during winter months via a system of underground tubes filled with liquid. Some systems can reverse the process during the hot summer months to sink heat into the ground. Geothermal systems require electricity for pumping, but since they rely on the earth to do most of the heating and cooling, they ultimately consume much less power than a comparable furnace and air-conditioner combination.

The problem with such systems, beyond their high initial costs, is that they are only truly useful for buildings that are surrounded by large lots where the required tubing can be buried. Large lots in turn require a suburban-style infrastructure of roads, utility networks, and cars, which means that household geothermal systems are only slight improvements on what is a terrifically energy-intensive pattern of living overall. When builders cluster houses into walkable and bikeable neighborhoods or combine them into efficient multiunit urban apartment and condo buildings, geothermal systems become all but unworkable.

Larger geothermal plants hold more potential, but they only make economic sense in a few select locations where the planet's crust is unusually hot or flows with heated springs. Large geothermal systems are ideal for communities in these locations, but hardly a solution for everyone else on the planet. If energy companies are willing to risk drilling deeper, engineered geothermal systems (EGS) can function almost anywhere on the planet. But they cause earthquakes. In fact, they require manufactured tremors to function properly. Basel, Switzerland, shuttered its pricey geothermal plant when it triggered a 3.4 magnitude earthquake in 2006. Furthermore, these operations can churn up underground radioactive compounds such as radium–226 and radium–228, producing radioactive water and steam within geothermal facilities and leaving behind solid radioactive waste.[5] And even red-hot geothermal zones can degrade over time, sometimes unpredictably.

Taken together, these restrictions will limit the rise of geothermal power. Even large geothermal corporate players agree. For instance, Dita Bronicki, CEO of the geothermal company Ormat, admits "Geothermal is never going to produce 10 percent of the world's electricity, but the way we look at it is that if it reaches 2 percent in 20 to 50 years, then that is a lot."[6] As with hydropower, the best way to increase the share of geothermal power may be to use the resulting energy to do more.

Cold Fusion

When two light atomic nuclei fuse into one, they release a great deal of heat. Such reactions power the sun. Nevertheless, facilitating this reaction on a smaller scale at a lower temperature to capture the power would be an impressive feat. In fact, there's no plausible explanation for electrochemical cold fusion within the existing laws of physics, but that's only one limitation of this energy generation scheme, and perhaps not the most problematic.

The larger hurdle facing cold-fusion researchers is economic. Cold-fusion research funding dried up following embarrassing hoax discoveries in the field, which led remaining researchers to rebrand their work as "low-energy nuclear reactions." The small lingering stream of cold-fusion funding now trickles down to laboratories outside the mainstream. There are a couple more plausible high-temperature fusion proposals—hydrogen bombs work on this principle—but efforts to scale them down for utilities are dreadfully far from commercialization. An international collaborative team, named ITER, originally planned to complete a prototype reactor core in 2016, but the project is running over budget and its organizers have pushed back the deadline multiple times. Once completed, scientists plan to take twenty years to study ITER's fusion behavior, safety requirements, material characteristics, and related issues before venturing to build a functioning power plant.[7]

Fusion comes with a myth, too, if somewhat tedious: There's no need to worry about energy consumption or its side effects since fusion will eventually deliver plentiful energy without the headaches. Fusion proponents cough up various time frames for their optimistic scenarios, which usually pool around the thirty-year mark. Unfortunately, expectations for a fusion-powered planet have been thirty years away for quite some time. First in the 1950s, then in the 1960s, the 1970s, the 1980s, the 1990s, the 2000s, and well . . . today fusion power is still about thirty years away, as its most staunch supporters will affirm with a straight face. So as a general rule, whenever anyone attempts to defend the existing energy establishment by playing the fusion card, you might choose to disregard anything that thereafter happens to fall out of their mouth.

Fusion may be an idea worth researching further, but it is most certainly not a basis for public energy policy today. Anyone who suggests that it is, probably has something up their sleeve. If by some stroke of luck fusion should ever prove to become the cheap and available energy source its proponents insist it will become, you are invited to pull this book from your shelf and drop it in the trash—along with the rest of your nonfiction books. A world with inexpensive and scalable fusion would initiate a reset on every level of human and environmental interaction (along with some unintended consequences to be sure). But for now, please carry on.

Concentrating Solar Photovoltaic

Concentrating solar photovoltaic systems employ lenses to focus broader swathes of radiant energy onto solar cells. It's a lot cheaper to build a plastic lens than a larger solar cell. However, concentrating solar systems only work with direct sunlight—they can't take advantage of diffused rays on cloudy days. Also, because lenses force the solar cells to work harder and hotter, they don't last as long. So photovoltaic replacement costs largely offset the benefits. Solar advocates argue over whether concentrating solar strategies are a bit better or a bit worse than standard solar photovoltaics. Either way, "a bit" likely won't be enough to propel these systems into larger roles.

Solar Thermal

High-temperature solar thermal systems employ mirrors to superheat oil, salts, or steam for electrical generation. Of all the solar energy systems, solar thermal best accommodates grid demand because the hot fluids remain hot when clouds pass over or even into the night if stored in reservoirs. And since these plants ultimately run on steam-driven turbines, engineers can easily integrate fossil-fuel backup to pick up the slack on cloudy days. Generous government subsidies are fueling their growth and they arguably hold greater power generation potential than solar photovoltaics. Still, ecologists warn of the dangers massive mirror and lens arrays pose to desert ecosystems. And solar thermal power plants generally consume millions of gallons of water annually, a big problem anywhere, but one that is especially agonizing in deserts. Finally, their utility drops significantly outside hot and arid locales since they require expensive transmission lines to get their energy to market.

Other solar thermal strategies hold widespread potential, but these are frequently overlooked in the fanfare surrounding high-tech solar devices. As mentioned earlier, the most basic solar thermal strategies use dark tubing and simple optical devices to capture the sun's rays to heat water—coiled swimming pool heaters work through this principle. These boring heaters are cost-effective and generally advantageous accompaniments to standard hot-water heaters and boilers. And solar hot water can also power *thermal coolers* to chill buildings during the summer.[8]

Heat pumps are another option to draw upon the sun's energy. Essentially, they concentrate the tiny bit of warmth in cool air to heat buildings. Like liquid thermal coolers, they can also work in reverse during the summer. These thermal solar strategies can start to take on characteristics of efficiency strategies, which we will consider later.

Natural Gas

The natural gas industry has rather successfully reframed and marketed its product as a bridge fuel, in effect placing itself in a role that is both necessary and environmentally legitimate. But a bridge between what exactly? Between a "dirty" fossil-fuel present and a gleaming alternative-energy tomorrow? Is natural gas a bridge to nowhere?

Even though natural gas produces less CO_2 when burned than other fossil fuels, its extraction operations induce corresponding harms. For instance, one of the largest reserves of natural gas in the United States lies under the Marcellus Shale watershed, which provides drinking water to more than fifteen million people. Until recently, analysts considered these gas deposits unrecoverable since they are thinly distributed in tiny pockets of impermeable rock a mile beneath the earth's surface in a formation that extends from the northern edge of Tennessee to Syracuse, New York.[9] However, a newer technique called "fracking" is changing that. Energy firms combine underground horizontal drilling with high-pressure chemical slurries to fracture rock formations, thus releasing the entombed gas. This geologically violent process leaves behind copious amounts of injection effluent and permanently alters geological formations surrounding one of the nation's most prized fresh-water resources. Fracking also releases enough methane to more than cancel out any greenhouse-gas benefits that natural gas promoters so aggressively tout.

Many locals are peeved. An upstate New York group of investigators sifted through the Department of Environmental Conservation's database of hazardous-substance spills from the last three decades and found that natural gas operations also bring a variety of toxic chemicals and petroleum compounds to the surface. These effluents are often, incidentally, radioactive.[10] Walter Hang, president of the group, recounts one story from the investigation:

> A Vietnam vet living in Candor, New York, had discovered that, even though he had lived in the same house since 1962, his water started to release gas, and he discovered that you could light it. . . . He complained to the Department of Environmental Conservation. . . . The incredibly shameful thing is that the Department of Environmental Conservation did not even come to look at this situation. They simply told this disabled vet, Mr. Mayer, "Don't drink the water." And that was it.[11]

Meanwhile, Hang himself points out that the Finger Lakes region of western New York is facing difficult choices:

> These communities are just desperate for jobs. And so, it sounds so good: we're going to get this gas out, we're going to make tons of money, communities are going to benefit, the state of New York is going to benefit. . . . What happens when hundreds and hundreds of these hundred-thousand-ton trucks start pounding these structurally deficient bridges that have been neglected for decades into pieces? Who's going to pay for that? What about

Illustration 8: Flaring tap A still from the documentary film *Gasland* shows tap water bursting into flames when lit, allegedly following natural-gas hydrofrac extraction operations near this home. (Still courtesy of Josh Fox)

the roadways that are going to get destroyed? What are we going to do with all of this toxic wastewater? . . . You have these upland reservoirs, hundreds of miles away from the city, and the water flows completely under gravity through these giant tunnels. It's so pure it doesn't even need to be filtered. And so, this is a jewel. Any city in the world would give anything to have this water. That's why it has to be safeguarded. It has to be protected. Once it's polluted, then the city would have to treat that water at gargantuan cost.[12]

As with clean coal, natural gas can only seem clean if we first sharply narrow our focus (to one CO_2 output metric) and then proceed to disregard absolutely every other well-established and documented side effect, limitation, and long-term risk of deploying the fuel.

Hybrids and Electric Cars

My first appearance on national television, twenty years ago, was followed by an unexpected realization. The CNN segment featured a hybrid car that I had designed and built. The small two-seater hybrid (a plug-in electric and natural-gas hybrid with regenerative braking) used far less fossil fuel than its contemporaries. It was also scaled for city driving, measuring just an inch longer than a golf cart. I thought it was an especially beneficial solution to our environmental challenges. I was wrong.

First of all, what counts as an alternative-energy vehicle and what doesn't is hardly a straightforward reckoning. Such definitions are social contrivances, adeptly evolved to serve a variety of purposes. For instance, is an electric car a true alternative if its drivetrain is ultimately powered by coal, nuclear power, and lithium mines rather than petroleum?[13] Perhaps, but it's not a clear-cut calculation. According to Richard Pike, chief of the UK Royal Society of Chemistry, fully adopting electric cars would only reduce Britain's CO_2 emissions by 2 percent due to the country's electric utility fuel mix.[14] When we start to exchange one set of side effects for another, the exchange rates become confusing. This opens a space for PR firms, news pundits, environmentalists, and others to step in and define the terms of exchange to their liking.

For instance, during the hype for the launch of its Volt electric vehicle, General Motors claimed that customers could fill up for the cost of less than four cents per mile.[15] The California utility company PG&E came to a similar conclusion.[16] So did the DOE's National Renewable Energy Laboratory.[17] These calculations assume a retail electricity rate of ten cents per kilowatt-hour, which is roughly the national average. In academic research, government reports, environmental reports, and journalistic accounts of electric vehicles, these two persuasive figures run the show. Four cents per mile. Ten cents per kilowatt-hour. Given the apparent thrift of these two figures, it's stunning that electric vehicles hadn't caught on earlier. Was it a Big Oil conspiracy? Did someone kill the electric car? Or do these two honest-looking figures have something they're not telling the jury?

In reality, when analysts use these two celebrated numbers to calculate the fuel costs of an electric vehicle they arrive rather quickly at one remarkable problem. If car buyers intend to drive their electric car farther than the extension cord from their garage extends, they won't be able to take advantage of that cheap ten-cent electricity. They'll have to rely on a battery—a battery they can only recharge a finite number of times before it must be replaced, at considerable expense. The battery step, not the "fuel" step, is the expensive part of driving an electric vehicle.

We don't usually think about the storage costs of fuel in our cars because the gas tank is comparatively inexpensive. We can recharge it over and over (to varying degrees) and it generally outlives the life of the car. Not so for batteries. Batteries are many times more expensive than the electricity required to fill them. The better the battery, the more expensive it is. A sixty-watt-hour laptop battery with a seven-hundred-recharge lifespan that costs $130 to replace will ultimately cost about $3 per kilowatt-hour to operate—so expensive that the ten-cent-per-kilowatt-hour "fuel" cost to charge it becomes negligible.[18] And even though electric vehicles are moving to cheaper batteries, the costs of exhuming their required minerals extend far beyond simple dollars and cents.

Environmentalists generally stand against battery-powered devices and for good reason: batteries require mined minerals, involve manufacturing processes that leak toxins into local ecosystems, and leave behind an even worse trail of side effects upon disposal. Though when it comes to the largest mass-produced battery-powered electrical gadget ever created—the electric car—mainstream environmental groups cannot jump from their seats fast enough to applaud it.[19]

Even at current battery-production levels, mining activities draw fire from local environmental and human rights organizations that are on the ground to witness the worst of the atrocities. An analysis by the National Research Council concludes that the environmental damage stemming from grid-dependent hybrids and electric vehicles will be greater than that of traditional gasoline-driven cars until at least 2030, given expected technological advances.[20] But even if mining companies clean up their operations (which at least will require much stricter international regulations) and engineers increase battery storage capacity (which they will, very slowly) there is still a bigger problem looming on the horizon: Alternative-fuel vehicles stand to define and spread patterns of "sustainable living" that cannot be easily sustained without cars. Cars enable people to spread out into patterns of suburban development, which induces ecological consequences beyond the side effects of the vehicle itself.

Even the most efficient hybrid or electric cars can't resolve the larger ecological impacts of sprawl. In fact, their green badges of honor might even help them fuel it. For a time, this may not pose a problem, but eventually it will. Over time, a car's odometer may be more environmentally telling than its fuel gauge.[21] Sprawl has positive and negative effects on Americans, but its intensification is clearly at odds with the long-term ideals of the environmental movement. The future to which alternative-vehicle proponents aspire may place more people at risk of resource and energy volatility swings in the future. Alternative-fuel vehicles up the bets in an already heated game of energy threats dealt by what sociologist Ulrich Beck calls our "risk society.[22]

In cities, alternative-fuel vehicles bring some distinct environmental benefits—less noise and smog for instance. But if these vehicles become less expensive, as their proponents claim they will, then people who once relied on cycling or walking might purchase them—potentially negating some or all of the presumed environmental benefits. Furthermore, battery-powered vehicles tend to be heavy. Drivers generally state they feel safer in a larger, heavier car, and sometimes (but not always) they are. However, heavy vehicles place pedestrians, bicyclists, and those in lighter vehicles in greater danger. Therefore, the risk of injury or death has not been reduced, but rather transferred to others.

Upon closer inspection, the benefits of green automobiles start to appear synonymous with the benefits of smoking low-tar cigarettes. Both seem healthy only when compared to something framed as being worse. And on some level, I suspect people are tuning in to this. Even with all of the hype surrounding hybrid and electric vehicles, these machines are becoming somewhat of a cliché in some circles. In my experience, even those who most willingly embrace alternative vehicles still smell something funny in the air as they drive by—a kind of ephemeral suspicion that's hard to pin down. Hybrid and electric vehicles may offer partial solutions within certain contexts, but those contexts are frightfully limited. Nevertheless, marketers will continue to flash their green credentials in order to sell car culture to a greater number of people worldwide.

It isn't acceptable for doctors to promote low-tar cigarettes. Should environmentalists promote alternatively fueled automobiles? What about alternative fossil fuels, nuclear power, or alternative energy more broadly?

We might expect mainstream environmental groups, with their concerns about resource scarcity, to identify alternative-fuel vehicles as a drawn-out delay before the inevitable crunch—a crunch that would presumably be far worse if even more livelihoods were placed at risk over the intervening years. But they don't. In fact, there is little critical engagement with the whole array of dubious low-tar energy options perched on the wall behind the cashier.

Why?

That's the intriguing question to which we shall now turn.

Part II
From Here to There

8. The Alternative-Energy Fetish

The society which rests on modern industry is
not accidentally or superficially spectacular, it is
fundamentally *spectaclist.* –Guy Debord, *Society
of the Spectacle*

Perhaps it's all too easy for us to miss the limitations of alternative energy as we drop to our
knees at the foot of the clean energy spectacle,
gasping in rapture. The spectacle has become a
divine deity around which duty-bound citizens
gravitate to chant objectives without always re-
flecting upon fundamental goals. This oracle
conveys a ready-made creed of ideals, objectives, and concepts that are convenient to re-
cite. And so these handy notions inevitably become the content of environmental discourse. In
a process of self-fashioning, environmentalists
offer their arms to the productivist tattoo artist to embroider wind, solar, and biofuels into
the subcutaneous flesh of the movement. These
novelties come to define what it *means* to be an
environmentalist.[1]

Environmentalists aren't the only ones lin-
ing up for ink.

Peer pressure is a formidable power, and
there's no reason to assume that rational adults

are above its dealings. Every news article, environmental protest, congressional committee hearing, textbook entry, and bumper sticker creates an occasion for the visibility of solar cells, wind power, and other productivist technologies. Numerous actors draw upon these moments of visibility to articulate paths these technologies ought to follow.[2]

First, diverse groups draw upon flexible clean-energy definitions to attract support. Then they roughly sculpt energy options into more appealing promises—not through *experimentation*, but by planning, rehearsing, and staging *demonstrations*. Next, lobbyists, strategic planners, and PR teams transfer the promises into legislative and legal frameworks and eventually into necessities for engineers to pursue. A consequence of this visibility-making is the necessary invisibility of other options. There's only so much room on the stage.

Productivist Porn

During the rise in oil prices through the first years of the twenty-first century, I remained safely in the library. I studied a corpus of thousands of articles, environmental essays, and political speeches staged around energy through those years.[3] I found that most writers fell into a predictable flight pattern, confidently landing their conclusions atop a gleaming airstrip of alternative energy and offering a sense that alternative technologies are all it will take to cure our energy troubles. The way to solve our energy production problems is to produce more energy.

Why do the options of wind, solar, and biofuels flow from our minds so freely as solutions to our various energy dilemmas while conservation and walkable neighborhoods do not? Why do we seem to have a predisposition for preferring *production* over energy *reduction*? The answer is neither straightforward nor immediately apparent. Some claim that modern conceptions of prosperity, progress, and vitality structure our preference for production. Evolutionary biologists point to physical characteristics of our brain. Other theorists argue that the productivist spirit rose from Christian values as people abandoned holistic pantheism to worship a creator they understood to be separate from creation. As the natural world was desacralized, it was left exposed to investigation, definition, and scientific manipulation.[4]

Some intellectuals maintain that the productivist drive should be linked not to Christianity but to philosophical developments in the fifteenth and sixteenth centuries. This is when the mind and body came to be understood as separate, largely due to Descartes' dualistic view of the self and Kant's distinction between active subjects and passive natures. Perhaps Newtonian physics also played a role by divesting the material world of spiritual value—depicting it as a gear set devoid of spiritual value— to be sacrificed and exploited without moral consequence. Was it here, nestled in the bosom of the Reformation and scientific achievement, from whence the productivist penchant hath been spawned?[5]

The genesis of productivism may forever remain a complex mystery. Religious, philosophical, scientific, and capitalist traditions did not develop in separate vacuum chambers, but in orchestration and entanglement with one another.[6] Regardless of the particulars of this codevelopment, one outcome is clear. Productivist leanings have effectively nudged nature to the sidelines of Western consciousness. Whether we are considering energy production, human procreation, the work ethic, or any other productivist pursuit, there endures a common theme: that which is produced is good and those who produce should be rewarded. These values till the fertile soil of an almost religious growthism with invasive roots of techno-scientific salvation.[7] Many voices of influence willfully hoist up figures such as Buddha, Jesus, Thoreau, and Gandhi onto pillars, even while trampling over their conceptions of simplicity, humility, community, and service, which they instead characterize as radical and naïve.[8]

Perhaps these values are naïve when viewed from a hyperindividualist and hyperconsumerist perspective—one where families not only have their own home theater, their own power tools, their own heating system, and their own laundry facilities but have also settled into a concept of living wherein such hyperindependence is normal and unproblematic. Is our perceived independence the residue of the suburban diaspora? Manifest Destiny? Perhaps it arose from the same crib as the productivist spirit. In any case, if we have come to understand the human condition as narrowly arising from the individual, rather than as individuals arising from within the relationships of others, then it demands some attention. For as the poetic philosopher Martin Buber realized, "All real living is meeting."[9] We are born creatures and become human through community.

The Energy Pornographers

We don't just want our own energy. We want to create it ourselves. It is perhaps not so shocking that President Obama's opponents mocked him during his initial run for the presidency when he advocated for proper tire inflation in the face of those who were thirsting to drill for oil in the Alaskan wilderness. It is likewise understandable that other politicians, corporations, and many environmental groups steer their platforms away from energy *reduction* strategies, driving instead toward the glowing symbolism of green energy *production*. If they include energy reduction strategies in the program, they remain as side acts in the larger alternative-energy big top. Clean energy steals the show. A *USA Today* journalist goes so far as to claim that alternative-energy production is the *only* option, insisting that reducing greenhouse-gas emissions "requires" that we immediately deploy "other sources of energy, such as windmills and nuclear plants."[10]

Hogwash.

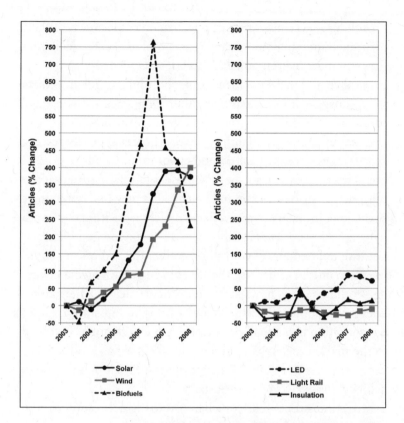

Figure 8: Media activity during oil shock Gas prices more than doubled between 2003 and 2008, but media coverage of solar, wind, and biomass energy shot up faster, averaging a 400 percent increase. Meanwhile, media coverage of energy reduction strategies remained low over the entire period, averaging just a 25 percent increase. (Data from author's study of fifty thousand articles)

The fact that the national media routinely discharges such gratuitous streams of productivist sewage without objection from environmental groups is a testament to how impoverished environmental movements in the United States have become. As pundits lock their sights on alternative energy, people are not only fooled into ignoring far more promising solutions but also charmed into overlooking how these technologies generate greenhouse-gas emissions, instigate negative side effects of their own, encounter limitations, and most problematically, act to further stimulate overall power demand.

Massive multinational firms control the bulk of media operations. These multinationals own other companies in diverse sectors such as defense, logging, real estate, oil, agriculture, banking, manufacturing, and utilities. Media boards typically reserve spaces at their weighty hardwood tables for these business leaders. Start to see the issue here? These relations assure that the majority of stories, as fair and balanced as they may sometimes seem, are ultimately conceived and developed within the womb of corporate, not *public*, influence. This may be good, bad, or neutral, depending on your perspective. Regardless, it should hardly come as a surprise that the aggregate of media coverage contains more news segments and articles on alternative-energy technologies, which can be bought and sold, than on conservation and simplicity efforts, which are not as involved in the market mechanism and could in some cases threaten to reduce the very consumption patterns multinationals rely on to achieve quarterly sales forecasts.[11]

Media industry executives will retort to this accusation with an emphatic, "Squash!" They argue that corporate owners place no pressure on editors to align stories with corporate interests—conglomerates even keep editorial offices separate from their other businesses. But this defense presumes it is possible to maintain a clear separation in the first place. It isn't. Boardrooms don't send edicts down to editors—productivist influences on media are typically far more subtle than that. They arise from interstitial forces within the operating procedures of newsrooms themselves. It starts with a tradition that people tend to hold in high esteem: objectivity.

Flirting with Truth

Objectivity in journalism is frequently, yet mistakenly, understood as truth. Facts are slippery things, and news organizations understand that attempting to sell them directly would be sheer folly. Instead, news organizations operate through proxy. Journalistic objectivity is not so much a rendering of truth as much as it is an attempt to accurately convey what others believe to be true. In order to achieve this rendering, experienced journalists instruct young journalists to keep their own beliefs and evaluations to themselves through a conscious depersonalization. Second, mentors instruct them to aim for balance, or field "both sides" of a controversial subject without showing favor to one side or the other. The news industry generally accepts this framework as the best way to go about reporting on issues and events. It's certainly a lot better than some of the alternatives. Nevertheless, this truth-making strategy carries certain peculiarities.

For example, news editors tend to judge stories supporting the status quo as more neutral than stories challenging it, which they understand as having a point of view, containing bias, or being opinion laden. Investigations that present empirical evidence and consider unfamiliar alternatives are not as valued as the familiar "balance of opinions." As a result, journalists reduce energy debates to a contest between alternative-energy technologies and conventional fossil fuels. We have all witnessed these pit fights: wind versus coal for electrical production, ethanol versus petroleum for vehicular fuels. Pitting production against production effectively sidelines energy *reduction* options, as if productivist methods are the only choices available. Have you ever seen news segments that pit solar cells against energy-efficient lighting or that toss biofuels in the ring with walkable communities? Probably not. I have so far come across only a handful of examples out of thousands of reports.

Pitting production versus production seems natural, but it leads to some unintended effects. First, these debates set a low bar for alternative-energy technologies; it's not difficult to look good when you are being compared to the perfectly dismal practices of mining, distributing, and burning oil, gas, and coal. Imagine if wine critics judged every Bordeaux against a big bottle of acidic vinegar that's been sitting in grandpa's cupboard for two decades; it would be difficult for a winemaker to perform poorly in such a contest. Secondly, journalistic dichotomies reduce apparent options to an emaciated choice between Technology A and Technology B. This leaves little space for nontechnical alternatives. It also misses negative effects that both Technologies A and B have in common. Finally, pitting alternative-energy technologies against fossil fuel gives the impression that increasing alternative-energy flows will correspondingly decrease fossil-fuel consumption. It won't—at least not in America's current socioeconomic system—as we shall consider in the next chapter.

Implants

Time pressures and streamlining media operations force journalists to increasingly rely on quotes and comments from a short list of contacts, usually government, industry, public figures, or other sources that viewers see as credible. Professor Sharon Beder claims that this "gives powerful people guaranteed access to the media no matter how flimsy their argument or how self-interested."[12] In the effort to provide credibility, journalists may unknowingly give equal voice to views that are blatantly exaggerated, have already been widely discredited, or are given little credence by those more familiar with the topic.

For instance, in their book *Merchants of Doubt*, historians Naomi Oreskes and Erik Conway show how over a period spanning decades, oil and industry groups effectively convinced the public that a scientific controversy surrounded climate change when, in fact, there was little disagreement.[13] Consensus among climatologists actually began to solidify in the 1970s. In 1988, researchers organized the Intergovernmental Panel on Climate Change (IPCC) to assess the risks associated with human-induced climate change. That same year, NASA reported to Congress that climate change was occurring and that it was caused by humans. After years of research, the IPCC stepped forward to agree with NASA scientists.[14]

Feeling threatened, several oil companies and other large corporations joined forces to fund advertising campaigns, foundations, and organizations such as the American Enterprise Institute, the Global Climate Coalition, and the George Marshall Institute, in order to attack the credibility of scientists studying climate change and to frame climate change as a scientific "dispute" rather than a consensus. These organizations hired many of the same public relations and legal consultants who had earlier ridiculed doctors for warning about the risks of cigarette smoke.

In the early 1990s, these skeptics organized test markets to ascertain the most effective ways of producing "attitude change." When they discovered that people tended to believe scientists over politicians or corporations, they test-marketed names of scientific front organizations. Once set up, these front organizations would produce reports that questioned climate change. They distributed their arguments via pamphlets, mass media, and the Internet, rather than publishing in peer-reviewed journals. Internal documents from these organizations reveal that they found radio ads to be the best way to influence "older, less educated males." For "younger, low-income women," they selected magazine ads.[15] They even test-marketed the spokespeople for their believability. By the early 1990s, these organizations had launched a full-fledged public relations tour to frame climate change as both a controversy and a topic that required more research before consensus could be reached. They ensured that journalists would have ample opportunity to "balance" the views of climatologists with those of the skeptics, even if the naysayers could not speak with scientific authority themselves.

The public relations campaign proved a magnificent success. It swayed public opinion, greatly influenced media coverage, and delayed policy to mitigate the effects of climate change. In 2006, a *Time Magazine* poll showed that a majority of Americans believed global temperatures were rising. Yet 64 percent also believed that scientists were still busy making up their minds on the matter, when in fact the scientific consensus had already been gathering dust for a decade.[16] In 2010 a FOX news employee leaked an internal email from the Washington bureau chief that instructed, "Refrain from asserting that the planet has warmed (or cooled) in any given period without IMMEDIATELY pointing out that such theories are based upon data that critics have called into question. It is not our place as journalists to assert such notions as facts, especially as this debate intensifies."[17] By 2012 half of Americans felt climate change probably wasn't caused by humans and its potential risks were exaggerated.[18]

Beyond depicting controversy in science where none actually exists (as with smoking and climate change), corporations also use front groups to propose superficial solutions to important problems in order to divert attention away from more serious policy or regulatory changes. For example, a program for the Keep America Beautiful Campaign frames the problem of waste simply as a case of individual irresponsibility. If only people would just be more responsible, they say, perhaps the problem of waste would go away. This frame distracts public attention away from regulatory mechanisms addressing packaging design, recycling, waste legislation, and corporate ethics.[19] Similarly, voices for energy reduction strategies are only faintly audible in a room full of deafeningly loud productivists auctioning off their new, refashioned, or camouflaged technological widgets.

Spreading It Mouth-to-Mouth

About twenty years ago, the Pew Research Center found that 33 percent of local journalists felt that business pressures affected their reporting.[20] When Pew Research checked in more recently, they found the number had more than doubled to 68 percent. Journalists now cite financial concerns as the foremost challenge facing their craft—more influential than the quality of coverage, loss of credibility, ethics, and concern for standards combined.[21]

Understaffed news rooms increasingly fall back on source journalism—initiating stories using material distributed by public relations firms and corporations. (This contrasts with the more time-consuming practice of investigative journalism.) Today about half of news stories arise from press releases. This helps explain the nauseating barrage of articles touting new green gadgets, which are simply rewritten press releases from companies promoting their products or researchers eager to attract attention (and funding) for their often half-baked schemes. Readers and viewers have a hard time distinguishing between these rewritten PR scripts and traditional journalism. Reported uncritically and replicated in bulk quantities, these pieces toke news users on the kind of consumerist high typically achieved only through infomercials.[22]

Commercial news conglomerates aren't necessarily focused on the public interest as much as on keeping readers and viewers entertained so that they stay tuned for the next round of advertising. Short and exciting stories with captivating visuals attract and hold appropriate audiences for advertisers. Therefore, firms typically provide journalists with videos and computer renderings. The drive for entertainment leaves little space to cover background, contextual fundamentals, or the structural origins of increasing energy consumption. As Robert Pratt from the Kendall Foundation observes, "One of the things that is a difficulty with energy efficiency is that it's kind of fundamentally boring. Photovoltaic cells, solar cells, or wind turbines are much more interesting to people."[23] It's perhaps no surprise that this form of journalism generally doesn't spur minds to engage critically with the social, political, economic, and cultural complexities of our energy system. In fact, a report by the Woodrow Wilson Center contends that people who rely primarily on television for their news are among "the least-informed members of the public."[24]

A decade before BP's Gulf oil spill, the company put these techniques to work in rebranding British Petroleum to "Beyond Petroleum" as it installed solar panels on filling stations throughout the world. The company swamped newsrooms and television stations with press releases, complete with ready-made renderings and photographs of the new panels. BP crafted "Plug into the Sun" promotions to show the greener side of filling stations with copy like, "We can fill you up by sunshine," even though the gasoline in the pumps was still the same.[25] Journalists mechanistically rephrased the press releases, attached the handy visuals, and sent them to press. During the media campaign, journalists rarely offered any context for readers to assess how much solar energy the tiny arrays would produce or how this affected the company's expanding oil exploitations.

The campaign worked. In fact, out of the collection of articles I studied from this period, I could not find a single one pointing out what might seem an obvious design flaw—some of the solar cells appeared to aim not at the sun, but down toward the asphalt, presumably so incoming drivers could see them as they rolled in. Though perhaps it wasn't a design "flaw" after all, considering that the company allegedly spent more money on its marketing campaigns than on the solar panels themselves.[26]

Luckily, some might say, the Internet challenges such greenwashing, as online news sources democratize media and create voices for thousands of niche publications even as traditional journalism declines. At least, that is how the story goes. Unfortunately, the reality is more complicated than this weary story line indicates.[27] Despite the many sources of Web-based media, just ten news organizations (which incidentally are associated with legacy media corporations) form an oligarchy attracting most of the eyeballs. Compared with print journalists, Internet journalists are far more likely to indicate that corporate owners and advertisers "have substantial influence on news coverage."[28] Two-thirds of Internet journalists claim that bottom-line pressures are hurting news coverage as they shift from being reporters to recoders of information.

The ultimate result? Churnalists pump out information on the newest energy gizmos in uncritical articles, which allow us to forget about the complex challenges of our energy systems and instead bask in the warm glow of artificial techno-illumination. As those ideas fade, others will surely show up to warm our hearts soon after. Eight in ten journalists agree. They claim that the scope of news coverage is too narrow and that news rooms dedicate insufficient attention to complex issues such as global energy production, use, and related side effects. When Pew Research asked journalists what media are doing well, just a scant 15 percent pointed to their watchdog role of investigative journalism.[29] Media consolidation, with its narrow focus on the bottom line, has fundamentally altered journalistic practice and lubricated passageways for corporate views to slide into the public realm, ensuring that their technophilic, consumerist, and productivist spirit is immediately salient to audiences around the world.[30]

Power Tools

Unsurprisingly, profit motives likely induce much of the gravitational field surrounding productivist energy solutions. For the most part, knowledge elites can patent or otherwise control productivist technologies—manufacturing, marketing, and selling them for a profit (or at least federal handouts). On the other hand, many energy reduction strategies are *not* patentable because they are based on age-old wisdom and common sense. Solar photovoltaic circuitry, wind turbine modulators, nuclear processes, and even biomass crops are all patentable and commodifiable in a way that passive solar strategies and walkable neighborhoods are not. The profit motive of this ilk is a chronic theme in America; we are a country that values drug research (commodifiable) over preventative health (not commodifiable); most of our soybean fields are planted with corporate-issue genetically modified plants (patentable) rather than seed saved from last year's crop (not patentable). The debate about whether profits are a noble or a corrupted motivation is a political matter to be argued over a pint of beer, not here in these pages. I aim only to shed the humblest flicker of light on the illusion that the world of alternative energy operates within some virtuous form of economics. It doesn't. The global economic system rewards the commoditization of knowledge and resources for profit—why would we expect it to be any different for the field of alternative energy?

It's not just our economic system that offers virulence to our productivist inclinations; our political system does as well. The politics of production are far more palatable than the politics of restraint, as President Jimmy Carter learned in the 1970s. When he asked Americans to turn down the thermostat and put on a sweater, he received a boost in the polls. But voters ultimately turned to label him a pedantic president of limits. "No one has yet won an election in the United States by lecturing Americans about limits, even if common sense suggests such homilies may be overdue," remarks historian Simon Schama. "Each time the United States has experienced an unaccustomed sense of claustrophobia, new versions of frontier reinvigoration have been sold to the electors as national tonic."[31]

Clean energy is the tonic of choice for the discerning environmentalist. Over recent decades, flows of power within America and other parts of the world began pooling around alternative-energy technologies. Mainstream environmental organizations took a technological turn, which gained momentum during the

1970s and became especially palpable in the 1980s when the Brundtland Commission brought the idea of sustainable development into the spotlight. The commission passed over societal programs to instead underline technology as the central focus of sustainable development policy.[32] The commission's 1987 report, *Our Common Future*, stated, "New and emerging technologies offer enormous opportunities for raising productivity and living standards, for improving health, and for conserving the natural resource base."[33] This faith in the ability of technologies to deliver sustainable forms of development evolved during a period of public euphoria surrounding information technology, agricultural efficiency through petrochemicals, management technology, and genetic engineering. As in other periods throughout American history, there was a sense that if nature came up short, the wellspring of good ol' American know-how would take up the slack.[34]

Mainstream environmental organizations were all too eager to fill the pews of this newly energized church of technological sustainability, which they themselves had helped to consecrate. For instance, a World Resources Institute publication declared in 1991, "Technological change has contributed most to the expansion of wealth and productivity. Properly channeled, it could hold the key to environmental sustainability as well."[35] The next year the United Nations developed a sustainable development action plan called Agenda 21, which charged technological development with alleviating harmful impacts of growth. As the new centerpiece of social policy, there was little debate around technology, other than how to implement it. During the 1980s and '90s, environmental organizations began to disengage from the dominant 1960s ideals, which centered on the earth's limits to growth. They shifted to embrace technological interventions that might act to continually push such limits back, making room for so-called sustainable development. The former enthusiasm for stringent government regulation waned as environmental organizations expanded the roles for "corporate responsibility" and "voluntary restrictions." As a result, legislators pushed aside public environmental stewardship and filled the gap with corporate techniques such as triple-bottom-line accounting and closed-loop production systems, which purported to be good for the environment and good for profits.[36] In 2002, breaking with past mandates, the United Nations World Summit on Sustainable Development's Plan of Implementation narrowed its assessments by assuming that technological sustainability would require "little if any political and cultural negotiation about modern lifestyles, or about the global systems of production, information, and finance on which they rest."[37] And by 2004, Australia Research Council Fellow Aidan Davison observed that "the instrumentalist representation of technologies as unquestioned loyal servants" had come to fully dominate sustainable development policy.[38]

The so-called limits-to-growth theories of the 1960s certainly had limits of their own as effective conceptual tools for change. Yet the mass exodus away from these guiding concepts and toward an overwhelming reliance on technological fixes may have overshot the realm of reasoned and vital inquiry into the diverse causes of our unsustainable energy system. It has also erected a formidable fortress of interests.

Objects of Affection

Solar power means different things to different people, yet the notion's heartiness manages to sustain a common identity across various disciplines. Solar cells are "boundary objects," described by Susan Leigh Star and Jim Griesemer as concepts "both plastic enough to adapt to local needs and the constraints of the several parties employing them, yet robust enough to maintain a common identity across sites." These objects of affection "have different meanings in different social worlds but their structure is common enough to more than one world to make them recognizable, a means of translation."[39] While it may seem peculiar to categorize solar power as a translator, it can certainly be understood as serving this function.[40] Beyond its obvious service in producing electrical power, solar power plays less apparent social roles in politics, industry, academia, and in the public sphere where various groups employ solar for their own purposes. Let's consider an example.

If solar energy were a puppy, it would be a super cute little tike—a happy puppy trotting along with all of society's dog walkers. The academic dog walker likes Solar because walks with Solar bring benefits such as recognition and grant money. Industry likes walking Solar because it offers tax breaks, production opportunities, and good public relations for the company. Government enjoys being out with Solar as well. The public sees Government in a good light for walking Solar. And when elections come around, all the more reason to take Solar out for a stroll! Media likes Solar too—Solar offers exciting graphics, news stories, and segments people will watch, which brings in more advertising dollars. Public is always out on the walks; Solar makes Public feel happy, responsible, and successful in combating environmental challenges. Sometimes, being just a puppy, Solar gets tired, but there is always someone around to pick Solar up and keep going.

On walks with Solar, the various dog walkers all get along fairly well. Occasionally, they even release Solar from the leash to run ahead—"Go get 'em, Solar, go on!" Their interests in a healthy and happy Solar generally run in parallel or complement one another, making these walks a friendly outing even if they often just go to the park and walk in circles.

Now let's consider the walk with Energy Reduction. Reduction is a huge dog with lots of potential, though walks with Reduction differ from walks with Solar. Reduction's walks include many more stops to pee. Academia likes Reduction and gets some government funding for going on these walks. Industry, on the other hand, sees Reduction as more of a nuisance, sometimes saving some money, but more often getting caught up in regulations along the way, bringing the walk to a stop at times. Usually, there is some bickering between Industry and Government, but they usually work things out since Government doesn't have much to gain; Government's constituents don't pay much attention to Reduction. Media finds the walks with Reduction to be tiresome—not much new news here. Public agrees, seeing walks with Reduction as a decrease in living standards. Why go for a walk with Reduction, when the walks with Solar are so much more fun?

Poor Reduction doesn't get taken out for walks nearly as much as is healthy for a dog of Reduction's size and age. The dog walkers seem, at best, ambivalent, and at worst, plainly disgruntled. As policymakers, students, environmentalists, and concerned citizens, we need to understand how to organize walks with Reduction that interest more of the dog walkers. We'll come to that challenge in the next chapter.

Erecting the Clean-Energy Spectacle

Since progrowth ideals are well-funded, politically powerful, connected with media, and pervasive in public thought, it's no surprise that most of us have come to accept many progrowth premises as self-evident truth. Together they form a formidable force within local and international polity and economy. We expect companies to increase their earnings, labor to expand, and material wealth to increase throughout the world until every last child is fed, clothed, educated, and prosperous. This story line is conceivable only if we are willing to delude ourselves into believing there are enough resources on the planet for many more inhabitants of the future to consume, eat, play, and work at the standards that wealthy citizens enjoy. I've come across little convincing research to support the possibility that this is physically viable today, let alone in a more heavily populated and re-source-depleted planet of the future.[41]

These progrowth ideals act to structure future energy invest-ments. For instance, the International Energy Agency (IEA), as do other governmental agencies, crafts long-term predictions of world energy use, primarily extrapolating from past trends in population growth, consumption, efficiency, and other fac-tors. Subsequently, large energy firms evoke these predictions in their business plans in efforts to prod governments and in-vestors to support drilling, exploration, pipeline construction, and other productivist undertakings. Alternative-energy compa-nies have historically done the same. Once firms translate these predictions into investment, and investment into new energy supply, energy becomes more affordable and available. Energy consumption increases and the original predictions come true.

Numerous actors and factors hold the self-fulfilling prophecy together. Powerful energy lobbies promote their productivist in-clinations in the halls of government. Industry and a consumer-driven public sop up any excess supply with a corresponding in-crease in demand.[42] And since side effects are often hidden or displaced, the beneficiaries can continue at the expense of oth-ers who are less politically powerful, or who have not yet been born. For all practical purposes these side effects must remain hidden in order for the process to continue.[43]

Experts have developed a language to determine what is counted and what is not.[44] For instance, an influential congres-sional report from the National Research Council, entitled *The Hidden Costs of Energy*, explicates numerous disadvantages, limi-tations, and side effects of energy production and use. But it spe-cifically excludes some of the most horrible of these—includ-ing deaths and injuries from energy-related activities as well as food price increases stemming from biofuel production.[45] The authors dedicate several pages and even a clumsy appendix to convincing readers that such factors needn't be interrogated because they don't meet the economic definition of "externalities." Here their tightly scripted definition comes to run the show. It stands in for human judgment to decide what gets counted and what doesn't. Within this code, a whole world of side effects needn't be interrogated if they don't fit neatly within the confines of a definition.[46] In a moment of trained incapacity, the authors miss that it's the definition itself that requires interrogation.[47] It's not particularly shocking that this could happen in a formal policy report. It happens all the time. What's shocking is that a report featuring such glaring omissions could attract the sign-off of over one hundred of the nation's most influential scientific advisers. There are some oversights it takes a PhD to make.[48]

Since we live on a finite planet with finite resources, the system of ever-increasing expectations, translated into ever-increasing demand, and resulting in again increased expectations will someday come to an end, at least within the physical rules of the natural world as we understand them. Whether that end is due to an intervention in the cycle that humanity plans and executes or a more unpredictable and perhaps cataclysmic end that comes unexpectedly in the night is a decision that may ultimately be made by the generations of people alive today.

Perhaps we should find the courage to do more than simply extrapolate recent trends into the future and instead develop predictions for a future we would like to inhabit. These are, after all, the aspirations that will become the basis for policy, investment, technological development, and ultimately the future state of the planet and its occupants.[49]

The immediate problem, it seems, is not that we will run out of fossil-fuel sources any time soon, but that the places we tap for these resources—tar sands, deep seabeds, and wildlife preserves—will constitute a much dirtier, more unstable, and far more expensive portfolio of fossil-fuel choices in the future. Certainly alternative-energy technologies seem an alluring solution to this challenge. Set against the backdrop of a clear blue sky, alternative-energy technologies shimmer with hope for a cleaner, better future. Understandably, we like that. Alternative-energy technologies are already generating a small, yet enticing, impact on our energy system, making it easier for us to envision solar-powered transporters flying around gleaming spires of the future metropolis. And while this is a pristine and alluring vision, the sad fact is that alternative-energy technologies have no such great potential within the context that Americans have created for them. An impact, yes, perhaps even a meaningful one someday in an alternate milieu. However, little convincing evidence supports the fantasy that alternative-energy technologies could equitably fulfill our current energy consumption, let alone an even larger human population living at higher standards of living.

This isn't to say there aren't solutions. There are. We've just been looking for them in the wrong place.

9. The First Step

Probable impossibilities are to be preferred to
improbable possibilities. —Aristotle

If we were gunslingers, we'd be in trouble.
Several sinister energy challenges are staring
us down, but the productivists are asking us to
choose our weapon from a rack of toy guns. The
alternative-energy project's fundamental weak-
ness lies in its failure to engage with obvious cultural factors such as consumerism, corporatism,
and middle-class desires. Instead, we allow pundits to frame energy challenges as technological
problems requiring a technological fix. Every
day, media troupes relay news snippets touting the latest bio-eco-green energy sources—
all designed to jury-rig a mode of life that is not
optimal, desirable, or even affordable for most
of the world's communities. The "energy crisis" is more cultural than technological in nature and the failure to recognize this has led to
policies that have brought us no closer to an alternative-energy future today than we were in
the 1960s when the notion was first envisaged.[1]

In fact, since the 1960s, humanity has become

quite adept at intensifying large-scale risks through a variety of productivist pursuits. We've built neighborhoods deep in forests that are bound to catch on fire, we've built our cities right up to the banks of constricted rivers prone to flooding, we've erected tall buildings atop triggered faults, and so it's really no surprise that we've constructed an energy system pressed right up against the very limits of power production.[2] Attempting to push these limits back by creating more power through alternative means is a futile endeavor, at least in the current sociopolitical environment of the United States. A growing population insisting on greater affluence will quickly fill any vacancy such maneuvers might pry open. This would not only expand overall energy risks but also increase the number of souls in danger when energy supplies inevitably waver again. This is what I call the boomerang effect.

Energy Boomerang Effect

A central project of this book is to interrogate the assumption that alternative energy is a viable path to prosperity. I have not only outlined the many side effects, drawbacks, risks, and limitations of alternative technologies but have also indicated that we cannot assume that shifting to them will lower our fossil-fuel use.

Alternative-energy production expands energy supplies, placing downward pressure on prices, which spurs demand, entrenches energy-intensive modes of living, and finally brings us right back to where we started: high demand and so-called insufficient supply.[3] In short, we create an energy boomerang—the harder we throw, the harder it will come back to hit us on the head. More efficient solar cells, taller wind turbines, and advanced biofuels are all just ways of throwing harder. Humans have been subject to the flight pattern of this boomerang for quite some time and there is no reason to suppose we have escaped its whirling trajectory today.

In the existing American context, increasing alternative-energy production will not displace fossil-fuel side effects but will instead simply add more side effects to the mix (and as we have seen, there are plenty of alternative-energy side effects to be wary of). So instead of a world with just the dreadful side effects of fossil fuels, we will enter into a future world with the dreadful side effects of fossil fuel *plus* the dreadful side effects of alternative-energy technologies—hardly a durable formula for community or environmental prosperity. If we had different political, legal, and economic structures and backstops to assure that alternative-energy production would directly offset fossil-fuel use, these technologies might make more sense. But it will take years to institute such vital changes. Focusing our efforts on alternative-energy production now only serves to distract us from the real job that needs to be done. Worse yet, if fundamental economic, social, and cultural upgrades are not instituted, the project of alternative energy is bound to fail, which would likely lead to crippling levels of public cynicism toward future efforts to produce cleaner forms of power. As it stands now, even if alternative-energy schemes were free, they might still be too expensive given their extreme social costs and striking inability to displace fossil-fuel use. But as it turns out, they aren't free at all—they're enormously expensive.

This affront may seem intimidating, however many of the first steps for dealing with it are rather straightforward. We'll soon discuss these options. But before we do, there's another rather grisly topic to deal with. It is becoming apparent that energy solutions both large and small are subject to the pernicious fangs of a menace that is well camouflaged. In fact, the most astute energy experts occasionally walk right by it without even noticing. Environmentalists rarely speak its name. Pundits and politicians don't acknowledge it. And researchers know little about it. This phantom goes by many names.

The Rebound Effect Phantom

The nineteenth century brought us a collection of ghoulish and chilling immortals—the headless horseman of Sleepy Hollow, Bram Stoker's Dracula, and even Abraham Lincoln's phantom train, which has been heard leaving Washington DC late at night on a circuitous funeral route toward Springfield, Illinois, where it never arrives. It was during this era, in 1865, that a man named William Stanley Jevons wrote a book about a similar sort of phantom. His book, entitled *The Coal Question*, started out innocently enough. Jevons documented how James Watt's introduction of the steam engine greatly improved efficiency. Seems nice. But this increase in efficiency in turn made steam engines more popular and ultimately drove coal use ever higher.[4] This rebound effect, also termed the "Jevons paradox," arises again and again in various incarnations throughout the history of energy use: *Increases in energy efficiency make energy services relatively cheaper, encouraging greater consumption.*

Energy efficiency can actually lead to negative environmental impacts unless regions institute taxes, caps, or regulations to prevent growing consumption patterns from smothering efficiency gains. As long as energy-efficiency strategies come with checks to prevent the rebound effect, efficiency proponents argue that they are highly effective. For instance, new refrigerators use just a fraction of the energy of models sold decades ago, yet because there is a limit to the amount of refrigeration space one can fit in a kitchen, the benefits of efficiency are usually not usurped by the rebound effect. Similarly, there's no indication that drivers of small cars, who achieve twice the gasoline efficiency of those driving large vehicles, tend to drive twice as much as a result. And based on numerous case studies of businesses, Rocky Mountain Institute researchers claim, "We have not seen evidence that radically more efficient commercial buildings cause people to leave the lights on all night and set their office thermostats five degrees lower. In fact, energy savings in everything from office towers to schools have often been higher than projected. People do not seem to change their behaviors simply because they have a more efficient building."[5]

That's nice, too. But it's not the whole story.

There's another problem. Even though energy consumers might not spend their efficiency savings to buy more energy, they may choose to spend these savings on other products or endeavors that still lead to energy consumption. In this case, energy-efficiency measures can unintentionally inspire other types of consumption, leaving overall energy footprints unchanged or even larger. This occurs at the macroeconomic level as well. In short, energy-efficiency savings frequently lead to larger profits, which spur more growth and thus higher energy consumption.

For instance, another Rocky Mountain Institute study shows that reducing drafts, increasing natural light, and otherwise making workplaces more efficient, can increase worker productivity by as much as 16 percent.[6] This higher productivity allows firms to grow, and the resulting labor cost savings can be spent on new machinery, buildings, or expansion. These rebound effects often dwarf the original energy-efficiency effects, leading to far greater overall energy consumption.[7] In fact, the authors of a central report on the rebound effect conclude, "While the promotion of energy efficiency has an important role to play in achieving a sustainable economy, it is unlikely to be sufficient while rich countries continue to pursue high levels of economic growth."[8] Thus, efficiency efforts will only prove effective as long as we institute contemporaneous reforms to move from a consumption-based economy to one grounded in sufficiency.

It all seems too complex to handle—the dirty secrets, the boomerang effect, displaced externalities, the phantoms! How could we possibly change a complex system with so many entrenched cultural and physical roots? It will require innovation on many fronts, a lot of coordination, and quite some time to be sure. But the first steps aren't too large at all. And even though they are quite simple, they could change the future of our nation and the world. As the British historian Simon Schama has observed, "However dire the outlook, it's impossible to think of the United States at a dead end. Americans roused can turn on a dime, abandon habits of a lifetime . . . convert indignation into action, and before you know it there's a whole new United States in the neighborhood."[9]

Indeed, the big environmental dilemmas looming over us today are really, *really* big—much larger than humanity has ever before been forced to confront. And with big dilemmas come big unknowns:

Is it possible for all of the world's nations to come together in agreement on large-scale energy and environmental regulations, practices, taxes, or caps?

If some nations regulate oil, gas, coal, and other dirty industries, won't multinational corporations simply move to regions with slacker regulations?

Even if rich nations were to dramatically reduce fossil-fuel use, won't increasingly affluent populations in China, India, and other parts of the world simply burn away the fossil fuels anyway?

If the rich world created the vast majority of global environmental damage and put the vast majority of the CO_2 into the atmosphere, why should poorer nations help clean it up?

Energies and economies are conjoined. They always have been. And that's one of the principal reasons that humans find it so difficult to share energy resources and the responsibilities that come with them. It's unlikely that citizens of the rich world will willingly part with their high standards of living. It's even less likely that the world's poor will cease pushing to increase their own. We find ourselves approaching not one impasse, but several.

If you have come here looking for definitive answers, I'm sorry, I can't help you. I'm just another member of the search party. If you're looking for certitude, you can close this book and move on.[10] But if you're willing to deal with something less decorated, we can proceed together, see what we find, and perhaps take a shot at those beasts with a slightly larger gun.

But first, I should come clean about something.

The Little Secret

Meager alternative-energy schemes won't topple the hulking environmental concerns standing before us. They probably won't even budge the beasts. In fact, the alternative-energy boomerang, with all of its side effects and limitations, may make matters worse. I don't suggest that this book will solve these problems, though I do hope it might help clear some undergrowth off an alternate conceptual path—one that will bring us to a point where we can approach the really big problems with a bit more leverage. After all, the future of environmentalism lies in paths, not destinations. That said, there remains the matter of a little secret—a twist if you will—that I have sidestepped until now. This simple confession may already be evident: *Someday, renewable energy will supply most of humanity's energy needs.*

Now before you slam down this book, feeling you've been disingenuously tricked into reading thus far, please bear with me. After all, the only thing I promised you on the cover was the dirty secrets of clean energy. I never vowed to dismiss it altogether. Renewable forms of energy fueled humanity before the age of fossil fuels, and so they will after the fossil fuels are gone. A problem remains, however. There likely won't be enough of the precious renewable energy to go around.[11]

Previously, I maintained that alternative-energy technologies are only as durable as the contexts we create for them. I argue that it's the contexts, not the technologies that require the most development. If in some future age human societies are to operate on cleaner forms of energy, it follows that humans will have less overall energy to work with. A lot less. But even given the enormous sums of energy available today, we have issues with sharing (to put it mildly). Our backs are already up against the wall. And there is little reason to believe that calls for more alternative-energy production or pleading for citizens to drive more fuel-efficient cars will be enough to take much of a bite out of the problems we face on a global scale.

With this small twist in the story, I shall proceed on. Ahead I'll argue that before renewable forms of energy can ever deliver meaningful proportions of supply, we must achieve specific structural reductions in global energy consumption. This in turn will require democratic reimagining of certain cultural goals. As knotty as it may be, I'll argue that nations can move toward success by shifting their *measures* of success from material abundance to abundant communities, and from frivolity to utility.[12] The best way to get precious renewable energies to meet our needs is to simply *need less*—a chore that will be more fun than we might think.

Sacrificing Sacrifice

Two decades ago, antitobacco campaigners spent millions to educate teens on the cancer risks of cigarettes using a simple and important message:

"Sacrifice now, and you'll live longer."

Teens ignored them—living longer was for old people. So campaigners changed their tactics. Instead of linking cigarette smoking to cancer, they linked it to something much less frightening. That is, unless you are a teenager.

The new ads featured suave teenagers coming in for a kiss, but just before impact, their mouths opened to expose a mouthful of smoldering cigarette butts. Teens got the message. The risk of smoker's breath left an impression because it appealed to teens' immediate concerns—their Friday night date. Since the rest of us are just grown-up teenagers, the same tricks happen to work with us. Unfortunately, for the most part the reverse is also true.

"Sacrifice now, and you can help prevent catastrophic climate change."

"Whatever."

Even though the long-term risks of climate change are widely acknowledged in public discourse, it's difficult for citizens to mobilize changes to their individual behavior in response to such nebulous concerns. Concerned citizens may lionize sacrifices for being noble, even virtuous, but as a society we unsympathetically ignore them in practice. It's not because we are bad people. No matter how well-intentioned we may be, we're often unaware or unable to accurately assess the impact our choices have on the larger world. Even when we do have some sense of our impact, we're often persuaded more by immediate interests than by the more abstract and less tangible "common good." Furthermore, individual sacrifices don't hold tremendous potential in the larger scheme of things since corporations, the government, and the military leave the largest energy footprints.[13]

Policies that rely on sacrifice, will power, or appeals to ethics, no matter how valid or insightful they may be, are doomed to be pummeled or even knocked out when pushed into the ring against human behavior. Policies to reduce energy consumption will have a much better chance of success if they generate tangible, upfront benefits such as cost savings, free time, and other valued attributes. I'm not arguing against sacrifice or restraint. I'm only pointing out their limitations as policy tools—especially where there are more capable options.

Developing Congruency

The alternative-energy fairy tale was not scripted by a few sleuth conspirators. Rather, it grew out of a particular alignment of interests between legislators, corporate board members, scientists, environmentalists, journalists, consumers, and many others. Everyone had something to gain from the story. We may not have tried or even expected to become progrowth productivists, supporting the projects of Shell Hydrogen, Exxon Biofuels, and BP

Solar, but it just so happens that after plinko-ing our way down the pegboard of established duties and rewards in society, we ended up in their slot. Our alternative-energy inclinations lined up with the ambitions of those around us. These in turn lined up with the currents of power flowing around the corporate energy sector, which is by its nature designed to consolidate wealth for silent shareholders whom we've never even met. Perhaps you can call that a conspiracy, but it's not the clandestine maneuvering of a few skilled operatives; it's far more subtle and pervasive than that. A web without a weaver.[14]

So we went to an Earth Day party and woke up the next morning spooning the Exxon PR director. What to do? Take the walk of shame back home and give up on environmentalism? Fold our hand and leave the table? Absolutely not! If a particular alignment of interests created the bind we are in, then perhaps it's time to align those interests in a new way.

There's a ready example in California. Decades ago the state "decoupled" energy production from utility company profits. In short, the *less* energy California customers use, the *more* money utilities stand to make. While utility companies in other states are planning their futures around additional capacity and higher production, California utilities are buying energy-efficient light bulbs for their customers, installing smart meters, and replacing inefficient machinery—anything they can think of to get their customers to use less of their energy commodities. This may seem counterintuitive, but only if approached from a productivist mindset. From a well-being mindset, it makes perfect sense. Profits should flow toward undertakings that are socially and ecologically beneficial, not those that are socially and ecologically destructive. California residents have shown that it's their game and they can set the rules as they like. Through a few painless tweaks to its energy system, California armed itself to fight the phantom of Jevon's paradox head on and is still winning on several fronts. The trick was just a simple realignment of interests.

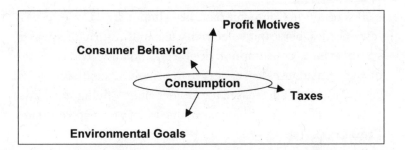

Figure 9: Incongruent power plays Various forces pull energy consumption in different directions. Here utility profit motives and consumer behavior overpower environmental goals.

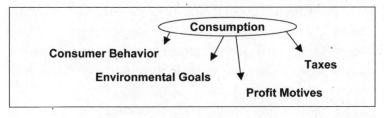

Figure 10: Congruent power plays In a decoupled energy system with stronger tax incentives, interests to reduce energy consumption align and work together to pull energy consumption down.

In regions *without* this congruency, things look less promising. Utility profit motives, low energy costs, and customer ambivalence generate an upward lift on energy consumption. This often overcomes environmental efforts to pull consumption down. It's difficult to lower consumption when muscular forces are pulling it elsewhere.

Meanwhile, decoupling brings these forces into congruence. Profit motives of utilities align with environmental goals, and customers choose to consume less energy when given incentives to use low-energy light bulbs and appliances. Additionally, California's cost penalties for heavy energy users induces a stronger downward pressure on consumption levels than just a simple flat tax rate.

The results in California have been impressive. Over the past few decades, per-capita electricity use in California remained steady even as consumption doubled nationally. Today, a pair of Californians uses less energy than a single Texan.[15] And lower energy use hasn't decreased the Golden State's living standards. In fact, Gallup ranks California among the top ten happiest states (Texas ranks twenty-one).[16]

Even more powerful are those energy strategies that become automatic, where performing the energy-saving task is built into daily life. For example, people who live in dense cities appreciate the convenience of walking down the street for groceries or to see friends. Even though they are using far less energy than their suburban counterparts, they wouldn't know it. These energy-saving conveniences become such an appreciated part of daily life that residents of walkable communities perform them unconsciously.

The environment is not an objective *thing*, sitting there in front of us, plain and obvious, but a complex cognitive construct, a hybrid between ecological states and social understandings. It follows that the energy and environmental solutions we develop will have to be equally enveloping forces, addressing individual behavior, social norms, institutional actions, and technological advancement—all at the same time.[17] Such pervasive changes are difficult, if not impossible, to orchestrate without considerable self-organization among various interest groups. Therefore, the policy paths we pioneer should be structured to appeal to human behavior and, most importantly, draw upon human creativity, an energy resource that too often goes untapped.

There are two primary paths to reducing the energy state of the economy: making the economy more efficient and shrinking the overall *material* economy. We'll have to do both. A green economy is a smaller economy. The options for realizing it would be laudable projects even if we weren't backed up against the proverbial energy wall.

In the following chapters, I'll introduce numerous durable proposals that hold the potential to both decrease energy use and increase human well-being. They don't require advanced technologies. They aren't expensive. And while some may require adjustments, they typically won't require sacrifice—their congruency makes them palatable to wide segments of the population and therefore realistically achievable.[18]

The First Step

I do not intend to pour the sweet syrup of a utopian grand narrative over the remaining pages of this book for you and other readers to passively lap up. Grand narratives have their place, but they too often come at the expense of practical and achievable first steps. Throughout this book, I have aimed to initiate an intervention—a reframing of certain taken-for-granted environmental story lines. I won't be scribbling a quick prescription to fix the mess I've created. Instead, I aim to acknowledge how complex it will be to thoughtfully trim the motivating roots of our energy challenges. And I'll ask for your help.

The future environmental movement won't offer itself up as a receptacle for energy firms, car companies, and product marketers to plug into. Nor will it focus narrowly on preaching at people to consume less. The environmental movement's greatest returns will come through building alluring sociocultural frames wherein citizens are offered the opportunity to consume less energy and can enjoy the benefits of doing so.

To start, I'll bring some first steps to the table. These first steps are not majestic solutions, but I argue that they might hold the potential to bring us to a place where grander possibilities become possible. I've selected each to be:

Achievable: They have already been instituted with success either on a small scale in the United States or on a large scale in another part of the world.

Congruent: They appeal to the self-interest of many people and groups, lending them the potential to catch on.

Meaningful: Instituting them will tangibly improve people's well-being and/or reduce the risks of human suffering.

We needn't chart the entire path toward a more socially prosperous energy system at the outset of our journey; we need only to take a much less dramatic step with solid footing down the right path. And I ask, Why not take those steps down a path that will accrue other benefits along the way? As we progress, gaining a better lay of the land, we can recalibrate our bearings and move on from there. Adjustments aren't a failure of the original plan but a tribute to the flexibility that it offers. The very constitutional foundations of our country have endured because they were designed to allow for updates.

The following chapters are admittedly incomplete, for writers far more qualified than I have written volumes on each of these themes. For now, I aim simply to introduce these topics, if through a slightly different lens. At the end of each chapter, I'll suggest some ideas that, hopefully, spur some thought into how we might practically move from material and energy consumption to more durable and meaningful forms of social growth and well-being. I am certain that you and other readers will have even better ideas for developing thoughtful energy policies that reach beyond conventional doctrine to overcome social, cultural, cognitive, organizational, and political barriers standing in the way of building a prosperous energy system. My goal is to instigate a shift in focus so that others might imagine solutions that are socially congruent, politically achievable, and decisively influential. If the following chapters assist in those efforts, they will have served their purpose. After all, we are the architects of reality. The question is, What kind of reality do we want to build?

Part III
The Future of Environmentalism

10. Women's Rights

Although the world is full of suffering, it is full
also of the overcoming of it. –Helen Keller

Women's rights, while a noble virtue in itself,
might not seem the likely extension of an argument concerning kilowatts and carbon dioxide. But in fact, one could argue that this unexpected key has greater potential for reducing
greenhouse gases, preventing resource conflicts,
shrinking energy consumption, and improving
human well-being than all of the solar cells, wind
turbines, and hybrid cars that we could possibly churn out of our manufacturing plants.
But is this right? Could empowering girls and
women to control their livelihoods and bodies
against the legislative, cultural, and economic
barriers that still stand in front of them actually
be a worthy *environmental* undertaking? We'll
come back to a rather contentious answer later
in this chapter. But to begin, let's take a look
at why this question is itself so controversial.

A subterranean rift is emerging between environmental advocates and women's rights advocates. And though it hasn't even touched the

political and media surface, the threat of its eruption is perhaps the greatest environmental risk that humanity faces. Here's the issue.

On one side of the rift, certain sectors of the environmental movement are characterizing human population growth as an unsustainable pandemic to blame for a host of environmental troubles. They are moving to embrace women's health and contraception programs in an effort to slow population growth and reduce humankind's impact on the planet. They indicate that such funding is presently inadequate and in some sectors falling. For instance, international women's health programs have markedly decreased as a percentage of the total health aid budget from about 30 percent in 1994 to 12 percent in recent years.[1]

On the other side, human rights advocates argue that this approach treats women simply as wombs and is suspect for ethical reasons. Focusing on contraception, they say, obstructs the benefits of supporting a comprehensive women's rights agenda, not to mention the benefits of a human rights project more broadly conceived. As the alleged carbon and ecological impacts of overpopulation gain attention within the power centers of the environmental movement, they fear we could witness a repeat of the dubious population control programs from the 1960s and '70s. And they point out that while population programs are most commonly associated with the "global South," indicators of women's and girls' well-being in the United States have fallen to among the most dreadful in the Western world. For instance, in 1950, the United States ranked fifth highest among the world's countries in female life expectancy at birth. It now ranks forty-sixth.[2]

It's true that women who are educated, economically engaged, and in control of their own bodies can enjoy the freedom of bearing children at their own pace, which happens to be a rate that is appropriate for the aggregate ecological endowment of our planet. But simply handing out condoms won't foster these requisite preconditions. There's much more to it than that.

In the near future, we won't be faced with deciding *whether* the environmental movement will engage with population concerns (it's already begun), but rather *what form* such engagement will take. It's therefore helpful to address the issue of population first, with all of its complexities and unknowns. Then we might better understand why some frame contraception programs as a solution and why others consider a wide-ranging program for women's rights to be superior. This is a knotty topic, to say the least. We don't have to settle here what is poised to be a large discussion in coming years, but instead simply bring some perspective to what is at stake in these decisions.

A Population Parable

Hundreds of years ago, something mysterious occurred on Rapa Nui, a Polynesian island formed by three extinct volcanoes that we know more commonly as Easter Island. Today, great monuments peering over the beaches echo a formerly thriving society of roughly twenty thousand, with a centralized government that once inhabited the island, but then suddenly and inexplicably collapsed. Archaeologists have attempted to extract a history of the island's enigmatic events by unearthing and analyzing the island's waste heaps. Their findings have become a topic of great interest and contention, perhaps because they have proven a not-so-subtle hint to the challenges we could face. As author Jared Diamond chillingly insists, "Easter Island is the earth writ small."[3]

The island's early residents seemed to have enjoyed a bounty of edible roots and fruits as well as abundant giant palms, which they felled and used to transport stone for the now famous monolithic moai statues. The islanders built canoes for hunting marine mammals and harvested plants for medicinal purposes. But as time went on, the daily lives of the islanders changed and their garbage piles changed along with them. Island natives seemed to be working their way down the food chain, from turtles, to shellfish, to grasses. The once prevalent porpoise bones and agricultural crops were absent from later waste heaps, indicating that deforestation may have prevented islanders from constructing canoes to hunt. It may also have triggered topsoil erosion. In the later piles, aquatic mammal bones were replaced by rodent bones and, eventually, in the later stages, human bones.[4]

If this was indeed the fate of the Easter Islanders, it was not the fate of their contemporaries in other parts of the world. For instance, when Japan faced deforestation in the 1600s, during a population and construction explosion following the 150-year civil war, the Japanese implemented several safeguards. They shifted construction practices away from heavy-timber and toward light-timber methods. They also employed more fuel-efficient stoves and heaters. Finally, they instituted a system of forest management, which has resulted in a modern-day Japan that, despite its high population density, is over 70 percent forested.[5] Throughout history, some societies have succeeded in managing certain shared resources in ways that securely pass those assets on to subsequent generations. As a global community, we have failed to do the same.

Perhaps the most infamous figure to argue the case of resource limits was Thomas Robert Malthus, an early nineteenth-century Englishman who claimed that while population can increase exponentially, food supply can only increase linearly. In other words, we are better at making babies than producing the food to feed them. This, argued Malthus, leads to S-curve population swings: periods of bountiful agriculture and increasing family sizes are followed by periods of high demand, increasing costs, and eventual famine. According to Malthus, this was a natural law that interfered with any kind of great improvements in society.

Malthus was not shy about translating his theory into social prescriptions. For instance, he railed against protections for the poor, arguing that poor laws and charity would simply remove natural checks to population without increasing the food supply, subsequently leading to a larger population. His contemporaries pleaded with him to incorporate contraception as a preventative check in subsequent volumes of his work. But through six published editions of his notorious treatise *An Essay on the Principle of Population*, he refused. Contraception and charity were against his moral philosophy. Understandably, he attracted numerous critics. Friedrich Engels called his theory "vile." Karl Marx lashed out at Malthus for being a "shameless sycophant of the ruling classes," and Charles Dickens developed a character based on Malthus that we know today as Scrooge.

Moral philosophy aside, the central inadequacy of Malthus's theories was that in the industrializing and urbanizing nineteenth-century Europe, they simply no longer applied. It appeared that Malthus's theories were obsolete as soon as he had developed them. This didn't stop aristocrats from adopting his theories to justify their own self-interested political agendas. Even today, pundits call upon Malthus's theories to promote everything from extreme anti-immigration measures to projects of eugenics and other initiatives that are not only naïvely simplistic but often racist or classist. These evocations of Malthus's work corrode his already less-than-shiny reputation in the history of demography.

Still, numerous academics, environmentalists, and social critics dare to view ecological constraints through a population lens. For instance, Professor Albert Bartlett from the University of Colorado, Boulder, is not afraid to get down and dirty with the physics of unconstrained resource consumption and population growth. He points out that even small growth rates of just a few percent per year, whether it be in material consumption, energy use, or population, can lead to enormous escalations in scale over modest periods of time. He claims that our inability to understand the exponential function is "the greatest shortcoming of the human race."[6] Using a local example in Boulder County, Colorado, which has fifteen times more open space than developed land, he shows that the seemingly modest population growth rates

suggested by city council members would cause the city to overflow the valley within a single human lifetime. What about "sustainable growth" and "smart growth"? He insists that at their
core these are still just the same pernicious forms of growth. If
we compare the pitfalls of unconstrained growth with the tragedy
of the Titanic, then "smart growth" is simply a first-class seat.

Bartlett extends this account to natural resources such as oil
and coal, showing that the growth rate of these resources will
inevitably drop to 0 percent growth—and perhaps sooner than
we think. For instance, coal advocates often claim that there is
enough coal to last hundreds of years, but only with the caveat
"at present consumption levels." However, if even a small population growth of just 1 percent annually is figured in, the length
of time it would take to burn through that coal drops dramatically.[7] Even at 1 percent growth, a population will double every seventy years. At 2 percent growth it will double in thirty-
five years (simply divide seventy by any rate of growth and
you'll get the doubling time). The New America Foundation,
a think tank, estimates that the U.S. population will grow from
310 million in 2010 to 500 million 2050, and a billion by 2100.
It's dubious to assume that "present consumption levels" are an
adequate measure of future demands, yet this is the statistical
monkey business that energy productivists employ when presenting their various commodities as the resources of the future.
This deception is not limited to those in the coal, gas, and oil
business—those in the business of selling wind, solar, and biofuels employ the same techniques.

Population critics also point to several factors that endanger
future food security. First, they warn that the petrochemicals fer-
tilizing the green revolution will eventually be in short supply.
Second, they argue that the nitrogen, phosphorus, pesticides, her-
bicides, and intensive-farming practices that advanced the revolution are mounting long-term risks such as superpests, dead
zones, and eroded topsoil.[8] Third, they point out that the natu-

ral process of photosynthesis is less effective above eighty-six degrees Fahrenheit, so expected regional temperature increases could stunt crop yields.[9] When a European heat wave captured headlines in 2003 for killing tens of thousands of people, it was not yet known that grain and fruit crop yields would fall between 20 and 36 percent that summer.[10] Climate scientists expect that the few degrees' rise in average temperature that year will be the norm by the end of the century.[11] The net result? They conclude that our future world will either contain more humans living on much less or fewer humans living on more. For instance, author Ronald Wright calls his readers to action at the end of his book, *A Short History of Progress*:

> Things are moving so fast that inaction itself is one of the biggest mistakes. The 10,000-year experiment of the settled life will stand or fall by what we do, and don't do, now. . . . We are now at the stage when the Easter Islanders could still have halted the senseless cutting and carving, could have gathered the last trees' seeds to plant out of reach of the rats. We have the tools and the means to share resources, clean up pollution, dispense basic health care and birth control, set economic limits in line with natural ones. If we don't do these things now, while we prosper, we will never be able to do them when times get hard. Our fate will twist out of our hands. And this new century will not grow very old before we enter an age of chaos and collapse that will dwarf all the dark ages in our past.[12]

Wright is not alone in extending the collapse of the Easter Islanders as a poignant reminder of what could happen to us. Such evaluations are tenuous at times because they compare simple, small-scale, premodern societies with our globally interconnected systems. Much is lost in translation. Others criticize these stories for offering only a convenient heuristic, not a robust model to account for the staggering risks that we wager today. Still, we shall never encounter such a robust model since there exists only one large-scale global economy that we know of—ours—and its story has not yet been completed. The best we can do is to craft a plan based on our ancestral wisdom, ethical bearings, technological capabilities, and historical understandings. If the population critics are correct, we have but one chance to get our future right because a failure will not be restricted to a Polynesian island, but to a planet upon which billions of souls rely. But before we ponder the utility of this parable, let's first take a look at the numbers.

The Population Dilemma in Sixty Seconds or Less

Our numbers are currently growing by about 1.5 million per week, equivalent to adding a fully populated San Francisco to the earth every eighty-six hours.[13] As recently as 1970, the world population was just under 3.7 billion; today it is nearly twice that, at over 7 billion. The U.S. Census Bureau expects the world population to reach 9 billion by 2043.[14]

Yes, the numbers are large and undeniably growing larger, but does that necessarily mean population growth is really such a big problem? The staunchest defenders of population growth enthusiastically point out that all of the world's inhabitants could easily fit within a landmass the size of Texas with enough space for each family to have a modest apartment. Journalists, politicians, and economists lazily allow this nifty estimation to parade through their population reports (as a quick search of the literature makes immediately evident), even though it entirely dodges a very basic observation: the land we physically occupy is far smaller than the landmass required to sustain us. There may be enough room for all of us in Texas, so long as we don't cook, shop, bathe, drive, use energy, or eat much of anything for the rest of our lives. Long before we start bumping elbows with the rest of the world's inhabitants, we will reach limitations on farmland, water, and other resources necessary for our survival. It's misleading to state that we could all fit in Texas with-

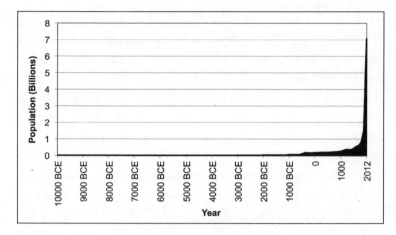

Figure 11: Global world population World population grew slowly and remained under one billion until the Industrial (coal) Revolution. (Data from the U.S. Census Bureau)

out acknowledging that the land and resources to feed, clothe, and otherwise support us could not.

The pertinent question, therefore, is not how many people can *fit* on the planet, but rather, how many people can *live* on the planet? Attempts to answer this question are raising hairs in some circles, with many theorists claiming we have already surpassed a sustainable population level and others insisting that such limits are far off or irrelevant. For instance, cornucopian superstars Julian Simon and Bjorn Lomborg have argued that through better technologies and by expanding mining and agricultural activities, we can easily provide for a growing population well into the future. They could be right, but others ask if that's really what we'd ideally choose to leave our grandchildren—a massive high-tech human feedlot surrounded by cavernous pit mines?

What's possible isn't always what's preferable.

Some technological optimists have gone so far as to insist that extraterrestrial colonization is the logical accommodation for our growing numbers. In fact, they've convinced 63 percent of the nation's citizens that the mass colonization of the moon is a possibility.[15] The fact that American science writers so frequently evoke extraterrestrial colonization as a reasonable option to deal with overpopulation or mass planetary resource depletion is disappointing and embarrassing. Even if it were an option, the United States doesn't even have the financial means to fly its population from Tampa to Tallahassee, let alone to another celestial body.

Down-to-earth population pundits point out that our numbers aren't the problem—modern capitalism is. If every person globally were to start consuming food, goods, and energy at the rate that Americans enjoy, there simply would not be enough superfluities to supply everyone for long, an uncomfortable detail that remains untarnished despite all of the food throwing among population theorists. In fact, sustaining American levels of consumption for every person on the planet would require the service of multiple planets like ours, each decked out with the most advanced systems for agriculture, mining, manufacturing, power generation, and water management. But we have just one planet and we can only maintain high levels of consumption so long as many others do not.

I don't point to these assessments to arouse guilt or lend them legitimacy, but to highlight a commonly posed ethical dilemma that's being lifted to the table: increasing our numbers necessitates that many others live with less, but placing limits on procreation trespasses on the sexual and reproductive autonomy of individuals in a free society.

What gives?

The Population-Consumption Impact

It appears that population growth among rich consumers cannot continue to expand at today's pace indefinitely. As echoed in the straightforward wisdom of economist Herb Stein, "If something cannot go on forever, it will stop."[16] Most demographers expect the human population to peak sometime after 2050, but there is much debate over what shape and timing of the population curve is most desirable. Furthermore, population forecasts are about as reliable as those predicting the weather. Just a small ripple in birth rates today can derail projections by billions tomorrow. This doesn't stop activists from calling upon these estimations to support a variety of population initiatives.

In the rare case that politicians address population growth, they usually frame it as an issue for the governments of poorer countries to deal with. Yet the average American accounts for more energy use in forty-eight hours than a Tanzanian will use throughout the entire calendar year, indicating that large populations may not be so problematic on their own. Growing populations become problematic when multiplied by Costco.[17] As Matthew Connelly points out in his book *Fatal Misconception: The Struggle to Control World Population*, population growth should not be taken out of its cultural context. He argues that if citizens of poor nations were to have on average of just 2.1 kids, but also drive cars, use air conditioners, and consume like citizens of rich nations, their environmental impact would be far greater than that of an additional billion subsistence farmers.[18]

Presumably, if there were fewer of us, the energy we use, the products we consume, and the waste we produce would not pose much of a problem, but the sheer magnitude of our escalating numbers dominates almost every consumption calculation. Take for instance Chris Jordan's artistic project *Running the Numbers*. His project depicts: one million plastic cups, the quantity airlines use every six hours; 426,000 cell phones, the number Americans retire every day; 1.14 million paper bags, the quantity Americans use every hour; 170,000 disposable Energizer batteries equivalent to fifteen minutes of production; and two million plastic bottles, the quantity Americans throw away every hour.

If the world population were extremely low, everyone could

drive large diesel rigs and the effect on the world's ecosystems would be negligible, but ten billion people driving electric cars would prompt an environmental disaster. Somewhere in between, presumably, lies an optimum level, where human residents of the earth can live in numbers plentiful enough to support the modern economies, societies, arts, and technologies to which we are accustomed while not depleting the ecosystems necessary for our survival. But what is it?

The Elusive Optimum Population

The Optimum Population Trust, a nonprofit group led by a wide array of educators, CEOs, and prominent naturalists, including Jane Goodall and David Attenborough, estimated a world population the planet can sustainably support while allowing for the possibility of every person to achieve a high standard of living. A decade ago, the organization's founding chairperson, David Willey, claimed that we can estimate the maximum carrying capacity of the planet by first considering the most basic share of energy resources, freshwater, agricultural land, and other factors allotted for each person. He adjusted this resulting maximum of about two to three billion people to account for additional niceties such as personal liberty, recreation, political representation, and all of the other features we'd like to maintain that are above and beyond our basic necessities. Finally, he maintained that too small a population could adversely affect technological innovation and economies of scale. Accordingly, Wiley determined an optimum population of one to two billion people should not drop below half a billion.[19] Incidentally, Dartmouth College's Jim Merkel, who identifies smaller families as "a solution with no losers," argues that if small families with an average of one child per woman became the norm throughout the world, within just one hundred years, the human population would naturally shrink to about one billion people, which would allow 80 percent of the earth's bioproductive land to go wild for the other twenty-five million or so other species on the planet.[20]

Others challenge such estimates. Joel Cohen, head of the Laboratory of Populations at Rockefeller University, points out that "no scientific estimates of sustainable human population size can be said to exist."[21] That's because it's not so simple to quantify factors such as human desire, material well-being, economic arrangements, and other social and demographic variables. Cohen collected over five dozen estimations of an ideal global population going back to 1679, with figures ranging from less than one billion to over one thousand billion.[22] He concluded that ideal population numbers are ultimately political contrivances, "intended to persuade people, one way or another: either that too many humans are already on Earth or that there is no problem with continuing rapid population growth."[23]

Even if nations could agree upon some optimum population, achieving it could be a messy affair, ethically and politically speaking—and not only because of humans' history with such projects. There's another hitch with easing population growth. Humankind's population throttle seems to be stuck.

Population's Sticky Throttle

Population isn't as easy to modify as critics sometimes suggest. Even if the current worldwide average birth rate of 2.6 children per couple were to somehow fall to a simple replacement rate of about 2.1, the world population would still continue to expand for seventy years before stabilizing at about thirteen billion people, due to several population momentum factors.[24] First, younger members of society will be here for a while and a disproportionately large 40 percent of humans are now in their teenage years or younger. In other words, all of the expected two billion senior citizens in 2050 have already been born. Second, we are lucky enough to be living longer than our ancestors did. The average lifespan during the first millennium AD was about twenty-five years; the average life expectancy in 1900 was just thirty years; today it is sixty-seven.[25] In rich countries, the average life expectancy is seventy-eight and could rise to the mid-nineties by 2050, according to a comprehensive study between researchers from the Max Plank Institute and Cambridge University.[26] Centenarians were once rare but the United States currently has over one hundred thousand inhabitants over one hundred years old—in fact, they are the fastest-growing segment of the population.[27]

A final factor, overlooked by most population forecasters, is the emerging field of life extension, encompassing biological interventions as well as rapamycin, sirtuin stimulants, and other experimental drugs to slow the aging process. The small firms that popped up to research these elixirs attracted rather incredulous glances from medical professionals until 2009, when Glaxo-SmithKline made news by swallowing one for a hefty $720 million. Now they're taken seriously. If successful, these various strategies could continue to draw life spans closer to, or some say beyond, the theoretical "Hayflick limit" of roughly 120 years, at least for the world's more affluent individuals.[28]

Talking about Population in Public

Given that our sheer numbers so momentously rock the calculations surrounding just about every consumptive activity, why do politicians, economists, and journalists so rarely approach the topic of population? One logjam is the staggering girth of the statistics—it's difficult for people to get their minds around the meaning of such large numbers. What's the difference between a stadium filled with ten million penguins versus one filled with one hundred million? It is challenging to appreciate the distinction. Similarly, it is challenging to isolate demographic scale from the many environmental issues we face today. For instance, it's difficult to separate *population* from *consumption*— either can be blamed for the impact the other presumably causes.

Furthermore, researchers are simply apprehensive about pursuing certain modes of population research for fear of being associated with campaigns of eugenics and sterilization. And then there is the matter of blame, as author Robert Engelman argues in his book *More*: "Who wants to be seen as implying that parents who have three or more children and want decent lives for them are somehow more at fault for our environmental problems than governments or corporations or drivers of sport utility vehicles?" He continues, "It's not that there's any compelling research absolving population growth as a long-term force in environmental degradation. It's just that researchers don't like to risk their reputations by appearing to hold prolific parents answerable for the sorry state of nature."[29]

Politicians are similarly prone to switching their stance whenever it is politically advantageous (it is a little-known fact that before recalibrating his bearings toward a run to become president, George W. Bush earned the nickname "Rubbers" for his interest in population reduction and his support of family planning initiatives).[30] Still, given our lengthening life spans, our high consumption rates, and the eventual resource limits to our growing population, we might wonder what official population policies our elected leaders have initiated to address these issues.

There aren't any.

In fact, a United States government report to the World Summit on Sustainable Development in Johannesburg clearly states: "The U.S. does not have an official population policy, in part because population density is low in the United States and large regions of the country are sparsely populated"—a mentality more recently evoked by presidential economic adviser Lawrence Summers.[31] In other words, a bureaucratic reprise of the we-could-all-fit-in-Texas tale.

Even though the United States does not have an explicit population policy, critics argue that it does have de facto population policies built into its tax and legal systems that culminate in a field favorable for continued population growth.[32] These invisible population reinforcements aren't surprising given that population growth has been historically associated with economic growth and prosperity. Corporate sales and production forecasts rest on expectations that there will be more Americans to consume their products next year than last year. Since population growth adds labor, multiplies customers, and lifts profits, what politician would dare mess with it?

Growth isn't just an American phenomenon; capitalism, socialism, and communism are by many measures all under the same umbrella of growthism. Nevertheless, economic growth in the sense that we have come to understand it—wealthy populations continuously escalating demand to be satisfied by supply cycles ultimately fueled by the exploitation of natural resources and human labor—will *necessarily* come to an end at some point given the finite limit of our natural surroundings. Economies can grow exponentially—the planet cannot. Mainstream economic thinking bars entry to this simple and obvious physical limitation. The rest of the field has been slow to take up those realizations as well. There's a saying among economists that their discipline progresses one funeral at a time.

Economic expansion and growth are so entrenched in our psyches as positive factors that it's difficult to conceive of a form of prosperity without them. Progrowth economists stress the importance of a growing population in order to create consumer demand, supply an ample workforce, and provide for older generations. A shrinking population ignites challenges to the progrowth model of prosperity. A feature section in *The Economist* portends: "As more people retire and fewer younger ones take their place, the labour force will shrink, so output growth will drop unless productivity increases faster. Since the remaining workers will be older, they may actually be less productive . . . a reasonable supply of younger people is needed to counterbalance—and fund the pensions of—a growing number of older folk."[33]

Others are more strident in framing population growth as a Ponzi scheme—one where nations expect ever-greater numbers to support older populations in a cyclical progression toward unmitigated growth that inches precariously upward like a house of cards. Population critics argue that we don't have time to wait for funerals in the field of economics. For instance, Paul Ehrlich, author of the controversial 1968 book *The Population Bomb*, and Anne Ehrlich, coauthor of numerous subsequent works, are spearheading a renewed interest in overpopulation, affirming their central thesis that "encouraging more population growth now just delays the inevitable. Sooner or later the age composition change must occur, and any disruptions it will cause will need to be dealt with by our decedents in an even more overpopulated, resource-depleted, and environmentally devastated world."[34] Ultimately, today's younger people may not find much utility in the brand of economic thinking their parents' generation bequeathed them.

Fears of a Shrinking Population

We are accustomed to imagining our lives as providing momentum for future generations, but if future populations are set to decline, does that in turn make our lives less meaningful? In a *New York Times* op-ed, columnist David Brooks pondered a certain form of population reduction by asking: "What would happen if a freak solar event sterilized the people on the half of the earth that happened to be facing the sun?"

His response stopped just short of a total Armageddon:

Without posterity, there are no grand designs. There are no high ambitions. Politics becomes insignificant. Even words like justice lose meaning because everything gets reduced to the narrow qualities of the here and now. If people knew that their nation, group, and family were doomed to perish, they would build no lasting buildings. They would not strive to start new companies. They wouldn't concern themselves with the preservation of the environment. They wouldn't save or invest. . . . Within weeks, in other words, everything would break down and society would be unrecognizable. The scenario is unrelievedly grim.[35]

Or unrelievedly nutty. Brooks's proposal that when faced with imminent decline, we will neither adapt, nor adopt children, but rather just throw in the towel "within weeks" seems unreservedly ahistorical. We might be tempted to write off Brooks's tirade as nothing more than a heteronormative and xenophobic anxiety fest. But his particular episode of population panic may prove indicative of broader posterity angst.

Meanwhile, philosopher David Benatar is raising hairs through his book *Better Never to Have Been: The Harm of Coming into Existence*. He argues that coming into existence is a serious harm, procreation is always bad, and that all would be better had we simply never been born in the first place (though he is a gentleman to maintain that once born, our lives should be valued and protected). He sums up his thoughts on procreation by stating:

> Creating new people, by having babies, is so much a part of human life that it is rarely thought even to require a justification. Indeed, most people do not even think about whether they should or should not make a baby. They just make one. In other words, procreation is usually the consequence of sex rather than the result of a decision to bring people into existence. Those who do indeed *decide* to have a child might do so for any number of reasons, but among these reasons cannot be the interests of the potential child. One can never have a child for that child's sake. That much should be apparent to everybody . . . not only does one not benefit people by bringing them into existence, but one *always* harms them.[36]

Whether or not we agree with Benatar's allusion to reproduction as a selfish project of living beings rather than one evolving from some enduring good for posterity, he certainly does, and rather unsympathetically so, split open posterity as a contested notion, one held together by the bindings of a particular social imagination.

Hardly a topic for polite dinner conversation, Dr. Benatar's thesis is unlikely to accumulate heaps of backers anytime soon. More likely, the thought of a dwindling population will rouse fear, not relief. For this reason, population critics are seeking to balance the challenges of shrinking our numbers with the corresponding benefits of smaller families and a smaller global population.

Risks and Benefits of a Shrinking Population

Economies will have to mature to accommodate lower birth rates; a ballooning number of seniors will be expensive for younger generations to support. A decade ago, Japan had four workers supporting every retiree; in a decade it will have just two. With a birth rate of just 1.4 children per woman, demographers expect Japan's population to drop from 126 million to well under 100 million by 2050. Countries such as Japan, Germany, and Italy, with birth rates far below the replacement value of two children per couple, are already facing the challenges that aging populations pose, including threats to pension and health-care commitments. How are they holding up? In a smack to the face of finger-waving economists, these countries remain standing, at least so far. Perhaps age begets wisdom; their levels of overall well-being are far less dreary than many had prophesied. Yet the road ahead looks to be far more rocky and rutted, as a glut of recent literature explicates in detail. These will be the countries to watch, as they struggle to develop modes of prosperity that are not tied to the convenient growth that expanding populations offer. Their experience with aging populations anticipates imminent adjustments elsewhere, making their plight especially relevant to future environmentalists.[37]

These challenges are at least partly offset by some less-publicized and intriguing benefits. For instance, these regions have fewer children to birth, cloth, bathe, house, and educate. Families spend less on formal childcare and companies spend less to replace the pay of caregiving parents. Since younger people perpetrate most crime, crime rates tend to fall as populations age—so do national expenditures for policing, incarceration, antiterrorism, and related security activities. The Norwegian demographer Henrik Urdal determined that for every 1 percent increase in youth population, a country's risk of armed conflict rises a staggering four percentage points.[38] The correlation between youth and violence holds in autocracies as well as democracies. In a demographic context where younger adults are in short supply, countries may think twice before committing their precious few youth to military service and consequently may be more proactive in preventing regional disagreements from boiling over into armed conflict.[39]

Aging nations could presumably channel cost savings from childcare, policing, military, incarceration, and other such activities into programs to serve elderly citizens. But this will require forms of governance that focus on the well-being of people over established special interests, another matter that shrinking populations might transform. Some theorists argue that overpopulation dilutes democracy. For instance, early in American history, each House member represented fewer than thirty thousand constituents but today each representative stands for over six hundred thousand, a dilution of representative democracy by a factor of twenty. Similar effects occur in state governments, town halls, and school boards across the nation.[40]

Some argue that this dilution affects our perceptions of the value of life more broadly. When journalist Bill Moyers asked Isaac Asimov how population growth will affect our concept of dignity, he replied, "It will be completely destroyed." He explained:

I like to use what I call my bathroom metaphor: If two people live in an apartment, and there are two bathrooms, then both have freedom of the bathroom. You can go to the bathroom anytime you want to and stay as long as you want to for whatever you need. And everyone believes in the freedom of the bathroom; it should be right there in the Constitution. But if you have twenty people in the apartment and two bathrooms, no matter how much every person believes in freedom of the bathroom, there is no such thing. You have to set up times for each person, you have to bang at the door: "Aren't you through yet?" and so on. In the same way, democracy cannot survive overpopulation. Human dignity cannot survive it. Convenience and decency cannot survive it. As you put more and more people onto the world, the value of life not only declines, it disappears. It doesn't matter if someone dies. The more people there are, the less one individual matters.[41]

Berkeley researchers Malcolm Potts and Martha Campbell back up Asimov's point with an economic correlate. They argue that "low birthrates aren't a consequence of national wealth, rather, they're needed to create it. . . . Democratic success stories such as Brazil, Chile, Indonesia, and South Korea only came after populations in these countries stabilized."[42]

Finally, parents that opt for smaller families may also find themselves financially better off. Parents pay on average $8,000 for childbirth and over $12,000 a year on food, shelter, and other necessities, totaling over $200,000 by the child's eighteenth birthday.[43] College is extra—a two-year degree costs $24,000; a four-year degree from a public university averages $72,000; and a degree from a private college, $148,000.[44]

Professor Bartlett goes so far as to pose the following question: "Can you think of any problem in any area of human endeavor on any scale, from microscopic to global, whose long-term solution is in any demonstrable way aided, assisted, or advanced by further increases in population, locally, nationally, or globally?"[45]

The Push for Contraception

In order to meaningfully reduce the world population in the midterm future, humans would have to shrink the average number of children per couple to well below two. Nevertheless, many of us shift in our seats when the topic of population reduction arrives to the table—and for good reason. In addition to the unsavory list of famines and disasters that Mother Nature keeps, humankind has engineered all sorts of frightful schemes for population control under the flag of various social, ideological, and political causes. Columbia University historian Matthew Connelly asserts: "The great tragedy of population control, the fatal misconception, was to think that one could know other people's interests better than they knew it themselves."[46] This tragedy is exemplified by the one-child policy in China, a failure by many measures, including its purported goal. Chinese fertility still exceeds rates in many parts of the world where no such authoritarian regulations exist.

But out of the ashes of the widely disgraced population control movement is rising a reenergized cohort of activists who insist that preventing population growth needn't be unjust and coercive. In fact, they argue that most of the world's couples already practice some form of procreative control voluntarily. Humans have sex far more frequently than is required for desired procreation (indicating there might just be something else people enjoy about sex). Therefore, fertility rates are centrally dictated by the ability of people to enjoy sex while avoiding the procreative part.

Throughout a wide variety of settings, demand for contraception increases as it becomes more available, especially if accompanied by fertility education. Contraception advocates point to Costa Rica, Iran, Sri Lanka, and Thailand, countries that cut their overall fertility rates in half by providing basic fertility education and contraception choices. When Iran introduced fertility counseling for newlyweds and broader contraception availability, the country's overall fertility rate dropped from an average of 5.5 children per couple to under 2.0 within fifteen years; the fertility rate in Tehran is now 1.5.[47] Analysts attribute the program's outcomes to its cooperative design, which most notably garnered support from local religious leaders.

Still, many women worldwide are denied access to even basic contraception and denied empowerment over their own bodies by husbands, mothers-in-law, religious authorities, and even their medical providers. In response, contraception proponents advocate for education to correct misinformation and fear surrounding contraception.[48] Researchers from University College London, John Guillebaud and Pip Hays, insist: "As doctors, we must help to eradicate the many myths and non-evidence-based medical rules that often deny women access to family planning. We should advocate for it to be supplied only wisely and compassionately and for increased investment, which is currently just 10 percent of that recommended at the UN's Population Conference in Cairo."[49] Indeed, the UN asserts that its first Millennium Development goal of eradicating poverty and hunger will be a difficult or even impossible task without more attention and funding for family planning.

But even while women's advocates agree that poor access to contraception is a problem that must be addressed, they argue that a focus on contraception is obscuring a much larger problem looming over the world's women.

Moving beyond Contraception

Women have led many initiatives to fight environmental injustice worldwide as they often endure disproportionate environmental risks. For instance, women not only suffer the ecological injustices stemming from energy production themselves, but also disproportionately care for the young and elderly who are adversely impacted. Population control campaigns of the 1960s and '70s framed women as wombs and in the most troubling cases led to a litany of injustices including the sterilization of women without their full consent.[50]

Environmental dictates have sometimes made life harder on women. In an effort to comply with international carbon-trading schemes, large corporate energy users have cordoned off sections of rainforests and set up monoculture tree plantations to act as carbon offsets to their activities, making it difficult or even impossible for locals to gather firewood, a chore often assigned to women and girls. In the worst cases, this reallocation of resources forces locals to leave their established communities altogether. During periods of dislocation women are especially vulnerable to sexual and domestic violence. They are also at greater risk of malnutrition in societies where scarce foodstuffs and medical care are divided along gendered hierarchies. Climatologists expect local environmental risks to multiply as poor populations absorb climate hardships directly resulting from entrenched fossil-fuel use in the rich world.[51]

So to blame mothers for producing too many children, or to place the responsibility of population reduction on women and girls of the global South, or to believe that simply prescribing contraceptives will solve the population crisis is clearly problematic. Yet this is exactly how much environmental thought on the subject has historically been framed.

The outcome for women, their families, and the environment will likely be more favorable if nations premise policy initiatives on women's rights rather than technocratic fertility programs.[52] When organizational bureaucrats measure success in terms of raw birthrates and other statistical contrivances, considerations for quality of care and freedom of contraceptive choice are shoved to the sidelines in the name of "efficiency"—a crude approach that in effect says, "Let them eat contraceptives."[53] Rights advocates maintain that this is a concept of technocratic efficiency not unlike the kind historically evoked to draw support for eugenic campaigns, sterilization schemes, and other insalubrious human undertakings. Contributors to a special report in the *Bulletin of the World Health Organization* insist that when it comes to family planning programs, individual rights should come first, with environmental and population benefits following as welcome and important secondary benefits. The authors argue, "Using the need to reduce climate change as a justification for curbing the fertility of individual women at best provokes controversy and, at worst, provides a mandate to suppress individual freedoms."[54]

Future environmentalists will interpret high fertility rates not as a problems in themselves but instead as a symptoms of broader economic and gender inequities.[55] Widespread changes in the risk biosphere will force environmentalists to increasingly address human adaptation to climate events. Broadly conceived human rights are central to addressing these extended challenges. Therefore, it may be helpful to conceive of reproductive rights as part of a comprehensive endeavor that includes HIV/AIDS treatment and prevention as well as literacy, education, and other programs to give women agency over their bodies and lives. As Betsy Hartmann, director of the Population and Development Program at Hampshire College, asserts: "Take care of the population and population growth will come down. In fact, the great irony is that in most cases population growth comes down the less you focus on it as a policy priority, and the more you focus on women's rights and basic human needs."[56]

Resolving the Dilemma

It is now apparent that the ethical dilemma I posed earlier, between leaving others with less by increasing our population versus limiting procreative rights, is in fact a false choice. Rights advocates and environmentalists may discover some common ground—at once avoiding resource risks and bypassing suspect population tactics—by concentrating on the individual rights of women. This will deliver economic and social benefits to the world's citizenry, reinforce democratic governance, prepare populations for climate hardships, and eventually bring the world population down to levels appropriate for our planet's carrying capacity.

There are still some ideological divides that may never be reconciled, but as Frances Kissling, founder of Catholics for Choice, points out, they need not be entirely resolved to move forward. She argues:

> For one set of organizations, whose central goal is achieving women's sexual and reproductive health and rights, there is no reason to include environmentalism or population stabilization advocacy in their agenda. In fact, there are good reasons to avoid these issues. The social transformation needed for women's reproductive rights to be fully accepted as fundamental human rights is in process, but it is not complete. Some groups must continue to work single-mindedly for that transformation in culture and politics by insisting that women's rights are an end in themselves and not a means to a better life for children, men, and society at large.

But she continues:

> At the same time, there is no need for sexual and reproductive health and rights groups to attempt to prohibit all organizations from making links between population, environment, development, and reproductive health or to offer blanket public criticism of such efforts as unethical or unfounded. We have become extremely sensitive to the efforts of the right to ignore or subvert evidence and science in service of ideology. We would fall prey to the same dishonesty were we to insist that these links cannot

be explored. And to claim that they do not exist at all would be intellectually dishonest.[57]

It is tempting to turn away from such difficult issues, but doing so would pass on great risks. From an energy perspective, if our numbers stay high, eventually there won't be enough fossil fuel to go around. Energy pundits bicker about the timing of the crunch but eventually the natural resources upon which we have built our societies, our jobs, our families, and our livelihoods will become scarce and therefore too expensive for most of the world's inhabitants. If meager alternative-energy supplies are not on hand to take up the slack (and they won't be up to the task any time soon, given the scale of global energy use) the ramifications could be disastrous. As traditional fuels stretch thin, nations will shift to low-grade coal and shale oil to fuel their economic activity. As heating costs rise, the world's forests will understandably become an irresistible resource to exploit for fuel. The natural gas and petroleum-based fertilizers that cultivated the green revolution will become too expensive for many of the world's farmers right at the time when crops for biofuels will be in highest demand. Nations may impose food export bans as they did following the 2008 and 2011 food price shocks. Others may use food aid as a weapon, as Henry Kissinger once suggested the United States might do.[58] As the costs to exhume fossil fuels rise, the invisible hand of the market will go right for our throats.

By increasing our numbers, we are ratcheting up bets on human prosperity that cannot be so quickly taken down. Population shifts require generations to run their course. But communities will feel the numerous social, economic, environmental, and civil benefits of broadly prioritizing women's rights within months and years, making this struggle the most important environmental issue of our time, regardless of the degree to which it is overlooked by energy productivists obsessed with technological gadgetry.

Creating a biosphere where populations renew themselves sustainably will take much more than just family planning. It will require sound governance, education, human rights, health care, improved consumption patterns, civil rights, workers' rights, HIV/AIDS prevention and treatment, and a host of other local variables that are themselves tied to international polity and economy. Well-established movements are pursuing these agendas and even though they are not today considered to be environmental issues, they will be.

First Steps: Approaching Population Concerns of Poor Regions

In a widely acclaimed photograph by Stephanie Sinclair, an eleven-year-old girl casts an apprehensive glance in the direction of an unfamiliar gray-bearded man who will soon be her husband. In her country of Afghanistan, girls frequently separated from their families and removed from school were to marry, resulting in a dreadfully high female illiteracy rate of 82 percent.[59] Since the ouster of the Taliban, the outlook for Afghan girls has vastly improved. But the practice of offering young girls for marriage in order to settle disputes or raise money is not limited to Afghanistan; throughout the large Indian states of Rajasthan and Uttar Pradesh, the Amhara region of Ethiopia, northwest Nigeria, and Bangladesh, roughly a third to a half of all girls are married off at the age of fourteen or younger.[60] In fact, the U.S. Department of State determined that child marriage is still a burden in sixty-four countries, affecting about sixty million girls worldwide—some as young as eight years old.[61] Exposed to the reality of this phenomenon, journalist Barry Bearak recounts, "Rather than a willing union between a man and woman, marriage is frequently a transaction among families, and the younger the bride, the higher the price she may fetch."[62] Husbands frequently expect the girls to engage in sexual intercourse before their first menstruation and once sexually mature, preteen or young teen girls are vulnerable to hazardous pregnancies. For an inordinately large proportion of girls, painful death follows.[63]

Family planning and contraception initiatives simply are not enough to confront these and other injustices. And even if they were, they shouldn't be—quick fixes won't do here. Population pressures are symptoms of a profound social malaise. They arise in the context of weak civil rights, anxiety about the future, poor access to health care, illiteracy, disease, lack of resources, lack of education, disregard for the welfare of women, and other complex factors. If future environmentalists and aid organizations relegate themselves to addressing population symptoms rather than the causes and risks identified by those on the ground, then they are bound for failure if history is any guide.[64]

As environmentalists, rights advocates, or both, we should approach the project of women's rights as an end in itself. First, on ethical grounds, and more pragmatically, because women's rights, health, and education can be approached through policy in a way that population cannot. For instance, Congresswoman Betty McCollum (D-Minnesota), Senator Richard Durbin (D-Illinois), and Senator Olympia Snowe (R-Maine) crossed party lines to cosponsor the International Protecting Girls by Preventing Child Marriage Act. It authorizes U.S. foreign assistance programs to combat child marriage and offer educational and economic opportunities for girls in poorer regions of the world. The nonprofit Population Media Center produces radio and television dramas to educate people on a positive array of social, education, and health initiatives that viewers have measurably replicated in their own lives. These are just a couple of the small steps possible toward bettering the global welfare of women exposed to violence upon their bodies and psyches.

Future environmentalists will advocate to expand education opportunities, abolish laws and customs that exclude daughters from property inheritance, and create opportunities for women in public life and public service (see the resources in the back of this book and at greenillusions.org). The United States could start by fulfilling existing international commitments and leveraging its clout as a large trading partner to mandate measurable progress in supporting these rights internationally. The enormous domestic benefits that accrue from supporting women's issues will make this a win-win negotiation. The United Nations Population Fund prioritizes five focus areas:

- Gender-based violence
- Reproductive health inequities
- Economic and education discrimination
- Harmful traditional practices
- Armed conflict[65]

First Steps: Approaching Women's Welfare in America

Forecasters expect the largest population increases in poor nations, yet some argue that slowing population growth in *rich* nations would have a much larger impact on greenhouse gases and energy consumption since affluent children are born into a lifestyle where a greater number of goods and services are allotted to them. Population forecasters have consistently underestimated U.S. population growth. In 1984 the U.S. Census Bureau predicted that the population of the United States would reach 309 million by 2050, but we've already exceeded that today.[66] Recent estimates range from 420 to 500 million by 2050 and up to a billion by 2100.[67] Immigration induces part of the expansion but the bulk comes from births. And many American births come from an agonizingly underserved subset of the population: teenagers.

The United States has the highest rate of teenage pregnancy in the industrialized world, about four times the European average and eleven times the average in Japan, even though rates of sexual activity across these regions are similar.[68]

Teen pregnancy reduction efforts could gain broad political

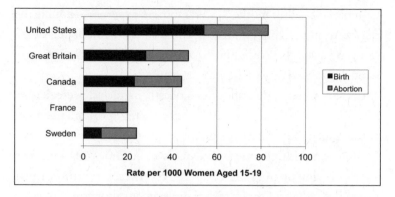

Figure 12: Differences in teen pregnancy and abortion American teens experience far higher pregnancy rates, birthrates, and abortion rates than teens in other wealthy nations. (Data from the Guttmacher Institute)

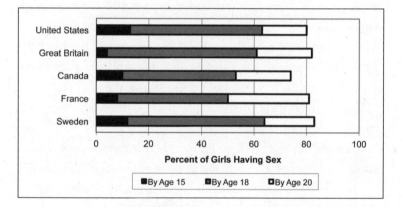

Figure 13: Similarity in first sexual experience Teenage sexual activity differences across wealthy nations are small. Shown are the percentages of women aged twenty through twenty-four who report having had sex in their teen years. (Data from the Guttmacher Institute)

support since risks to young mothers are widely accepted.[69] Researchers link teen childbearing to numerous disadvantages in later life such as lower education attainment, limited economic activity, mental health problems, and a propensity toward postnatal depression three times the rate that older mothers suffer.[70]

Compared to adult mothers, teen mothers experience double the risk of dying during childbirth, three times the neonatal death rate, and a range of other health complications, according to an American Academy of Pediatrics study.[71]

In the United States, about 440,000 teenage and preteen girls give birth every year (some of these young mothers even become impregnated before age ten).[72] A girl growing up in the United States is five times more likely to become pregnant than a girl growing up in France. And an American girl is three times more likely than a French girl to undergo an abortion.[73] Among industrialized nations, America's teens are least likely to use contraceptives or long-acting reversible hormonal methods such as the pill because of the stigma of teen sexual relationships, little motivation to delay motherhood, scarce availability, and high costs.[74] Lowering American teen pregnancy rates just to the level experienced in other industrialized nations would greatly benefit the well-being of the nation's girls and would initiate a host of positive side effects for the nation as a whole.

Almost all teenage pregnancies in the United States are unplanned but so are about half of other pregnancies.[75] Worldwide, there are about eighty million unintended pregnancies every year, which lead to enough offspring to fully populate a Chicago every month.[76] And as shocking as this number may seem, it may very well be a low figure. Couples often don't admit to having an unplanned pregnancy for fear of being seen as irresponsible or careless (their fears seem to be justified; well over half of Americans associate unplanned pregnancies with poverty, a poor education, and a decline in moral values).[77] Professor John Guillebaud, author of the report *Youthquake*, argues, "From an environmental perspective, the fact that so many births result from unintended conception and then, among teenagers, cause so much grief is plainly absurd."[78] It may be equally absurd from an economic perspective, according to James Trussell, director of Princeton University's Office of Population Research, who points out that "preventing pregnancies is far cheaper than the medical costs associated with unintended pregnancies."[79] In the United States, direct medical costs from unintended pregnancies add up to about $5 billion per year.

Some environmentalists are stridently drawing connections between unplanned pregnancies and the nation's carbon footprint or overall energy bill.[80] According to one particularly controversial study led by statistician Paul Murtaugh of Oregon State University, just one American mother with two kids creates a carbon legacy equivalent to that of 136 Bangladeshi mothers with their 337 kids.[81] Murtaugh claims:

> An American child born today adds an average 10,407 tons of carbon dioxide to the carbon legacy of her mother. That's almost six times more CO_2 than the mother's own lifetime emissions. Furthermore, the ecological costs of that child and her children far outweigh even the combined energy-saving choices from all a mother's other good decisions, like buying a fuel-efficient car, recycling, using energy-saving appliances and light bulbs. The carbon legacy of one American child and her offspring is 20 times greater than all those other sustainable maternal choices combined.[82]

Just lowering teen births alone to European levels would prevent the need to generate over thirty billion kilowatt-hours of energy annually (the amount required to support this growth in population).[83] Incidentally, generating that same energy using rooftop solar cells would cost over $500 billion *per year*.[84] I don't aim to validate such admittedly crude comparisons. Rather, I bring them up to introduce a point that is aligned with the topic of this book: *Alternative-energy technologies are hopelessly inadequate and inappropriate tools to address the persistent social ills that create and recreate the nation's energy-intensive cultural patterns.*

While environmentalists march on Washington for solar cells, wind turbines, and biofuels, teen health initiatives remain under-funded and underappreciated. Nevertheless, as we shall consider in coming chapters, we've managed to overlook more than just the well-being of the nation's teens in the rush to create more green energy.

First Steps: Addressing Complex Population Challenges

Teens too will need more than just condoms. According to a wealth of social science research, sexual health is intertwined with psychological pressures and social expectations as well as economic and educational variables.[85] These findings become most salient in cross-cultural studies between the United States and other countries where girls experience far lower rates of teenage pregnancy, birth, and abortion. According to the Gutt-macher Institute, one of the largest aggregators of international reproductive health data, "Countries with low levels of adoles-cent pregnancy, childbearing, and STDs [sexually transmitted dis-eases] are characterized by societal acceptance of adolescent sex-ual relationships, combined with comprehensive and balanced information about sexuality and clear expectations about com-mitment and prevention of pregnancy and STDs within these relationships."[86]

Based on extensive interviews with both Dutch and Ameri-can youth, Professor Amy Schalet, from the University of Mas-sachusetts—Amherst, determined that

> American white middle-class parents "dramatize" adolescent
> sexuality, conceiving of it as involving perilous struggles be-
> tween a young person and his or her difficult-to-control hor-
> monal and emotional urges, between the sexes, and between
> parents and children; while Dutch white middle-class parents
> view teenage sexuality as a phenomenon that can and should
> be "normalized": that is, be subject to self-determination and
> to self-regulation and embedded in relationships that are nego-

tiated with parents as well as integrated into the household. . . .
In the American families, girls are required to bifurcate between
their roles as good daughters and sexual actors because of the
assumed antinomy between the two; while in the Dutch families, parents and daughters negotiate similar tensions in such a
way that girls can integrate sexual maturation into their relationships with parents.[87]

These and other international comparisons suggest that improving the welfare of adolescents will require advancement beyond
harmful customs, as well as contraception and education. This is
hardly a straightforward undertaking, but one worth acknowledging. Other helpful steps are more clear-cut.

For instance, a universal health-care system would not only
ensure that reproductive health and education are available to
everyone but would also ease pressures to bear children arising
from anxieties regarding lack of care later in life. Correspondingly, if there will be fewer children in the future to support a
growing populous of elderly, we should offer our children the
best opportunities to succeed. According to a report by the international aid organization Save the Children, Sweden provides
the most supportive environment for young children, meeting
all ten of the organization's benchmarks for child development.
The United States meets just three. These benchmarks are neither new nor shocking; they're just embarrassingly underde-
ployed. To name a few: provide health care for mothers and
children, coach first-time parents, train early childcare providers, and invest in childhood development programs.[88]

There can be little argument, then, that advancing the rights
of women and girls is an important ethical goal in its own right
that could also dramatically lower the incidence of abortion,
save billions of dollars in health-care costs, increase well-
being, strengthen communities, and handily offset more fossil-
fuel consumption than all of the nation's existing and planned

solar cells, wind turbines, and biofuels *combined*. And unlike expensive alternative-energy programs, which falter during periods of economic hardship, rights, education, empowerment, and acceptance of teen sexuality will remain relevant and achievable even if stock markets fall. Citizens and organizations keen to help solve the world's energy and environmental problems will do the greatest good by directing their attention beyond the shiny eco-gadgets on the stage and toward the injustices behind the curtain.

11. Improving Consumption

And he puzzled three hours, till his puzzler was sore. Then the Grinch thought of something he hadn't before! "Maybe Christmas," he thought, "doesn't come from a store. Maybe Christmas perhaps means a little bit more!" –Dr. Seuss, *How The Grinch Stole Christmas*

Across former meadows, rippling plumes of heat levitate above sticky carpets of black asphalt rolled out as patchwork in front of chain stores standing infinitely shoulder-to-shoulder. Their slightly faded Tupperware claddings are mediated by painted courses of concrete block and punctuated by glazed doors whose scarred hand-pulls open to a fluorescent parallel universe selling everything television has to offer. Inside, chrome and gold flicker like Christmas above a sea of manicured marble tiles that appear to nourish the roots of dubious ficus trees whose bark and leaves blur the boundary between life and a plasticized version of it. Fantasmatic arrays of street culture, sex, food, nature, sports, alcohol, and music are available for the commodified self where this season's lifestyles are on sale, prepackaged and shrink-wrapped.

Ecoconsumerism

Aisle 14: Juice and Soft Drinks——regiments of jellybean bottles bask under a fluorescent sun. Generously labeled "All Natural," they could easily be described as something much different: rehydrated food products with heavily processed syrups and stabilizers packaged in petro-plastic containers, wrapped in labels secured with toxic adhesives, printed with volatile ink compounds, and bounced many miles across the country above the wheels of multiple fossil-fuel-burning vehicles. (One study reports that a third of total energy input for food is mobilized to create sweets, snacks, and drinks with little nutritional value.)[1] Perhaps "natural" says less about the colored liquid and more about how we'd prefer to relate to the food we consume. Is it a yearning to pour something down our throats that is more grounded, stable, and pure for our supposedly polluted bodies? Or perhaps a momentary escape from our hectic and sometimes frustrating day-to-day grind? A teeny tiny revolt against the hypermanufactured landscape surrounding us? In any case, "natural" appears to have deep-seated roots in our psyche and the concept has attracted the most eager of promoters who happen to have enjoyed notable success in translating "natural" into "cha-ching."

And if "natural" isn't enough to get your wallet out, perhaps "sustainable," "green," "organic," "fair trade," or "local" will do the trick—the buzzwords have become so ubiquitous that they now refer to anything and nothing.[2] Builders label luxurious kitchen and bath remodels, costing tens or even hundreds of thousands of dollars, "sustainable" when they veneer the cabinets with bamboo. Auto-show cars receive a green stamp of approval if their seat cushions contain soy-based foam. In fact, today we can purchase almost anything with ecological clearance, so long as we're prepared to lay down a few extra dollars for our conscience.

The green marketing trend began in earnest during the 1970s. Indeed, some manufacturers developed new product lines and manufacturing processes much to the benefit of the planet. However, others simply relabeled, rebranded, and shipped the same products with green halos above their shiny new packaging, perhaps with a higher price tag, too.[3] The organized assault of green marketing screams aloud from even the most remote corners of our superstores, coaxing us to buy green, buy *more*. These campaigns rely not only on a highly stylized concept of nature but also on a consumer class that is unaware of the system-wide implications of mass consumption, or is at least willing to suspend such knowledge upon passing through the doors of their favorite shopping centers.

Manufacturers expect us to believe two things: First, that eco-friendly qualities are measurable and objective. And second, that green products have a neutral or even beneficial impact on the greater environment. Both are falsities.

Sure, there's a patchwork of standards for ecolabeled products but the handing out of green halos is fraught with ambiguity and uncertainty. For instance, a so-called biodegradable diaper will require eons to degrade when entombed within the compacted layers of a landfill without oxygen or sunlight. It may also require more energy to manufacture than a traditional disposable diaper or decompose into undesirable byproducts. The diaper wars of the 1990s between commercial backers of disposable plastic diapers and reusable cloth diapers left consumers with a choice between expanding landfills or releasing detergents into waterways. There's still no consensus on which is worse. At the same time, there is yet another consensus emerging among environmental researchers: there really is no such thing as a green product, no matter how many soybeans went in to making it.

In his book *The Light-Green Society*, historian Michael Bess aptly underscores this conundrum:

Nearly every consumer article, in one way or another, at some point in its life cycle, detracted from or adversely affected the earth's natural environment. A paint thinner that killed no fish when poured into a creek, for example, might only be producible by using the bark of a rare tree, or by adopting a chemical process that released terrible toxins into the air at the factory stage of the product cycle. Even the most seemingly benign and simple items, like an unpainted wooden toy or a "natural" textile, only very rarely made it through the complex life cycle of manufacture, packaging, transportation, distribution, marketing, selling, use, and disposal, without exerting at least a modest negative impact at one or more points along the way.[4]

Green labels permit purchasers to believe that the products within create some sort of positive impact on the environment. But at their absolute best, green labels act merely as an approximate guide to discerning which articles are *least harmful*—which product cycles have resulted in the lowest or perhaps most easily remedied negative effects on the biosphere. Whether we evaluate such bleak prospects as green, ecofriendly, or sustainable depends on how loosely we are willing to define these terms.

When author Joel Makower wrote *The Green Consumer* in 1990, he advanced the idea that we could help to alleviate many of the world's environmental problems though green consumerism: "By choosing carefully, you can have a positive impact on the environment without significantly compromising your way of life. That's what being a Green Consumer is all about."[5]

Today, he disagrees.

"I fought the good fight. Twenty years later, I'm thinking of waving the white flag," he laments. "Green consumerism, it seems, was one of those well-intentioned passing fancies, testament to Americans' never-ending quest for simple, quick, and efficient solutions to complex problems."[6] Makower's consternations do not stand alone.

As the swarm of green products buzzes with an increasingly

sharp pitch, a growing body of ecologists and environmental scientists are warning that it's mostly nothing but hot air. When authors of an investigative report to Congress scrutinized thousands of "green" products, they found that over 98 percent of them contained labeling information that was misleading or even outright deceitful. Some companies falsely claimed that their products earned the federal government's Energy Star rating, a designation reserved for the top 25 percent most energy-efficient appliances, when in fact they failed the test.[7] Green consumerism has calcified into a hard yet vacuous shell—a purely ornamental ideal—little more than a niche segment for marketers to exploit.

The *best* material consumption is *less* material consumption. But that has been a hard sell in its own right. Four decades have passed since Fritz Schumacher wrote *Small is Beautiful* and the subsequent pleading for people to consume less has had a relatively muted effect on ever-expanding consumption patterns among the rich of the world.[8] That's largely because the message of consuming less directly conflicts with the inclinations of a vast majority of people who have become rather comfortable with material consumption and generally believe it leads to greater levels of personal happiness. It does in many cases. Nevertheless, in a great many more, material consumption clearly degrades long-term personal satisfaction. So perhaps we should consume less of the things we'd be just fine without (and perhaps might even be happy to be done with) and replace them with other forms of consumption that are personally and ecologically gratifying. But how?

Let's start with a bit of psychology—but not the standard fare.

The Dark Side of the Davenport

When most of us think of psychologists, we imagine well-heeled professionals leaning back into padded chairs, notepads in tow, liberating patients from their insecurities, obsessions, and vulnerabilities. But many of the nation's most highly paid psychologists are actually in the business of facilitating just the opposite. They work for ad agencies. These psychologists draw upon their expertise to induce insecurity, obsessions, and vulnerability among their subjects.[9] And their most moldable target? America's trusting youth.

Young children are especially suggestible because they have not yet developed the ability to understand bias.[10] In a study of three-to-five-year-olds, 76 percent reported a preference for french fries in McDonald's packaging compared to the same fries in a different package.[11] In a related study young kids chose branded carrots more than twice as often as identical unbranded carrots. The more they watched television, the more they yearned for the branded items.[12] But ads structure just part of the complex associations that children form with brands. According to critical psychologist Aaron Norton, "Children associate the packaging with not only the consumption of food but also with the excitement of going out to dinner, getting a free toy, seeing a clown—all of these experiences become associated with the food. In this case, they're not just tasting the fries—they're tasting the whole package of experiences that has been sold to them."[13] Those experiences are almost always branded around unhealthy foods. The advertising pyramid for children looks like an inverted food pyramid. Given that young children are so wholly immersed in the excitement of purchasing certain brands of fast food, candy, and sugary cereals—which in some cases are biochemically ideal for triggering overconsumption—it's hardly surprising that so many American children endure the ill effects that arise from obesity.[14] What may be more surprising is that there are still children who don't.

The practice of marketing to children surfaced as soon as the modern concept of childhood was socially constructed, but the most recent decades of commercial promotion manifestly differ in intensity, reach, and strategy. Before the seventeenth century,

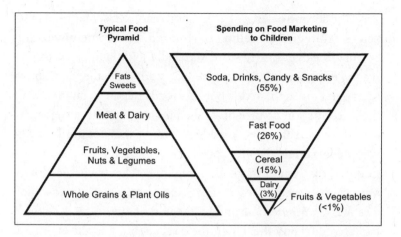

Figure 14: American food marketing to children A typical healthy food pyramid compared to spending on food marketing targeted to youth aged two through seventeen as reported by the industry. The Federal Trade Commission forcibly extracted this rare glimpse into secret budgets through a series of forty-four subpoenas. (Data from the Federal Trade Commission; special thanks to Clive Hamilton and Richard Denniss)

people didn't think of children as we do today. Parents did not recognize their birthdays and left their fallen children's graves unmarked. The cultural tradition of observing a child's age is only a couple hundred years old itself. The eighteenth century marked the beginning of the widespread recognition of modern childhood and everything that came with it, such as separate clothing, education, and games. Between 1750 and 1814, publishers released 2,400 children's books—but before this period they published almost none, even though printing presses had been churning out adult books for three centuries. In the late eighteenth and early nineteenth centuries, people began to recognize children as more than just underdeveloped adults. The government established itself as a guardian of the young, and when children committed crimes, courts gave them special consideration (formerly, guilty youth were simply flogged, chopped, and hanged according to the codes enforced for adults).[15]

As the accoutrements of childhood matured, so matured marketers' strategies for selling to children. In the late nineteenth century, the department store Marshall Fields launched a thirty-six-page toy catalog. As in other early advertisements, it spoke to parents more than kids but this focus began to change with the introduction of broadcast media. Starting in the 1930s, radio advertisers targeted kids. However, it was not until the development of television—specifically children's television programs—that youth marketing began to take its current form. Mattel advertised its toys during ABC's 1954 premier season of *The Mickey Mouse Club* and by the end of the 1950s, Kellogg's had created Tony the Tiger, the Trix rabbit, and the energetic trio Snap, Crackle, and Pop. Compared with the advertising of today, these ads were meek and understated, roughly cut from the same set of techniques marketers used to woo adults. Then something happened.

Corporations realized that little kids could bring in big profits. Firms began mobilizing resources and specialized research in order to promote youth consumerism, which child advocates now link to a multiplicity of social troubles. Nowhere did this shrill louder than in America. Here's the short list of a few of the well-honed strategies marketers use today:

- *Infiltration*: Marketers have partnered with groups such as the National Boys and Girls Club, the National Parent-Teacher Association, UNICEF, and other trusted organizations to gain access to children for marketing purposes.[16] For instance, starting in 1995, Girl Scouts instituted the ongoing "Fashion Adventure," with the large retailers Limited Too and Justice. Instead of scouting out natural habitats, wildlife, or waterways, girls embark on an overnight adventure beginning at the mall and ending with discount coupons for the supporting retail chains.[17]
- *Bro-ing*: Shoe and apparel marketers dress down and hang with the urban youth to discover what they can market as cool and

what they cannot. Insiders call this "bro-ing." Boston College sociologist Juliet Schor observes, "While the connection to inner-city life may sound like a contradiction with the idea that cool is exclusive and upscale, it is partially resolved by the fact that many of the inner-city ambassadors of products are wealthy, conspicuous consumers such as rap stars and athletes driving fancy cars and living luxurious lifestyles."[18] Starting in the late 1960s, Converse, Nike, and other shoe manufacturers aimed to associate their shoes with African American athletes, giving free samples to inner-city coaches and positioning their brands alongside the sociability of street athletics. Harvard University scholar Douglas Holt claims: "Street has proven to be a potent commodity because its aesthetic offers an authentic threatening edginess that is very attractive to both white and suburban kids who perpetually re-create radical youth culture in relation to their parents' conservative views about the ghetto, and to urban cultural elites for whom it becomes a form of cosmopolitan radical chic. . . . We now have the commodification of a virulent, dangerous 'other' lifestyle . . . Gangsta."[19] Schor and Holt maintain that many contemporary street creeds are not the homegrown and emulative grassroots phenomena they appear to be—they are first and foremost a product of mass marketing.

- *Spying*: A friend of mine recently purchased a doggie cam, a collar-mounted video camera, to record her St. Bernard's lumberings while she was at work. Marketers attach similar devices to children in focus groups before setting them free in toy departments, where their every glance and movement is recorded and analyzed so that marketers can more efficiently sell products to their peers. In order to gain insight into the deep intangibles of child behavior that psychologists cannot access through surveys and laboratory studies, researchers also observe kids in their bedrooms, bathrooms, and even while bathing.[20] During my research for this book, I visited a sterile

facility standing in a field adjacent to a West Coast highway where child psychologists hid behind a large one-way mirror to observe young children playing with toys in an eerie
windowless room humming with bright lights. I quickly got
the impression that monitoring children for marketing purposes might not be a particularly rewarding job. And, I thought,
if more parents were aware of this side of the industry, they
might be less ambivalent about its widespread exploitations.

- *Exploitation*: We'd be shocked to find a nine-year-old working a fast-food register, but bringing children into the service of ad firms is widespread even if it remains largely shield-
ed from public view. Advertisers rarely obtain consent for
videotaping children in retail or other public settings. Some
in-home researchers don't even employ release forms. Marketers frequently undertake research in schools (particularly
inner-city schools) assuming parental consent unless parents
step in to insist their child not be studied for marketing purposes. Dr. Schor claims that even when companies do offer
kids monetary compensation for their work, the amounts are
small: "When a couple of kids in Chicago gave Doyle Research the idea for a squeeze bottle for Heinz ketchup, the company made millions on it. The kids got the standard fee. . . .
Nike has it even better. When it goes 'bro-ing,' it has gotten
away with paying nothing at all. It's a form of financial exploitation we tolerate because it's kids, rather than adults, whose
ideas, creativity, and labor are being shortchanged."[21]

- *Neuromarketing*: For some time, the advertising industry has
employed eye-tracking devices to monitor and record the way
kids and adults view advertising in stores and on television.
Some marketing techniques employ brain-science research
based on neural MRI scans. I underwent an MRI study to see
what it was like. The researcher instructed me to lie down on
a cold plank, where he strapped my head to a foam cradle before
sliding me into a narrow cylindrical tube. The machine's hulk-

ing electromagnets buzzed, clicked, and tugged on my body's molecules as I viewed images flashed before me. Marketers use brain-imaging techniques to identify subconscious triggers that gain and hold people's attention. They have also optimized a "mind mapping" technique to use on children and teens without the MRI component. Which companies deploy these subconscious techniques? A representative of Bright House Institute for Thought Sciences in Atlanta answered: "We can't actually talk about the specific names of the companies. Right now, they would rather not be exposed. We have been kind of running under the radar with a lot of the breakthrough technology."[22]

So how does it feel to perform psychological special ops on the nation's youth? Not so good according to a study of child marketers. One child spy working for a New York advertising agency admitted, "At the end of the day, my job is to get people to buy things. . . . It's a horrible thing and I know it. . . . I am doing the most horrible thing in the world. . . . We are targeting kids too young with so many inappropriate things. It's not worth the almighty buck."[23] Other marketers described it as a double life—one life dedicated to manipulating children in the name of profit and the other at home shielding their own children from the corresponding messages. Some marketers quit their jobs after being exposed to standard industry practices and others refuse to market to children altogether.

One of my clients, whom I advise on his environmental philanthropy, was a well-known public figure as a teenager but now lays low, limiting his appearances to charity events. He shared with me his disillusionment with the promotion of kid stars and the advertising, co-branding, and cross-marketing that has come to define the industry:

"It's a big sales event, really. People don't see it for that but that's what it is. I didn't think that at first either but after a while you start to see it—you start to realize the only reason you're there is because you can sell shit."

He claimed that the industry is micromanaged by marketers concerned about placement, appeal, reach, and image—in his case, the object of the commercials, the print ads, the endorsements, and the figurines was to get kids to buy the "shit," or at least nag their parents to buy it for them. I had to ask, "Knowing what you know now, would you have done it again?"

He paused.

"I'd like to think no," he replied with a bit of hesitation.

"But I'm not so sure. I mean, I was so caught up in it. . . . It was such a rush and it happened so fast—it would never have crossed my mind to stop—to jump off the train. And the money. The money was, well, there was too much to keep track of. And you know, all of that money—at the end of the day it didn't have anything to do with talent," he chuckled.

"That is, except for the talent of *selling the shit*."[24]

Whether it's celebrity endorsements, advertising, or product placements, selling to kids works. The techniques that marketers have refined over the past few decades are especially effective means for turning children into consumers. Kids bond quickly to brands, drape their bodies in logos, adopt fashion trends mainstreamed by multinational conglomerates, alter their appearance with Botox injections, and don't seem to care if their idols are primarily marketing creations.[25] In one popularized study, 93 percent of teenage girls responded that their favorite pastime activity is shopping.[26]

The ad industry has derived a highly profitable formula but its effects on kids are far from beneficial. Dr. Schor studied the consumption habits and well-being of hundreds of children in and around Boston. Corroborating other research in the field, she implicates advertising, public relations, and product promotions in developing or worsening childhood depression, anxiety, low self-esteem, and psychosomatic issues: "Psychologically healthy children will be made worse off if they become more enmeshed in the culture of getting and spending. Children with emotional problems will be helped if they disengage from the worlds that corporations are constructing for them. . . . That is, less involvement in consumer culture leads to healthier kids, and more involvement leads kids' psychological well-being to deteriorate."[27]

Those millions of little emperors and empresses eventually each grow up to expect their own kingdom. Compared with children, the higher incomes and elevated status of adults makes them a far more potent consumer class to be reckoned with. Adult material consumption not only directly expands the nation's overall energy footprint but also provides a powerful channel for the social and cultural reproduction of materialist love fests (think birthdays, weddings, and Christmas for starters). Furthermore, it's no surprise that once grown up, the instilled materialistic values—the wanting and the needing—can perpetuate in adults the same psychological hazards evident in children. Affluenza.

Affluenza

In 1899 Thorstein Veblen coined the term "conspicuous consumption" to describe the impulse toward lavish or unnecessary spending in an effort to keep up with the Joneses.[28] Economists Clive Hamilton and Richard Denniss extend this concept to define *affluenza* as "a growing and unhealthy preoccupation with money and material things." They warn, "This illness is constantly reinforcing itself at both the individual and the social levels, constraining us to derive our identities and sense of place in the world through our consumption activity. It causes us to withdraw into a world of self-centered gratification— often at the expense of those around us."[29] If your small car, old house, outmoded furniture, outdated clothing, big nose, small boobs or anything else that you have neglected to upgrade in order to keep up with the Joneses (who are now available in widescreen high definition) should happen to make you feel inadequate, undervalued, or depressed, it certainly is not a problem with your society, but a problem with *you*. And you can buy a pill for that too.

It's no secret that America's increasing wealth has not led to the simple life and leisurely pursuits envisioned by futurists of the 1960s and '70s. What may be less immediately evident is that we have come closer to achieving just the opposite. We now spend more time in traffic to get to jobs where we work longer hours so we can stock our indebted homes with whatever it is that we saw on television during our abbreviated weekends—an existence that is neither simple nor leisurely. In fact, it more strongly corresponds with the traits of mental illness. And so it follows that the psychological fallout from elevated consumption levels is increasingly showing up on the therapist's couch. According to psychologist Tim Kasser, author of *The High Price of Materialism*, "When people and nations make progress in their materialistic ambitions, they may experience some temporary improvement of mood, but it is likely to be short-lived and superficial . . . some of the psychological dynamics related to the strong pursuit of materialistic goals (problems with self-esteem and discrepancies) keep individuals' well-being from improving as their wealth and status increase."[30] He claims that materialist preoccupations with wealth, status, and image directly work against the hallmarks of psychological health: close familial and interpersonal relationships and connections with others. In fact, once people reach a modest level of income of about double the poverty-line figure, higher income delivers little corresponding increase in satisfaction on average.

Perhaps more startling, studies show that merely *aspiring* for greater wealth or material possessions corresponds with higher levels of personal unhappiness. "People with strong materialistic values and desires report more symptoms of anxiety, are at greater risk for depression, and experience more frequent somatic irritations than those who are less materialistic," Kasser maintains. "They watch more television, use more alcohol and drugs, and have more impoverished personal relationships. Even in sleep, their dreams seem to be infected with anxiety and distress. Thus, insofar as people have adopted the 'American dream' of stuffing their pockets, they seem to that extent to be emptier of self and soul."[31] It certainly doesn't help that, unlike alcoholism or excessive gambling, compulsive shopping is socially sanctioned and even promoted despite its similarly detrimental effects, from financial hardship to psychological distress.

Cornell University economist Robert Frank, who studies consumption and happiness, points out that "satisfaction depends more on relative consumption than on absolute consumption. Many people, for example, recall being happy during their student days, even though they were living at a much lower material standard."[32] Why? Part of the reason has to do with the built environment of college campuses where students live, study, and work within a walkable network of friends and colleagues (We'll come back to this in the next chapter). But as working adults, a hefty amount of consumption consists of positional goods—things we feel we need simply because others have them. This induces the ratcheting effect of mass consumption—akin to everyone standing up for a better view at a concert only to discover that with everyone standing the view remains the same.

A primary driver of this phenomenon was anticipated forty years ago in the renowned book *Levittowners*, by Herbert Gans, who argued that middle-class families were no longer looking to their immediate coworkers, friends, and family for comparison of positional well-being, but to corporate managers, big-city counterparts, and celebrities, which mass media increasingly brought to the forefront of people's imagination. After an anthropological expedition through an American shopping mall, teeming with hip hop and hipster youth sporting the repli-ware of commercially honed celebrities, price tags a dangling, we might reckon Gans's assessment to have been strikingly premonitory. Mass media have fundamentally altered the way younger people compare themselves to others. Of the kids who dream of becoming celebrity musicians or professional athletes, few will succeed, but in our contemporary mass media and mass consumerist culture, they can at least purchase the outfits and jewelry to dress the part. For a price, the fulfillment of a dream can be had by proxy.

Among adults, this consumption arms race leaves fewer national resources for endeavors that would actually increase wellbeing, such as mass transit, safety research, preventative medicine, and simply spending more time with friends and family. Over the last three decades, Americans' time stuck in traffic has more than doubled. Vacation time has eroded by 28 percent as Americans work longer to pay off hefty mortgages on generously spaced homes, which incidentally have grown in size from an average of 1,600 square feet to an average of 2,500 square feet over the same period.[33] And whatever we can't fit in our distended homes, we store somewhere else. People once rented self-storage units as temporary spaces to assist with moving. Americans now primarily rent them to stockpile junk over the long term, even in the face of the economic crisis.[34] This unrelenting treadmill of buying—debting—working—hoarding is what social scientists term the work-spend cycle.

The Work-Spend Cycle

"I owe, I owe, it's off to work I go," sings the bumper sticker parody. It invites us to ponder what might happen if workaholics were to start not only consuming less, but also working less. They'd have less cash so they probably wouldn't buy as much stuff, they'd presumably live in smaller and more efficient homes to reduce their overhead costs, and their practice of working less could open up more opportunities for the unemployed to work. The proposition of working less might sound simply ludicrous to anyone settled into a growth-driven concept of prosperity but it isn't particularly remarkable if you travel outside the United States. After living and working for years throughout both the United States and Europe, I can personally attest to the very different mindset about work productivity between the regions. In Europe, I had a typical nine-to-five job, except that I only actually worked from nine to four——just thirty-five hours per week. I also received over eight weeks of annual paid vacation starting from day one. Such a work schedule might seem like a dream come true for many Americans, but it's hardly remarkable for Europeans. Nevertheless, it took me some time to adjust. After having become accustomed to my work schedule in the United States, working less seemed almost nefarious. I was likewise aghast to discover that if I neglected to take the allotted eight weeks of vacation, my employer would cut my pay and downgrade my performance reviews. So I adjusted. After the dust settled, all was fine, perhaps even better.

I frequently took long weekends, enrolled in cooking classes, spent more time with friends, and started to volunteer with a local charity. Sure, I didn't have a car or money to buy as much stuff. But there wasn't much room in my relatively small city apartment anyway. Overall, the quality of my lifestyle was at least equivalent to or greater than my previous life in the United States where I had a larger house, a car, more money, and worked longer hours. My experience wasn't unique. According to Dr. Frank, reducing absolute consumption levels doesn't negatively affect happiness or prosperity. "If everyone consumes a little less for a while, most people will adapt pretty quickly," he claims.[35]

On average, Europeans consume less (in most every category) when compared to Americans, yet they consistently rank higher on measures of happiness, as quirky as those ranking systems may be. A healthy majority of Europeans are satisfied with the basic social and governmental conventions that make their more egalitarian way of life possible. And there's not as much pressure to stockpile earnings for later in life since European nations provide health care and elder care to all citizens. I appreciated the European health care system while living there. My doctor stressed preventative health, and when I did feel ill, I was able to see her promptly (she even did house calls). I was actually surprised when I learned that per capita health care spending is lower in Europe than in the United States.

So why do Americans continue to work longer even though they would almost categorically prefer more vacation time? In addition to the work-spend cycle and affluenza, a number of institutional constraints have helped to calcify the forty-plus-hour workweek. It's difficult for employers to split jobs into part-time positions, allow employees to work fewer hours for less pay, or to implement job sharing because companies often cover health insurance for each employee. This means that offering unpaid time off or otherwise spreading work out to more people effectively increases health-insurance overhead costs. This is just one of many incongruities that policymakers and unions could address to encourage greater workforce flexibility.

Additionally, large proportions of the working population experience "deferred happiness syndrome," increasing their work hours and enduring more stressful work conditions in the belief that someday it will pay off. It's the work equivalent of hoarding. In one focus group, participants admitted that they sacrifice time with family and friends or forego activities that would make their lives more fulfilling, citing a desire to generate greater material wealth, anxiety about not having enough for retirement, or even fear about consequences if they were to change.[36]

It's perhaps worth noting that many people are surprised to find out how rich they actually are compared to others in the world. For instance, if you have a bank account with more than $100, then chances are you are among the richest 15 percent of the world's population already—if you make $25,000 per year or more, you're in the top 10 percent of earners.[37] And even though a majority of Americans are among the world's richest 10 percent, we tend more often to compare ourselves with the top few percent than with the poorer 90 percent. And just like other wealthy individuals, we manage to exhaust much of our hard-earned money on things that don't necessarily make us happier or healthier. In fact, we spend a sizable chunk of our income on junk. I'm not speaking figuratively of the infomercial relics collecting cobwebs in the rearward berths of corner cabinets. I am talking about junk in the most literal sense.

The Junk Business

My good friend occasionally wears an orange shirt that reads: "ADVERTISING *helps me decide*." The shirt tightens the eyes of onlookers into a perplexed squint for a few moments until their faces eventually offer up a bemused smirk. This sardonic slogan forces us to consciously think about something we usually don't—does advertising really help us at all? Or more broadly, what is the social function of advertising? What would happen if all advertising were to suddenly disappear? Probably sales figures for some companies would go down and the economy might "suffer." But it's possible we might be left with more money in our pockets, less debt on our credits cards, and perhaps even a bit more time to do what we'd like rather than sitting through ads, opening junk mail, and fretting over the fantastico-widgets that we never knew we needed.

For an institution without any clear-cut benefits, it does co-erce Americans into buying billions of dollars of extraneous gear that itself requires massive amounts of mining, processing, and energy resources to manufacture.[38] Eventually, we have to transport all of that stuff home, where we can unwrap, use, and eventually throw it away. The whole system requires energy inputs along the way. The junk-mail industry alone claims a hundred million trees every year, which producers must grow, cut, haul, process, roll, print, and ship to homes where they are usually immediately thrown away, hauled, processed, and finally dumped.[39] It is difficult to tally the total energy bill for this cycle but one critique of junk mail equates its carbon footprint to that of eleven coal-fired power plants running continuously at full tilt.[40] Moreover, the junk-mail lure induces unneeded purchases that come to clutter people's lives—so much clutter, in fact, that entire television series are predicated on showing viewers how to exorcise it from their lives. To make matters worse, another layer of junk clads most of those products: packaging.

Roughly a third of the waste in our trash bins is packaging.[41] Some of that packaging ends up in landfills, some gets recycled (which requires further energy inputs), and some makes its way into forests, parks, waterways, and oceans. In fact, many of humanity's discarded packages now float about a thousand miles off the coast of California in a slowly circulating island of debris covering about ten million square miles of the Pacific Ocean— at its current rate of growth, it may soon qualify as a continent.

Many packages serve an important function; they keep products safe from damage and they prevent food from spoiling. Other forms of packaging are excessively wasteful. Part of the reason comes back to advertising. Large consumer-product firms enter into contracts with big-box stores and other chains to reserve shelf space for their products. Marketers then aim to fill those spaces with large, colorful packages that they design (again with the help of psychologists) to grab the attention of shoppers as they walk down the aisle. These miniature billboards, replicated and wrapped around every product individually, are astonishingly wasteful—think of the times you've opened something containing more packaging than product. There are far less profligate ways of performing the same advertising function, say with fixed ads attached to the shelf instead of on every individual product.

Toward Improving Consumption

The roots of wealth are in the earth. We have become so technologized and globalized that we may sometimes forget that our affluence ultimately arises from material extraction—oil drilling, farming, mining, and so on. The service industry may seem an exception but it isn't—service payments come from excess extraction wealth. Filling the bottomless cornucopia will require a bottomless planet. And we haven't one. Eventually, Mother Nature will take away our credit cards, starting unjustly with the poor and disenfranchised. Humans will increasingly be left to draw on the planet's natural replenishment cycle, fueled by solar radiation and nuclear reactions within the earth's crust.

It is noble, yet all but pointless to lecture to several hundred million people about how they're consuming too much. It's especially futile when those millions are immersed in a socioeconomic hallucination whose very survival depends on endlessly filling the cornucopia. Yet this has been the tedious powerpoint presentation of the mainstream environmental movement for several decades.

Environmentalists are actually far more persuasive when they concentrate on techniques to make people's lives better and less energy intensive at the same time. For instance, most people would prefer to live with less junk mail and excess packaging—that's not a difficult reduction to sell. A healthy percentage of suburbanites report they'd be willing to live in a smaller flat or give up their car for the opportunity of living in a city that is walkable, culturally engaging, safe, clean, and green.[42] Not everyone will want to live in walkable communities, and that's fine—the goal shouldn't be to coerce people into less energy-intensive lifestyles. Instead, future environmentalists will concentrate on creating enticing opportunities so that people *prefer* lower-energy lifestyles.

A small but growing segment of the population claims they've found a way to start; they're leading happier lives—not despite their lower consumption, but because of it. They've shifted from a live-to-work mentality to a work-to-live mentality. They call themselves *downshifters*.

First Step: Enable Downshifting

In 2010 Karl Rabeder was a millionaire, a multimillionaire to be precise, but today he's essentially penniless. He sold his villa in the Alps, his old stone farmhouse in Provence overlooking the *arrière-pays*, his art collection, and the interior furnishings business that amassed his fortune. He gave it all to charity, but not because he was feeling generous or racked with guilt. He did it for his own good. Rabeder is part of a growing caucus of individuals who are voting to improve their consumption patterns by shifting from material consumption to other forms of spending their time, which are less expensive and more enjoyable. Karl Rabeder is in an extreme case, but anyone can adopt their strategies, especially if their nation enables such lifestyles.

Rabeder told the *Telegraph*, "It was the biggest shock in my life, when I realized how horrible, soulless, and without feeling the five-star lifestyle is." After a trip to Hawaii with his wife, spending all the money they wished, they felt as if they hadn't met a single genuine person. "The staff played the role of being friendly and the guests played the role of being important and nobody was real." He had the same experience during trips to South America and Africa, but with an additional convolution. "I increasingly got the sensation that there is a connection between our wealth and their poverty," he said. "More and more I heard the words: 'Stop what you are doing now—all this luxury and consumerism—and start your real life'—I had the feeling I was working as a slave for things that I did not wish for or need. I have the feeling that there are lots of people doing the same thing."[43] He's correct. Even former BP chief John Browne looks back at his twelve years of plotting takeovers, orchestrating consolidations, and making a lot of money for shareholders with "distaste and dissatisfaction."[44] Perhaps Socrates had a point when he claimed, "contentment is natural wealth, luxury, artificial poverty."

To varying degrees, a few people have found secret doors and passageways from their artificial poverty into new, more satisfying modes of living in which they spend fewer hours at work and in stores while giving and living more. They walk their kids to school, spend more time outdoors, take up new hobbies, become more politically engaged, and frequently report feeling uplifted and fit. In essence, they shift their consumption away from products and toward friendships, community, and family. This isn't a mode of consumption you'll see advertised on television, nor is it anything particularly new or shocking, but in many respects it's simply a fresh take on various tried and trusted wisdoms of life. Featuring downshifting as a more realistic option for Americans will take some work. Businesses, unions, and governments will have to imagine and create higher-quality, part-time work and more flexible work options beyond the typical forty-to-sixty-hour weekly grind. Universal health care and a trustworthy pension system will free Americans from the anxiety of each creating their own individual safety nets.[45]

Nevertheless, as downshifters move into smaller homes and shorten their workweeks, they frequently report not knowing at first what to do with all of the extra time they previously spent working, shopping, viewing advertisements, and maintaining their large homes and lawns. Quality work provides us with challenges, opportunities for personal enrichment, and purpose. It isn't just money that drives us, but "intrinsic rewards," according to Daniel Pink who studies Wikipedia, Firefox, Linux, and other successful forms of social entrepreneurship and social enterprise.[46] Unwilling to relinquish these intrinsic rewards, some downshifters swing their focus to hobbies, sports, or education. Others focus on another form of work. Volunteering.

First Step: Promote Volunteering

Volunteers improve not only the lives of others but their own lives as well. Community, school, and charitable projects can steal momentum from the work-spend cycle and direct it to more meaningful pursuits. Volunteering is a low-cost or no-cost activity that can replace consumption-related activities such as hanging out at the mall or watching television—and it delivers superior personal rewards. Furthermore, community involvement frequently exposes volunteers to groups of people from various socioeconomic backgrounds, making them less likely to view themselves at the bottom of the pecking order or feel as if they must impress others with material acquisitions. Volunteermatch.org is a nationwide Web site that lists thousands of volunteer opportunities searchable by theme, location, and time commitment.

As a cornerstone of citizenship, numerous countries expect their young adults to dedicate a cumulative year to public service. Even without a national mandate, American youths eagerly volunteer in droves—they'd be able to enjoy their pursuits even more if we didn't expect them to dedicate a cumulative year of their life to viewing advertisements.

First Step: Eliminate Advertising to Kids

The header above sums it up. This is one of the most important steps that future environmentalists can take to interrupt material consumerism and it happens to be wildly popular across the nation, beneficial to almost everyone, essentially free, simple to implement, and especially low risk given that numerous other countries have already accomplished it without incident. Professors Hamilton and Denniss insist that "advertising to children infects the next generation with Affluenza—and with a more virulent strain . . . it is no surprise that parents, teachers, and churches cannot compete. . . . One of the most valuable things parents can do for their children is to teach them to adopt a critical attitude towards marketers' attempts to influence them."[47]

Nevertheless, such strategies are only effective with older children since younger kids don't have a sufficient level of cognitive development to understand advertising even when it is explained.[48] Children's advertising so pervades American culture that asking parents to shield their children from ads is like asking them to keep their kids dry in a swimming pool. Of course, nationwide regulations would make parents' jobs a lot easier. Such restrictions are a popular success in other nations where parents and educators would be horrified if asked to submit their children to intense American-style marketing operations.

Regulating child advertising comes with some precedent in the United States, though it is weak. The Federal Trade Commission (FTC) limits the practice of "host selling," where a children's character endorses a product during the corresponding show. For instance, marketers cannot feature Fred Flintstone in an ad for Fruity Pebbles during an episode of *The Flintstones* cartoon. Any other time is fine, however, making this a relatively feeble regulation on its own. When both the Federal Communications Commission and the FTC attempted to outlaw advertising to children in the 1970s, marketers heavily lobbied Congress to block the proposed ban. Corporate lobbyists found a clever way to frame the threat. They claimed that halting ads to kids would in effect establish a "national nanny," which would eventually allow the state to dictate the lives of children—parents would be made obsolete. An absurd argument in retrospect, it was persuasive at the time.

Still, the FTC refused to back down, citing the then well-known ill effects of advertising on children. In fact, FTC regulators fought so hard for the nation's children that they placed the agency's very existence in jeopardy. Yet, fueled by corporate donations, ambivalence, and fear of being labeled a creator of the national nanny, congressional members ultimately voted to support corporate wishes. And they went further. To punish the FTC for even attempting to shield kids from marketers, Congress suspended the agency's funding. It reinstituted funding only after legislators had hollowed out the power of the FTC to enact protections for youth against the insurgence of advertising. Why would the advertising industry and some congressional leaders embrace such draconian subversions? Probably because they were terrified that such a ban might actually garner public support and pass.

Today American cartoons stand as the only English-language children's programs in the world to take breaks for advertising. Canada, Australia, Great Britain, and many other nations have long banned television ads to young children. Two decades ago Sweden banned not only television ads but also all other forms of advertising targeting children under the age of twelve.[49] Child psychologists hail the ban a success. One study curiously found that in letters to Santa, Swedish children requested "significantly fewer items" than their foreign counterparts exposed to advertising.[50]

An American Psychological Association (APA) study links children's television advertising to misperceptions of healthy nutritional habits, parental conflict, materialistic values, and more positive attitudes to tobacco and alcohol.[51] These effects are most pronounced among minority children, who view higher levels and more strident forms of corporate advertising than their peers of higher socioeconomic status. The APA strongly recommends restricting ads to children, which is particularly noteworthy given that some psychologists have actively worked to refine and intensify such practices.[52]

As we work to eliminate childhood ads, the advertising industry will claim that ads lead to certain benefits such as better products, as well as low-cost or free television and children's programming. To the contrary, network television is far from free—its costs are simply hidden. The public pays for television through larger price tags on advertised products. Second, advertising critics easily counter the concern that children's programming would die out without advertising revenue by simply pointing out that children's programming flourishes in countries with child advertising bans. Should child-programming gaps occur, regulators could expand the U.S. Children's Television Act, which obligates stations to air at least three hours of commercial-free educational programming per week in exchange for using public airways. Finally, advertisers will claim that ads promote competition, but they likely do the opposite—expensive advertising presents a barrier to entry for start-up firms, novel products, new research, and fresh ideas. For instance, advertising and public relations for the multibillion-dollar acne products industry effectively drowns out research published by the American Academy of Dermatology, Harvard School of Public Health, and other institutions showing that millions of teens could alleviate their acne simply by consuming less dairy (a product line that advertisers also sell to kids and parents).[53]

Here's how child welfare advocates recommend we begin to end childhood advertising:

- Legislate an outright ban: Ban advertising aimed at children under the age of twelve.
- Expose the spies: Require corporations to disclose who created each of their advertisements and who performed the market research for each ad directed at children under the age of twelve.
- Ensure commercial-free schools: Prohibit marketers from entering schools to pitch their products to schoolchildren and from using compulsory school laws to bypass parental oversight.
- Eliminate tax write-offs: Eliminate all federal subsidies and deductions for the costs of advertisements, market researchers, psychologists, ad agencies, and the like in campaigns aimed at children under twelve years of age.

- Tax advertising: Levy a tax on television, radio, film, print, and Internet advertising whenever more than 25 percent of the audience is under eighteen and direct the proceeds toward non-commercial children's media and programming.[54]

First Step: Social Enterprises for Youth

Prior to the beginning of the twentieth century, American children chose to spend much more time off the couch compared to their contemporaries. During the twentieth century, traffic laws kicked kids off the streets and into homes. As it became more dangerous to even be near the nation's roads, walking gave way to bus transport, which in time gave way to private vehicle transport. Suburbanization pushed wooded areas farther and farther from cities, making them unreachable for many kids.

Ensuring that our cities and roads are safe for kids will allow them to more easily find autonomy outside the bounds of a living room. Sixty years ago, Sweden began instituting reforms to make cities safer for kids by surrounding schools with webs of dedicated bike paths and walkways, lowering speed limits to levels designed to protect the youngest individuals of society, and instituting many of the walkable-city initiatives discussed in the next chapter. As a result, kids can freely navigate cities for social, athletic, cultural, and school activities at will.

American adolescents enjoy hanging out at the mall and related consumer pursuits because they are comprehensively social activities, so a large part of thwarting consumer culture will involve creating attractive social and educational alternatives. It's likely worth the investment—the HighScope longitudinal study of Michigan children found that for every dollar spent on self-directed learning for young children, the state saved seven tax dollars later on crime, unemployment, and welfare payments.[55] Self-directed learning can take many forms. A school in Berkeley, California, converted a vacant lot into an "Edible Schoolyard" project where students and community members volunteer to work on a one-acre fruit-and-vegetable farm complete with hens and even a kitchen where kids learn about food, health, and nutrition.[56] Other schools offer kids the chance to produce their own television or Internet news programs as an alternative to Channel One (a corporate "news" and advertising program wheeled into classrooms across the nation—the televisions are rigged so that teachers can't turn down the volume during ads). Primatologist Jane Goodall's thriving program, Roots and Shoots, combines naturalist education models with volunteering to engage youth in effecting positive change within their own communities. And in German cities, as well as rural areas, youth hang out in wildly popular *Jugendhaus* buildings rather than in shopping malls. Jugendhaus is a flexibly organized nonprofit assemblage of open-work cafes, discos, restaurants, farms, and so on, that adolescents organize, staff, and operate themselves. The German city of Stuttgart has forty-one Jugendhaus facilities, which youth independently reach via walking, biking, and public transit.

First Step: Shift Taxes from Income to Consumption

In some ways, our taxation policies motivate greater consumption. For instance, our tax system links school funding to nearby housing prices, meaning that parents have to move "up" if they want their children to get a better education. Changing our system of taxation will be a murky political process but there may be some short-term strategies to help shift taxes from income to consumption. To start, we could require retailers to post the full price of their products instead of just the before-tax amount, a simple switch that other countries already appreciate. Economists have detailed a variety of progressive consumption tax options that increase with a person's earned income or wealth. For instance, University of Delaware economist Larry Seidman suggests instituting a luxury tax on the highest levels of consumption, which would affect only the wealthiest 1 percent or so of the population. This, he claims, would make the tax not only easy to implement but also politically viable.[57]

An energy tax could proportionally increase the cost of goods to better reflect the *entire* cost of energy side effects. Food, toys, appliances, and other goods already carry a small energy surcharge—sticker prices include the fuel and generation costs for mining, processing, manufacturing, selling, and shipping. A tax on raw energy sales would similarly filter down into the price of goods, in effect creating a consumption tax appropriately adjusted for a product's energy footprint. This delivers much better signals to consumers than those achieved through green marketing campaigns; grocery shoppers could identify the almonds with the lowest energy footprint simply by checking the price tag.[58]

First Step: Smart Packaging

In the United States a full one-third of plastic production and a significant proportion of paper and pulp production ends up in packaging. Economizing this packaging not only lowers the energy inputs to mine, forest, process, and mold these encasements, but also initiates a *downward spiral* of associated energy inputs. First, shippers can more efficiently pack smaller packages into ocean freighters, rail cars, and trucks. Second, storeowners can fit more of these smaller packages on shelves, allowing them to reduce store size and refrigeration space (which initiates further downward spirals related to construction, heating, cooling, and urban planning benefits). The same holds for shoppers. It's easier to tote smaller packages (think biking and walking) and store them once we get home. Finally, with less packaging to throw away, municipalities spend less energy and money to haul, dispose, recycle, and build landfills for municipal waste. Walk into a European supermarket, where the floor plans, aisles, and even the shopping carts are smaller than their American counterparts, and you'll see this system in action. Several European governments require marketers to pay up front for the eventual recycling and disposal of their packaging.

Simple. Congruent. Potent. Ignored. Smarter packaging would cost nothing (it would actually save money). It could gain widespread support (who likes to fuss with excess packaging—and did I mention that sharp container edges, knives slipping off plastic encasements, and other packaging hazards send more than 300,000 Americans to the emergency room every year at a cost of over $15 billion annually).[59] Its effects would be felt immediately (while the long-term benefits could be far greater than those fabled by green-energy productivists). And it's already been successfully achieved elsewhere (meaning we can learn from the mistakes of other regions to avoid unintended consequences). If ever there was low-hanging fruit, this is it.

First Step: Introduce Junk-Mail Choice

Offering people a simple junk-mail choice will prompt a new mindset about the most egregiously wasteful forms of advertising. Why should society tolerate something with such questionable social worth and such clear social and environmental drawbacks? Under German regulations, postal customers may simply place a sticker on their mailboxes with a binding "no thanks" to all junk mail. In the United States, legally binding "no thanks to junk mail" stickers would deliver a greater energy impact than all the nation's existing and planned photovoltaics combined.[60] Perhaps junk mailings should be a right conferred only to locally owned businesses and charities. Or maybe we'd be just fine without them altogether. It should be our choice.

First Step: Ditch the GDP

As advertisers induce Americans to shop, the rest of our politico-economic system clears the way to the registers. If citizens slow their pace of spending, economists release dire warnings about consumer confidence, which they aim to adjust through policy interventions. And even though many forms of spending are wasteful, they absolutely define the success of the American economic system—not because waste is socially valuable, but simply because waste shows up as a positive indicator on the only miserable yardstick we use to measure success as a country: gross domestic product (GDP). Wasteful cycles of consumption, along with pollution, car accidents, fires, alcoholism, crime waves, and numerous other ills, all increase the national GDP, which economists, politicians, and policymakers in turn hold up as a testament to national prosperity. Meanwhile the GDP doesn't account for volunteering, open-source software, income disparities, quality of goods, and the negative externalities from harmful business practices.

Actually, the man who devised the GDP index, Simon Kuznets, never intended it to be an indicator of prosperity, as he plainly pointed out in a 1934 report to Congress: "The welfare of a nation [can] scarcely be inferred from a measure of national income."[61] He was most thoroughly ignored. We now use GDP to measure well-being when it could just as well represent the opposite. So why do we use it? GDP serves elite interests well. It's a powerful and easy-to-calculate (and manipulate) gauge to measure material wealth expansion. If citizens accept it as an indicator of prosperity—all the better. Indeed, over the past few hundred years there has been a somewhat positive correlation between economic activity and prosperity, so it's alluring to think that the same holds today. It doesn't. Over recent decades, GDP expanded enormously, yet measures of life satisfaction and happiness dropped.

Critics scorn the GDP for a variety of reasons—most notably because it doesn't offer a way of holding our elected officials accountable for our happiness, health, and safety. But in order to kick the GDP back to the sidelines of economics where it belongs, future environmentalists must advance a new mea-

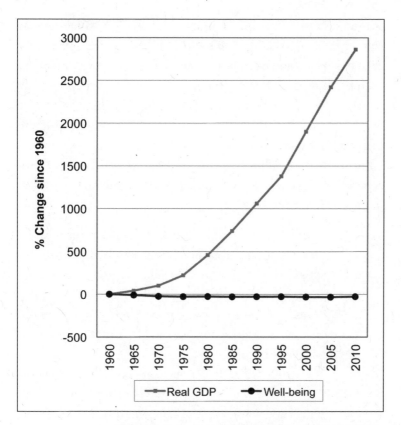

Figure 15: GDP versus well-being While real gross domestic product (GDP) in the United States increased dramatically between 1960 and 2010, well-being declined 25 percent, according to the New Economics Foundation. Gallup polls and other well-being indices support this discordance. (Data from U.S. Department of Commerce and New Economics Foundation)

sure of gross domestic *health* (GDH), based on health-care quality and availability, employment, education, crime, income distribution, consumer protection, environmental indicators, and related factors. Just like the GDP, a GDH could aggregate already available data. For example, a health-care metric could account for the number of people insured, vehicle fatalities, cancer rates, longevity, and teen pregnancies.

Since our politicians and corporate leaders have an interest in maintaining the GDP, the viability of a GDH rests solely on citizen organizations. The first big hurdle will be to convince local newspapers and newscasts to report the quarterly GDH along with the GDP, and then to convince national newspapers to do the same, with the eventual hope of nudging the GDP off the front pages altogether. The potential result? A better alignment of regulations, incentives, and policymaking with consumption patterns that are valuable to the health and well-being of citizens and the planet. Some promising GDH indicators are waiting for us in the wings:

- *Gross National Happiness*: A Bhutanese concept for measuring national happiness based on living standards, health, education, biodiversity, time use and balance, governance, community vitality, and psychological well-being.
- *Happy Planet Index*: A global comparison between nations based on happiness and environmental factors (find it at www.happyplanetindex.org).
- *Genuine Progress Indicator and Index of Sustainable Economic Welfare*: These indicators employ some GDP figures but adjust for volunteer and household work, income distribution, crime, and pollution.
- *Gini Coefficient of Wealth Disparity*: A 0.0 to 1.0 scale measuring the disparity of income between a nation's poorest and richest members with 0 corresponding to perfect equality and 1 corresponding to perfect inequality (one person gets everything).
- *Quality of Life Survey*: A European survey measuring personal satisfaction over time.

First Step: Shift Military Investment to Real Energy Security

When determining our national priorities, it is important to point out that of every federal tax dollar that Americans pay, less than seven cents goes toward education, environment, energy, and science combined. Meanwhile, the United States spends forty-one cents of every tax dollar on the military (we borrow much more beyond that for wars, which we'll have to repay in the future).[62] When Europeans wonder why a rich country like the United States can't seem to afford safe bike routes around schools and universal health care or why Indianapolis doesn't have forty-one Jugendhaus facilities, it's because of the nation's restricted concept of energy security. In reality, any of these initiatives could provide greater energy security than could a new fleet of bombers. However, vested corporate interests so tightly constrain notions of energy security that more meaningful, proven, and beneficial solutions go completely ignored. Ohio Congressman Dennis Kucinich rightly points out, "You can't have guns and butter at the same time in this country. We can't afford both anymore. We have to start focusing on what is the real security in America."[63] Peace is not an easy prospect—it requires greater bravery than does conflict.

I recently had dinner in Washington DC with a consultant assigned to reduce waste in the military. He estimates that the United States could cut the military budget in half without detriment.[64] In fact, leaner budgets might actually help the military focus on the types of conflict we might expect in the future rather than spending large sums of money on established and expensive war toys that are becoming increasingly obsolete. Incidentally, halving military spending would free up enough of the nation's resources to generously fund every first step in this book (including universal health care and solvent pensions) several times over with cash to spare.[65] But in reality, military cuts might be more difficult to orchestrate than any of the other first steps in this book. Washington's military contractors keep their lobbying and PR machines in high gear at all times. Their influence eclipses even that of Big Oil. That's because oil executives know we can't live without their product. Military contractors don't have that same comfort.

First Step: Vegetarianism

Note: Parts of this section have been censored due to legislation enacted in thirteen states that pose risks to investigative writers reporting on the food industry. So-called food disparagement laws (also known as "veggie libel laws") enable the food industry to sue journalists, writers, and other people who criticize their products, often placing the burden of proof on the defendant (guilty until proven innocent). Writers may be helpless, even if backed by scientists, since agribusiness plaintiffs need only convince a jury—a high-risk legal balancing act for even the most honest and dedicated investigator. In Colorado, breaking the law is not a civil but a criminal offense.

In the late 1990s, after being told that dead cows are routinely ground up and fed to other cows, which could risk spreading bovine spongiform encephalopathy (or mad cow disease), Oprah Winfrey commented, "It has just stopped me cold from eating another burger!" The beef industry promptly sued her under a Texas food libel law. Unlike Winfrey, I do not have the financial resources to defend myself in such a suit, and as a result you and other readers will be cheated out of the whole story.

I am a bit of a hobbyist chef and seek out local favorite restaurants when I travel. My favorite has been a local favorite for a couple hundred years, tucked down a narrow street within a now rapidly redeveloping section of Shanghai. I treated myself to a dinner of wood-fired pork and slow-roasted duck— they were perhaps the most succulent cuts of meat I have ever eaten—but they weren't meat at all. I accosted the waiter twice with my dull Mandarin, inquiring if these were truly vegetarian. He assured me that they were genuine fakes.

I searched out the restaurant because I was curious to try this ancient method of food preparation, originally intended for Buddhist monks, whose compassion for all creatures forbade them from consuming animal flesh. Over hundreds of years, chefs perfected alternatives using a mixture of bean curds for body, vegetable oils for richness, and innovative edibles such as silk threads for texture—the result, I can assure you, is transcendent.

Though it may seem an odd priority, subsidies for research into tasty meat alternatives might actually be far more promising than those for alternative energy when it comes to ameliorating harmful environmental impacts. Why? Because meat production, from feed to farting, instigates an extensive host of extreme ecological harms.[66]

but there is another reason

.

.

.

the resolution

.

.

.

.

.

there are three.

.

.

We are so frequently scolded about so many of our consumption practices that it is difficult to know where to start. In order to break through this overload, Michael Brower and Warron Leon from the Union of Concerned Scientists set out to identify the most harmful types of human consumption in terms of greenhouse-gas emissions, air pollution, water toxicity, water consumption, and habitat alteration. They developed a list of the seven most environmentally harmful consumer activities in the United States. Second on their list, after driving, is meat and poultry consumption. Meat production alone accounts for more land alteration than suburban sprawl and more water pollution than all household water and sewage systems combined.[67]

Cattle and hogs are big farters, generating tailpipe emissions of their own.[68] Animal emissions are especially problematic since they contain methane, an especially potent greenhouse gas.

together with the ████████████████. The UN Food and Agriculture Organization recently determined that livestock production alone leads to more human-linked greenhouse-gas emissions than all of the planet's cars, trucks, buses, trains, and airplanes, as well as the rest of the transportation infrastructure, *combined.*[69]

Vegetarianism or part-time vegetarianism is an obvious alternative and one that could offer other benefits as well. Vegetarian diets tend to contain high amounts of vitamins, minerals, and fiber as well as less saturated fat and cholesterol compared to meat-based diets, which may explain why vegetarians experience lower rates of obesity, lower cholesterol, lower blood pressure, and lower risk of dying from heart disease and strokes than nonvegetarians.[70] Some researchers even attempt to make links between vegetarianism and intelligence; it's purported that Albert Einstein was a vegetarian. Vegetarianism is gaining broader acceptance, but its appeal will be limited by cultural traditions that are difficult to modify. Perhaps the goal shouldn't be to guilt people into eating the veggie burger, but to create a genuine fake that carnivores prefer to the original.[71]

12. The Architecture of Community

We turn to science to free ourselves from fallible judgments of human experts, and we find that the scientific tests themselves require human interpretation. —Edward Dolnick, *The Forger's Spell*

Once a prairie, the landscape surrounding _____, USA, is now subdivided into standardized formations of standardized houses for standardized humans. Troops of satellite receivers stare wide-eyed at the southern sky. A drive, court, or way negotiates a serpentine path under the rubber soles of Fords, Toyotas, and Volkswagens, aimlessly twisting through the fertilized neon green of cookie-cutter plots only to end up exactly where it started—trapped in a temporal loop where every departure is a return to the beginning, each day a photocopy of the last.

This is the American suburb—described as a "formless human community . . . a richness of social surfaces and a monotonous poverty of social substance" by anthropologist Clifford Geertz in 1963.[1] Geertz is known for his concept of "involution," which he identified as "cultural patterns which, after having reached what would seem to be a definitive form, nonetheless fail either to stabilize or transform themselves into a new pattern but rather continue to develop by becoming internally more complicated." Geertz was describing Dutch colonies but he could just as well have been discussing a form of colonialism right here in America: suburbanization.

Muddied tracks of a drunken giant, winding blacktop patterns stamp onto farmlands, forests, and prairies surrounding urban America in a reconstituted and revitalized manifest destiny. We've all witnessed the carnage—long suburban drags of low-slung strip malls separated by asphalt sheets of parking that light up at night as if being prepped for surgery—hardly an inviting place to walk even in the rare case that there are sidewalks. Residents must negotiate their days by car, every interaction and every task mediated by a fuel pump. The modern energy system enables suburban expansion while it encourages demand for even more energy, which in turn breeds our most troublesome environmental problems. But beyond energy impacts, critics maintain that the proverbial sewage stemming directly or indirectly from suburban sprawl also pollutes family relations, communities, schools, and other social systems necessary for human welfare.[2] It's no shock that America's energy-intensive lifestyles, enculturations, designs, and technologies form through or in tandem with the suburban experience. What's shocking is the utter failure of the mainstream environmental movement to imagine a more alluring alternative.

To an arguable degree, the mainstream environmental response to suburbanization has been to (1) romanticize rural life while (2) hyping technological fixes. Their resulting models for sustainable living are unrealistic for the vast majority of people—even if droves of Americans moved into rural straw-bale homes powered by costly solar cells, the result might be a larger environmental catastrophe than today's suburbs. Spreading people in a thin layer across the countryside might seem sustainable, but it more frequently intensifies environmental harms and leaves the resulting impacts less visible and more difficult to address.[3]

In any case, if being an environmentalist means living off the grid in the middle of nowhere, not very many people are going to sign up. Worse yet, these rural imaginations—as charming as they may seem—often come at the expense of reimagining and improving the communities where we actually do live.

How We Got Here

America's city streetscapes of the 1800s looked much different than those of today; streets were engaging public spaces, offering room for playing children, business dealings, markets, chatting neighbors, and flirting teens. In fact, from the birth of the city until the end of the nineteenth century, walkways claimed the largest share of the public space between buildings (the *side-walk* is a modern contrivance). City dwellers liked it that way. New Yorkers resisted steam-powered trolleys in 1839. Philadelphia's residents followed suit in 1840. And in 1843 the Supreme Court of New York officially declared steam engines a public nuisance, in effect restricting rail systems to horse-drawn coaches. For the next two generations, citizens protected public walkways and gathering spaces from motorized trespasses.[4]

Then came the horseless carriage.

The automobile shifted America's conceptions about the spaces between buildings. What they once understood as a place of play and commerce, they would eventually see as a place to *drive*. The first open-air horseless carriages, chauffeured by the wealthy, rumbled through city streets in the late 1800s. They were a curiosity, at first. They eventually became a nuisance, and citizens initially resisted horseless carriages just as they had resisted the steam trolleys before them. Tensions in New York mounted in 1901 when a Wall Street chauffer ran down and killed a two-year-old playing in the street. Two years later, children playing in the streets of New York stoned a woman driver until she was unconscious.[5] Something had to be done.

In December of 1903, New York City officially split city

streets—part to be used for vehicular traffic and part for pedestrians. City hall's move not only codified the occupation of public space by automobiles but, more importantly, legitimated it. In just a few years—a remarkably short time—horseless carriages flourished. Congestion forced street vendors into buildings. Stables and mechanics took in cars while they were not in use, but these makeshift spots soon overflowed. Drivers needed parking.

New Yorkers struck down a proposed thirty-thousand-car parking lot in Central Park. Citizens also blocked plans to pave over large swaths of the National Mall in Washington DC. Other municipalities were not as fortunate. In Detroit, Dallas, Boston, Newark, and countless other cities, public officials gave their nod to pave over public squares, markets, and parks to accommodate the swelling car population. And with ample parking, the floodgates creaked open a bit more.

In rushed street congestion at levels never before imagined. In 1907 authors of an article published in *Municipal Journal and Engineer* observed that street improvements did nothing to reduce congestion.[6] Rather they produced the opposite effect as cars multiplied to overfill any new development. Nevertheless, cities widened their roads and when congestion increased, they widened them again, shoving sidewalks up against buildings and drawing them out into ever-narrower strips. In 1910 the *Saturday Evening Post* observed that the sheer volume of vehicles would bind up the whole system, leaving disgruntled drivers sitting idly in their cars! The writer called it a "traffic jam." Again cities responded by widening roads. And again, it did nothing to solve the problem. In 1925 a prominent book on street traffic control reiterated the hopelessness of the situation: "Any reasonable increase in street capacity . . . will not reduce the density of traffic."[7] A boomerang effect.

In the years before World War I, cars grew more robust and automobile owners ventured outside city limits, rumbling their way through fields, pastures, and forests. Landowners objected but growing satellite cities welcomed them, building tourist camps replete with free parking to lure the city's well-heeled to their remote enclaves. Eventually, these camps integrated cabins into the experience, which eventually evolved into motor courts, Holiday Inns, and then . . . well, you know the rest.

Today we don't travel back and forth to cities for holidays as much as we do for work, which takes us an average of 25.5 minutes each way. Americans spend an average of forty-five hours per month in their cars—more than a standard workweek of sunken time.[8] Automobile cabins expose drivers to benzene, carbon monoxide, formaldehyde, methyl tertiary butyl ether, and other toxins at concentrations up to ten times those outside their vehicles.[9] The worst of that pollution occurs in traffic.

Cars provide freedom until roads fill with them. Americans spend the equivalent of one additional workweek per year in heavy congestion, collectively wasting an extra 2.9 billion gallons of fuel and costing about $78.2 billion a year in lost time, according to a report by the National Academies.[10] As we sit there in a twelve-lane-wide slick of oozing traffic we often wonder, wouldn't this be faster if the road were just . . . *a little bit wider?*

And over the years, Americans continued to push their, local, state, and federal representatives to make the roads a little bit wider. Legislators willingly expanded funding for streets designed around cars, hiding the true costs of driving and selling sprawl at a perceived discount. Founder of *Adbusters* magazine Kalle Lasn insists that more than any other product, the automobile stands as an example of the need to account fully for the costs of production and operation:

> That doesn't just mean manufacturing cost, plus markup, plus oil, gas, and insurance. It means paying for the pollution, for building and maintaining the roads, for the medical costs of accidents and the noise and the aesthetic degradation caused by urban sprawl. It means paying for traffic policing and military

protection of oil fields and supply lines. The fossil-fuel-based automobile industry is being subsidized by unborn generations to the tune of hundreds of billions of dollars every year.[11]

The Addiction to an Ideal

In the United States, the land cultivated with turf grass exceeds the total irrigated cropland dedicated to wheat, corn, and soybeans combined.[12] Sustaining this colossal growing project costs upward of $40 billion a year and requires thirty-five million pounds of pesticides (including patently dangerous compounds such as the Agent Orange component 2,4–Dichlorophenoxyacetic acid), a stunning sum of petrochemical fertilizers, millions of gallons of gasoline to power lawn equipment, and a full third of the nation's residential water supply, which some regions process with energy-intensive desalination plants.[13] Americans dedicate all of these resources to supporting showy yards that, according to one recent study, families spend "negligible" time actually enjoying or even using.[14] In fact, the principal activity in America's lawns is maintenance. The ultimate purpose of the yard, it would seem, is self-justifying: its function is to be cared for.[15]

During suburban colonial expansion, developers slash, burn, and level fields and forests to make room for rolls of sod, which romanticize a landscape they actually smother. The suburban colonist's attempts to conjure up the romanticism of a countryside, whose intrigue has fallen into the jaws of a predictable cliché, were almost certainly calculated by well-intentioned committees. However, architecture and urban planning aren't so simple. They are organic wisdoms marked by a sense of location, weather, time, and space. Successful typologies arise in coordination with topography, history, building practices, and human ingenuity. The harmony of these components is lost by the suburban colonist whose mass-produced houses sit randomly on bulldozed lots. As with the imperial form, suburban colonialism arose not only from an addiction to profits but also from an addiction to an ideal.[16]

The German word for "suburb" makes this ideal visible: *Zwischenstadt*, meaning "in-between city." Thomas Sieverts from the University of Darmstadt comments, "Today's city is in an 'in between' state, a state between place and world, space and time, city and country."[17] The suburb is an attempt to merge the romanticism of the countryside with the convenience of the city, as imagined through "The Ideal," a 1927 poem by Kurt Tucholsky. He describes a magical house with a back door opening to a terrace nestled in the snowy Alps and a front door opening to the bustling city streets, only steps from the theater—the entire affair remaining, of course, "simple and modest." Tucholsky paints the struggle to reconcile yearnings for wilderness with the desire for urban culture and convenience. The late Amphicar, a German import from the 1960s, embodied this tension. It was a "Zwishen vehicle." Half car, half boat. It rolled off roadways into rural lakes with the grace of an overloaded golf cart. While the coupé enchanted onlookers with its ambidextrous charm, the Amphicar was neither a fine automobile nor a fine boat. Like the suburb, it was caught in between. But unlike the suburb, its new models fetched lower bids in the marketplace and it was abruptly discontinued in 1968.

Meanwhile, the suburb sustained its invasion through the language of a hush-hush war with traditional notions of community. Sieverts insists that the suburban façade is a counterfeit cosmopolitanism, lacking values inherent in the original. He argues that the true cosmopolitan experience evokes intellectual agility, openness, and curiosity about the world, though it's difficult initially to differentiate between cosmopolitanism and the physical constructions and commercialism surrounding it. Since we link cosmopolitan ideals in our mind with the hum of bustling commercial strips, a bait-and-switch is all too easy for suburban developers to deftly execute. With an eye to profit, they do just that. Nevertheless, their shopping malls have ended up being a reconstruction of the physical exclusively, a shell without the egg. Lacking the dexterous culture and economy interwoven through the fabric of cities, we might understand the manufactured suburban community as a forgery, a cultural prosthetic.

Suburban expansion ratchets up a host of risks and side effects. All that's required to drive the ratchet is wealth, an ideal of independence, and a healthy reluctance or outright incapacity to look too far ahead (a trap humans fall into despite our large brains). Once ratcheted up, the suburban cultural system is a messy knot to undo.

Suburbia's Ratchet Effect

Social cartographers began to chart the restless spirit of modern Americans almost as soon as there were modern Americans to chart. Alexis De Tocqueville noted how peculiar it seemed that Americans were "forever brooding over advantages they do not possess" in his famed nineteenth-century *Democracy in America*.[18] "In the United States a man builds a house in which to spend his old age, and he sells it before the roof is on; he plants a garden and rents it just as the trees are coming into bearing; he brings a field into tillage and leaves other men to gather the crops; he embraces a profession and gives it up; he settles in a place, which he soon afterwards leaves to carry his changeable longings elsewhere."[19] Tocqueville rightly acknowledged that the spectacle of a restless soul in the midst of abundance plays into a narrative as old as the world itself, yet the distinction for him was "to see a whole people furnish an exemplification of it."[20] In his book *The American Future: A History*, historian Simon Schama lends insight into Tocqueville's bemusement: "In the Old World you knew your place; in the New World you made it." Schama insists that "American liberty has always been the liberty to move on. Whatever ails you, whatever has failed; whenever calamity dogs your heels or your allotted patch feels too small for your dreams, there's always the wide blue yonder, the prairie just over the next hill, waiting for your cattle or your hoe."[21] And as it turns out, this state of mind has gotten us into a bit of a bind.

Suburban expansion begins with ready access to cheap land on the fringes of cities. Agricultural landholders are usually keen to sell their farmland at a premium to developers, and since remaining farmers expect they too will be bought out, they are leery to invest in improvements. As the land increases in monetary worth and agricultural yields plateau, the property becomes ripe for developers to pluck. Developers start with inexpensive housing patterns familiar to bankers interested in replication over risk. Distant investors squeeze out any space for substantive considerations of native landscape, water aquifers, pollution, erosion, wildlife, history, community, or fossil-fuel consumption unless mandated by law or shown to produce a market return in cash.[22]

As soon as the sod is tamped down, suburbia purrs a hypnotic Siren's song to the masses, wooing them to larger houses on spacious lots. Gated communities and good schools are the gems of suburbia, situated away from the chaos of the city in an open space where driving is unmitigated by traffic, parking is always free, and problems are never larger than twenty-seven inches diagonally. These inexpensive new developments lure urban dwellers into a drive-till-you-qualify swarm to the suburbs. Shopping centers arrive next, with office and industrial parks following a parallel pattern. But as the suburban cultural system ratchets forward, it compresses a spring of risks. Eventually that constricted spring starts to groan and jerk under the pressure.

After the clamoring developers have moved on, an astute ear can hear the ticking of a subterranean time bomb beneath the cracking parking lots that flank suburban strips. As restaurants and entertainment centers arrive, traffic and congestion start to intensify. Commutes to the city clog roads, increase driving times, diminish air quality, and lead to abbreviated evenings balancing on fast-food meals and microwave dinners as drivers grow more concerned about the grade of gas in their fuel tanks than the quality of the food they consume. Disconnected communities increasingly replace community sports with spectator sports. As traffic and crime intensify, parents feel kids are safer on the couch watching television than outside playing near the street. Since walking and biking are not practical, cellulite finds few impediments to growth on the superfluous rump of suburbia.[23]

Eventually, fully matured suburbs frequently acquire the crime, pollution, poverty, and costs of the city, along with the inconveniences of rural life, instead of the intended opposite. The final dénouement is an environment without open space, fresh air, unrestricted auto use, or safe communities for children—an environment residents must once again escape.[24]

A Tale of Two Economies

Profitable suburban development expands like a ring of brush fire, leaving behind a burnt-out urban core—a process that clumsily extends utility infrastructures, roads, and other services while abandoning perfectly functional ones. According to a report by the Urban Land Institute, the Minneapolis–St. Paul regional governments built 78 new schools in the suburbs between 1970 and 1990, even while they shuttered 162 schools in good shape located within city limits. Other municipalities face budget problems as they are forced to divert funding away from education and toward expensive reorganization of school buildings.[25]

A case in point is Detroit. As in most U.S. cities, new developments expanded outward while the valuable city infrastructure rotted. Detroit, however, is a notable case because it was both a perpetrator and a victim of urban flight.[26]

In the early 1900s, Detroit was a frothy metropolis with a roaring economy, stylish mansions, and an ornamented skyline of terra-cotta-clad towers. Judson Welliver vibrated in a 1919 *Munsey's* article: "Skyscrapers are everywhere, magnificent shops occupy palatial quarters, and the idea is that whatever is good enough for Detroit must be a little better than anything else in its class."[27] But the verve was already waning. Rents and congestion increased while electrical grids and telephones enabled people to move from Detroit proper, marking the beginning of a mass transfer of resources to the suburbs. Ford moved its headquarters to suburban Dearborn, and Detroit stumbled.

Finally, following the industrial high of World War II, Detroit tumbled headfirst down the staircase of prosperity. Today the scene is apocalyptic. Neither the city nor the suburbs of Detroit are particularly sought-after places to reside despite being geographically central, adjacent to an international border, and situated on the Great Lakes, which might have developed into a vibrant port and waterfront scene.[28] Detroit contains miles of abandoned streets, water supply lines, sewers, and power grids. As in many other cities, a dualism forms. One economy expands outward, chomping up cheap farmland and forests to fuel its profligate spawning of strip malls, industrial parks, and housing developments. Another stagnates, checked by burdensome infrastructure costs, poverty, and crime.

Detroit's urban residents helped pay the costs of the suburban expansion that eventually decimated their city. Even today, relatively efficient city dwellers end up subsidizing suburban road construction, power lines, sewers, and water mains—at a cost of up to $13,426 per suburbanite.[29] Suburban living seems inexpensive in part because others absorb many of the associated costs and risks.

Inevitable Stagnation

New urbanist Amanda Rees points out that "alienating everyone already living in post–World War II suburbia by simply labeling their physical, social, and cultural environment as 'bad' does little to persuade people . . . a more rigorous analysis of the ideological underpinnings of the movement is certainly required."[30] This is a valid and constructive point. Must social critics always characterize suburbia as a steaming vomit gushing over the American landscape rather than simply a different and equally valid form of social interaction? Hasn't suburbia matured?

Certainly suburbia has changed since its original Levittown beginnings. Developers now eagerly point to "forward-thinking" design elements such as solar panels to ease the environmental burden of their photocopied blueprints. And a few recent books render a portrayal of suburban maturation.[31] But *maturation* implies growth toward a defining form. Conversely, suburban development is reactionary and complicating; it has no future apotheosis to which it aspires. Its growth does not display the intentionality of design but the dogma of ideology. Solar cells and numerous other green construction gimmicks do not come from within but from without—reactionary modifications to address the consumer whim or psycho-political fetish of the season.

It is here that we revisit a defining element of Geertz's concept of involution, the inability of a cultural form to transcend to a new pattern from a former one. Each architectural revolution, like revolutions in the sciences, arts, and literature, was culturally authentic—a pattern capable of standing on its own merits. Could any of these progressions have been realized by simply slapping solar cells onto a preexisting blueprint and calling it a day? Absolutely not. The question alone is enough to turn a true architect's stomach.

Solar cells, roof-mounted wind turbines, and geothermal heat systems cannot be solutions to the suburban cultural system because they offer no fundamental transformative power. High-tech add-ons act to perpetuate the existing system, offering a bewitching promise of romantic utility even while delivering an uncalculated hypermediocrity of returns. Rather than bettering the human condition, productivist impulses too often gush with an unyielding complication of the status quo, another trip around the cul-de-sac and a stagnation of the progressive imag-ination that makes us human.[32]

Suburban sprawl did more than just reorganize the spaces around Americans. Sprawl reorganized American conscious-ness. A repli-tecture of fully disengaged homes and megastores linked together with wide streets and highways unleashed modes of operation within time and space that were fundamentally dis-tinct from previous ways of life. These physical displacements prompted a mass deployment of energy resources that would have been unthinkable just a generation before their formation. It's within suburbia's psychosocial spaces that Americans grew to understand extreme energy waste as perfectly normal. Dis-placed externalities erase consequences of this waste from pub-lic perception. If we want to see the harms our energy-intensive lifestyles instigate, we have to watch them on television. This brings contemporary American critics to argue that we have become numb to the broader impacts of our lifestyles, anesthe-tized by a frightened and self-directed culture—but perhaps it's not that simple. Perhaps it's because we are not exposed to bet-ter options.

The Village Model

Author David Owen opens his book *Green Metropolis* with a personal story:

My wife and I got married right out of college, in 1978. We were young and naïve and unashamedly idealistic, and we decided to make our first home in a utopian environmentalist community in New York State. For seven years we lived quite contentedly in circumstances that would strike most Americans as austere in the extreme: our living space measured just seven hundred square feet, and we didn't have a lawn, a clothes dryer, or a car. We did our grocery shopping on foot and when we needed to travel lon-ger distances, we used public transportation. Because space at home was scarce, we seldom acquired new possessions of significant size. Our electric bill worked out to about a dollar a day.[33]

The ecotopia that Owen describes is Manhattan. Yes, New York City, the most densely populated landmass in North America, also happens to rank first in public transit usage and walking and last in per-capita greenhouse-gas emissions (by a wide margin). Manhattan's population density is roughly thirty times that of Los Angeles, and Owen argues that "placing one and a half million people on a twenty-three-square-mile island sharply reduces their opportunities to be wasteful, enables most of them to get by without owning cars, encourages them to keep their families small, and forces the majority to live in some of the most inherently energy-efficient residential structures in the world: apartment buildings."[34] Many people think of cities as ecological disaster zones, yet city dwellers in dense villages of various sizes consume far less oil, electricity, water, and material goods than those living in suburban America.

Villages with the highest degrees of embodied efficiency also tend to be the most enjoyable to live in. There is a reason many of us reflect fondly upon our years in school or university. Chats between classes with friends—lunches in the park—movie nights—and walking home after a night out with friends—it's not only because we were young, it is certainly not because we were rich, but because we lived in close-knit communities that were walkable, multifunctional, and accessible. When we graduated, most of us got jobs, started making money, and decided to move to a big house with a big TV so we could watch streams of shows about celebrities, whom we came to know better than the people next door. Yet according to a University of Michigan and Cornell study, roughly 70 percent of people report they'd be happier with lower salaries if it meant living closer to their friends.[35]

Fifty years ago Jane Jacobs published *The Death and Life of Great American Cities*, which argued that the quality of urban life arises from the connections people make when living very near one another in compact communities with a haphazard mix of stores, residential buildings, and small green spaces.[36] At first the urban planning establishment ignored her. Later they ridiculed her work, even underscoring it for students as a dangerous example. Why was her work so threatening? Principally, it defied established ideology, which mandated strict zoning regulations to separate residential neighborhoods from large centralized shopping centers via long blocks of broad streets and open spaces. Conversely, Jacobs argued for short blocks with narrow streets containing a mixture of businesses and residences. She observed that pedestrians avoid parks and plazas in favor of bustling streets and cozier green spaces—akin to the kitchen effect at parties, where everyone crowds into the kitchen and nearby halls, leaving the large open rooms comparatively vacant. She also argued against centralized low-income housing projects (which were still in vogue at the time), instead envisioning a mixing of socioeconomic classes as benefiting everyone.

American urban planning departments oversaw construction of most of the nation's existing built environment by employing the *antithesis* of Jacobs's wisdom. But after half a century of reflection, it is becoming more apparent that she was absolutely right. Dense neighborhoods bring all sorts of people together in unexpected ways. This disco ball of continuously rotating social reflections leads to safe streets, engaging arts, close friends, excellent restaurants, and ultimately a satisfying life for residents. Creativity thrives where its roots are crowded.

And although her book wasn't sold as an environmental work, it could just as well have been billed as such. Moving people closer together doesn't just make them more interesting, it makes them greener too.[37] Density enables people to navigate their daily lives by foot for many trips. And even though life in a city brings certain frictions—every commute turns into a treadmill and every staircase, a Stairmaster—such frictions can lead to benefits; on average New Yorkers live healthier and longer lives than the rest of Americans.[38]

Bikeable Cities

During the restructuring of General Motors, when nobody knew quite what to do with the large firm as its financial sheets imploded, I proposed to Senators Carl Levin, Chris Dodd, and Debbie Stabenow the idea of a lightweight class of commuter vehicles ranging from human powered to electric assist, complemented by urban policy initiatives to promote their use. Bicycles. Improving bicycling facilities in the United States would provide urban mobility, bring health benefits, and enhance public welfare—all at a cost far lower than we pay to support new automotive transport infrastructure.[39]

Bicycles are the principal vehicle of the human race, numbering roughly two billion worldwide, and have forever changed the riders that have embraced them. "Tens of thousands who could never afford to own, feed, and stable a horse, had by this bright invention enjoyed the swiftness of motion which is perhaps the most fascinating feature of material life," wrote the nineteenth-century educator and activist Frances Willard.[40] Historians cite the interplay between society and bicycling as constitutive of monumental social changes including women's suffrage and feminist movements. (It is perhaps more than a historical curiosity that, even today, nations with the highest levels of gender equity also tend to have more bikes on their streets.)

Numerous global cities reserve select roadway lanes, pathways, or even entire streets for bicycles. The result? Cleaner, healthier, quieter, and safer neighborhoods. Amsterdam reserves a network of separated lanes and streets for bicycles and ultralight commuter vehicles. These narrow roadways leave more precious space for pedestrian sidewalks while allowing high densities of people to travel through the historic city (a street can hold eight bikes within the footprint of one car). At first glance, it might seem unsafe, but in practice it's safer than the conventional car-based alternative. Because the vehicles in these pathways are all lightweight, collisions tend to be more of a stand-up-and-brush-off affair rather than one that involves insurance companies, police reports, ambulances, and traffic backups. From my time working on energy research at the University of Amsterdam, one of my fondest memories was of zooming through the compact city by bike on my daily commute. "Who would ever have thought commuting could be so fun?" I would often think. This, unfortunately, is a simple pleasure unavailable to most Americans.

Future environmentalists will direct their sights toward increasing the number of trips made by bike rather than focusing on cars driven by alternative means. When compared to automobiles, cycling and walking necessitate only a fraction of the energy outlays, raw materials, infrastructure costs, and associated maintenance. Nevertheless, implementing bike lanes in America will differ from the European context.

Bicycling critics claim that American cities are more spread out so bicycling is not a realistic transportation option for policymakers to pursue. The first part of this argument is correct—European cities are certainly more compact. Yet over a quarter of the trips Americans make are shorter than one mile and over 40 percent are less than two miles.[41] These distances would be well suited for bike travel, yet Americans overwhelmingly opt to drive; Americans jump in the car for 90 percent of trips between a mile and two miles. Overall, Americans make less than 1 percent of trips by bike. Even Canadians, with their chilly weather and similarly dispersed urban configuration, bike at twice the rate. In some parts of Europe, the percentage of trips by bike is over twenty times higher than in the United States, indicating that there's plenty of opportunity for Americans to increasingly enjoy their neighborhoods atop two wheels in the open air.[42]

As a daily routine, bicycling is also an affordable and con-

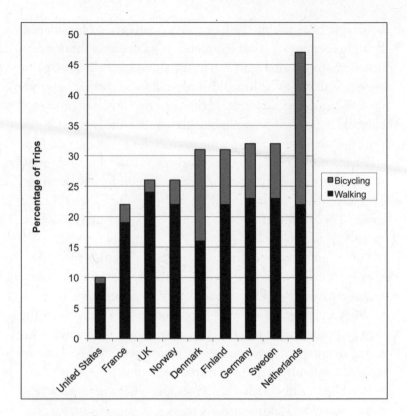

Figure 16: Trips by walking and bicycling Americans make far fewer trips by foot or bicycle compared to Europeans. (Data from David R. Bassett et al., "Walking, Cycling, and Obesity Rates in Europe, North America, and Australia," *Journal of Physical Activity and Health* 5, no. 6 [2008])

venient form of exercise. For instance, Steven Miller, from the Harvard School of Public Health, points out, "Our own good intentions are seldom enough to get us to change our habits. Few of us have the time, or the money, to spend time in a gym or going skiing or for multi-day bike rides. . . . What we as a nation need are ways to integrate physical activity into the everyday patterns of our life—walking to the local store, cycling to visit a friend, taking the trolley or bus to work and walking to the station, riding a bike to work."[43]

He estimates that people can lose or prevent ten pounds of fat accumulation per year just from biking two miles each way on their daily commute, not to mention the heart and mental benefits. Bicycling leads to longer, healthier life-spans and offers seniors independent mobility, which greatly increases their quality of life.[44] German and Dutch seniors make roughly half of their trips by walking or biking compared to just 6 percent of American seniors.[45] Many older Americans fear and eventually abhor the day they lose their driver's license because it is a virtual sentencing to house arrest if they live in a car-centric suburb. On the other hand, seniors living in dense villages, both large and small, still enjoy a host of mobility options—walking, cycling, public transit, and taxis.

Unfortunately, walking and biking are inconvenient in the United States for the elderly and nonelderly alike because most Americans live in a physical, legal, economic, and social terrain specifically designed over a period of many decades to accommodate motor vehicles above all else, making friendlier forms of transportation unpleasant, inconvenient, and even unsafe. The cost of owning a car in the United States is half what it is in Europe, and generous American subsidies for road construction, maintenance, and parking are inequitably absorbed by *all* taxpayers regardless of whether they're walkers, bicyclists, or drivers.

Furthermore, bicycling in the United States is more dangerous than in nations that prioritize bicycle safety. Germany reduced overall bicycle deaths by 68 percent over the last few decades, even in the midst of a bicycling boom, during which the number of trips by bike doubled.[46] By contrast, American bike fatalities dropped just 24 percent, and even that had little to do with safety efforts—the drop was merely a reflection of the decline of bicycling, most prominently among children.[47]

Many of the nation's schools stand behind a barricade of freeways, busy roads, and rushed drivers—hardly a safe environment

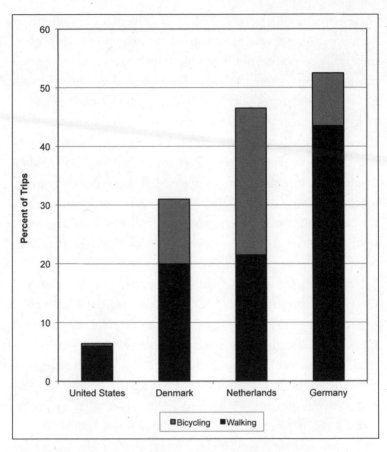

Figure 17: Walking and bicycling among seniors In the Netherlands, with dense cities and comprehensive bicycle infrastructure, seniors make 25 percent of their trips by bike. American seniors make less than one-half of 1 percent of their trips by bicycle. (Data from the U.S. Department of Transportation, Danish Ministry of Transport, Statistics Netherlands, and German Ministry of Transport)

for students to bike or walk to class. A student environmental group at Bridgewater-Raritan High School raised money for a bike rack only to have their principal reject it, citing safety risks. Similarly, a principal at Island Park Elementary School in Mercer Island, Washington, an avid bicycler herself, vetoed a proposed bike route, pointing out that a fifth-grader had recently been killed while walking his bike through a street crossing.

Stories such as these are all too familiar to Deb Hubsmith, director of the Safe Routes to School (SRTS) National Partnership, which institutes programs across the country to make walking and biking to school safer and more practical for students and educators. Testifying to Congress about an SRTS pilot program, Hubsmith stated, "In only two years, we documented a 64 percent increase in the number of children walking, a 114 percent increase in the number of students biking, a 91 percent increase in the number of students carpooling, and a 39 percent decrease in the number of children arriving by private car carrying only one student."[48] Nevertheless, even though children represent over 12 percent of pedestrian fatalities, and bicycle-related injuries send over a quarter million children to hospitals annually, the SRTS won just 0.2 percent of the U.S. Department of Transportation's safety budget.[49] And even though safe routes are a far more effective challenge to fossil-fuel consumption than solar cells, legislators overwhelmingly direct more money into the solar pot. In California, for every dollar spent on safe routes, well over ten dollars has flowed to solar cells during every budget year from 2007 thru today.

Given the clear and far-ranging benefits of walking and biking to school, the fact that communities hold bake sales to finance bike racks and safe thruways for students while the fetishized solar-cell industry bathes itself in billions of public funds is an inglorious national embarrassment. There is no secret to designing safe and convenient bikeable and walkable communities. The strategies are flexible to a wide array of neighborhood layouts, simple to institute, and return rapid paybacks in terms of public safety, quality of life, energy footprints, and long-term infrastructure maintenance costs. Ultimately, the success of bikeable neighborhoods hinges on a community's ability to establish a bicycling culture, where bicycling and walking stand as legitimate and esteemed modes of transportation. A coordinated study between Rutgers University and the European Commission identifies six policy priorities:

1. Better facilities for walking and cycling
2. Urban design sensitive to the needs of nonmotorists
3. Traffic calming of residential neighborhoods
4. Restrictions on motor vehicle use in cities
5. Rigorous traffic education of both motorists and nonmotorists
6. Strict enforcement of traffic regulations protecting pedestrians and bicyclists[50]

The Limits of Public Transit

For longer trips that are not convenient by bike or by foot, mass transit can be a powerful mobility supplement for people of all ages and abilities. But it won't work everywhere. The largest predictor of success for a public transit system isn't the fare cost, the number of seats, or even the frequency of service, but simply the *population density* of the locale it services. High-density housing and business districts make mass transit economical and practical. It doesn't take much rail to connect a hundred points of interest in San Francisco. However, connecting a hundred points in the sprawling suburbs of Kansas City would require massive investments and subject riders to long boring trips. One study indicates that regions must contain at least seven dwellings per acre to successfully support a bus line. Lower-density regions require significant public subsidies in order to maintain regular service. Jeffrey Zupan, one of the study's authors, points this out.

People often say they want public transit but then aren't willing to zone their area to support it. Then they complain to the government or the bus operator, "How come you're not running a bus in my neighborhood?" Well, they've got half-acre lots—what do they expect? Or they build their office building five hundred feet back from the road and put a parking lot next to it so that a worker in a car can park by the front door, while anyone getting off a bus has to negotiate a muddy lawn, maybe not even with sidewalks, just to get to the building. Under such conditions, who's going to use transit? You can't have a successful transit if you create an environment that doesn't support it.[51]

Communities throughout the country are fighting to establish light-rail lines though their neighborhoods in the hopes that when built, people will eagerly flock to ride them, but they might be in for a surprise. Experience shows that people are less willing to take public transit in areas where it is easier or cheaper to hop in a car and drive. Public transit can actually propagate car-dependant suburban growth when it connects previously uninhabited expanses of land lying between the outer legs of a city's mass transit system. In fact, that's exactly what occurred north of New York City, in the outskirts of Atlanta, and even in regions surrounding the celebrated bus system in Curitiba, Brazil.[52]

Regional officials frequently spoil rail and bus initiatives by supporting roadway construction, often with more gusto, in the name of fighting traffic, although road construction spurs developers to build new strip malls and housing developments, which eventually draw out automotive dependency. It's no coincidence that residents most strongly embrace mass transit, bicycling, and walking in areas where it's inconvenient to own and operate a car—metropolises such as London, New York, and Tokyo as well as smaller cities such as Amsterdam, Copenhagen, and Barcelona. In Denmark, cities charge for parking; cars carry a 180-percent sales tax; and gas costs twice as much as it does in the United States, making car ownership far less desirable. Owning and operating a car in most American locales isn't just convenient, it's practically mandatory, thus positioning automobiles at the top of the transportation pecking order.

Imagining Better Villages

In 1951 the California Division of Highways began building a grid of highways through San Francisco. The monstrous web of viaducts, ramps, and elevated highways started to bifurcate neighborhoods and decimate property values.[53] Public protests eventually halted construction but it was already too late for some. Workers had already erected the sky-darkening, double-decker Embarcadero Freeway, which lay supine like a fallen skyscraper across the city's northern waterfront. Even as crews were still assembling the monolith, residents pled to have it torn down. Proponents of the freeway maintained it was necessary for the city's growing traffic patterns and warned that if it were leveled, traffic chaos and economic hardship would follow. The 1989 Loma Prieta earthquake put their claims to the test.

After the Embarcadero Freeway's partial collapse, traffic snarled just as predicted—but it didn't last. Within weeks, the traffic self-organized into alternate patterns, which some deemed an overall improvement. San Francisco's mayor, Art Agnos, faced a choice: repair the collapsed section or tear the entire freeway down. His controversial nod to the wrecking ball in 1991 barely survived a six-to-five vote by the Board of Supervisors but a decade later he would claim that knocking down the freeway was the single best decision he made while in office.[54] Today the waterfront teems with activity, property values have recovered, and residents would never think of inviting the freeway back. The Embarcadero neighborhood now features a smaller road flanked by vintage electric streetcars (many brought in from cities that dismantled their trolley networks), broad pedestrian areas, and small parks. The current streetscape, with its skateboarders, pedestrians, tourists, farmers market, and even an acrobatic cabaret called Teatro Zinzanni, is again as functional as it was in the early 1900s. Other cities are exploring the benefits of taking back public space from automobiles and voluntarily exorcising freeways from their downtowns, even without the earthquake impetus.

Aside from highways, cities have their well-known disadvantages: noise, fumes, crime, bad schools, and overheated subway trips in the summer to name a few. And then there's the obvious critique: most people can't afford the extreme cost of housing in places like Manhattan (America's walkable communities are expensive because they're in exceptionally short supply).[55] These factors scare young couples to the suburbs to raise their families. Nonetheless, it's possible for cities to lessen or eliminate these challenges. For instance, the city of Amsterdam is incredibly dense, yet it isn't overcome by car fumes; public transit is comfortable, fast, and reliable; and the city center is quiet enough at night to hear distant frogs and crickets, perhaps interspersed with the occasional rustle of a bicycle. In other parts of the world, residents from across the wealth spectrum enjoy walkable and bikeable communities because such neighborhoods are abundant. Transforming our urban cores into friendlier places to live will be a better project for environmentalists to pursue than affixing solar cells to suburban McMansions or building disconnected "eco" forts far from grocery stores and other necessities.

In the coming years, growing numbers of the world's inhabitants will be living in cities, so it is especially vital that we draw upon strategies to make them cleaner, safer, inviting, affordable, and more energy-efficient places to live. Consequently, future environmentalists will be concerned about a lot of issues people don't label as environmental work today. They'll promote quality-of-life issues such as cleanliness, crime reduction, comfortable public transit, and street noise. They'll be developing entertainment and recreational facilities, senior programs, civic events, educational infrastructure, homeless services, film festivals, artist exhibitions, public health campaigns, and other projects that make cities more desirable places to live. If such accoutrements make dense cities more livable and alluring, then art galleries, cafés, and comfy bus seats are key environmental assets, whether they fit the standard model or not.

First Step: From Cars to Cafés

Americans who travel to Europe delight in the numerous cafés teetering out to the edges of sidewalks, where locals converse with friends, read books, and spar in political arguments. Vibrant sidewalks make neighborhoods safer and more inviting. Of course, another way to make cities more attractive is to physically *make them more attractive*. Widening sidewalks to accommodate trees and other plantings can change a blank road into an urban oasis by cooling the sidewalk in the summer and providing both physical and visual narrowing of the road. This in turn encourages drivers to slow down and heightens their awareness that they are traveling through a zone intended for people.

First Step: From Parking to Parks

American parking policies are among the most perverse and baffling public subsidies ever to be overlooked by a populous. We reserve and maintain some of the most valuable and potentially useful real estate in the nation in the anticipation that drivers might store their empty cars there—and we charge them mere pocket change, or even nothing, for the privilege. A growing group of artists, activists, and citizens are challenging this blatant misuse of public space—one day a year they temporarily transform metered parking spots into PARK(ing) spaces complete with trees, grass, picnic tables, and art installations. The movement has spread to over one hundred cities worldwide (ParkingDay.org).

Still, it would take a dangerously high dose of optimism to expect everyone in walkable and bikeable neighborhoods to simply abandon their cars.[56] That is, unless a superior option were to evolve.

Illustration 9: Reclaiming streets This "parklet" in San Francisco returns several automobile parking spaces back to pedestrian use by incorporating bicycle parking, benches, and plantings at sidewalk height. This is a temporary installation that city planners are testing in anticipation of a more permanent takeover. (Photograph courtesy of Aaron Norton)

First Step: Car Sharing

Car sharing offers many benefits over car ownership. Members don't have to pay for a car payment, finance charges, insurance, gas, oil changes, maintenance, or parking. In addition, they needn't stand in line at their local Department of Motor Vehicles to secure registration and vehicle tags or to pay transfer taxes. Instead, members simply pay a small yearly fee of about $50 for a universal keycard allowing them access to any car in the fleet for about $8 per hour—a scheme that ends up saving the average user $1,800 to $5,000 per year.[57] Since every car links to the Internet, members can view availability or reserve a car in advance. Worldwide, car-sharing membership is currently doubling every few years.[58] City CarShare, a California

The Architecture of Community   289

Bay Area nonprofit, enables locals the flexibility of choosing a pickup truck for moving furniture one day and a sporty car to pick up a date on another. Other outfits allow members to pick up and drop off cars at any designated lot without reservations or time limits. The largest car-sharing network, Zipcar, has over half a million users sharing nine thousand vehicles.

Car sharing prompts system-wide benefits as well. Since members have to pay at the moment they use the vehicle, they more frequently link trips together, drive with friends, or consider driving alternatives altogether. Car sharers report a 47 percent increase in trips by public transit, a 26 percent increase in walking, and a 10 percent increase in cycling.[59] Second, since car-share vehicles park at the same neighborhood hubs, rather than throughout dozens of garages, they are better positioned to accommodate alternative-fueling requirements. Third, with fewer car owners, cities can reclaim roadside parking spaces for sidewalk cafés and bike lanes. Urban homeowners can transform their garages into artist studios, student flats, or other uses, so long as they don't live in cities with outmoded zoning that prevents it. Finally, the book *What's Mine Is Yours: The Rise of Collaborative Consumption* projects the flexibility and economy of sharing onto a larger screen. If people find it more convenient and cost efficient to share and reuse items instead of buying new ones, their prudence could give new life to a now-faded virtue of building products to last.

First Step: Congestion Pricing

At first glance, congestion pricing sounds like a good idea. London drivers pay a steep toll to enter the city center, and other cities are proposing similar schemes to lessen rush-hour traffic. Reducing the number of cars in the city prevents gridlock and leaves fewer idling cars in traffic. However, congestion pricing is not without drawbacks. First, it delineates traffic reductions along class lines, essentially reserving streets for wealthy individuals or those with company cars and expense accounts (who don't have to pay the toll themselves) while most acutely restricting street use for middle- and lower-income workers. Second, drivers may rat-run around city borders to avoid the fees, offsetting or even adding to the overall number of car-miles driven. Finally, as city congestion eases, traffic speeds tend to increase. Just small speed increases greatly reduce the chances of a pedestrian or bicyclist surviving a collision and tens of thousands of such collisions occur every year in the United States. David Owen claims that traffic jams aren't environmental problems at all. They're a driving problem.

> If reducing congestion merely makes life easier for those who drive, then the improved traffic flow actually increases the environmental damage done by cars by raising overall traffic volume, encouraging sprawl and long commutes, and reducing the disincentives that make drivers think twice about getting into their cars. Traffic jams are actually beneficial, environmentally, if they reduce the willingness of drivers to drive and, in doing so, turn car pools, buses, trains, bicycles, walking, and urban apartments into attractive options.[60]

Still, congestion pricing has some virtues if planners anticipate and interrupt the potential negative consequences. Most valuably, as congestion eases, city dwellers can simultaneously reclaim their roads by expanding sidewalks into former traffic lanes, replacing parking spaces with corner or street cafés, and planting trees. Furthermore, legislators who value equality could ensure that congestion fees support public transit service. An organization founded by New York labor lawyer and activist Ted Kheel advocates reducing or eliminating bus fares in New York by instituting a variable congestion fee of about six dollars and charging higher bridge tolls and parking fees.[61] If implemented as part of a larger project for reclaiming streets, congestion pricing can work in large cities—residents of small to medium-sized cities may be better served through traffic calming techniques.

First Step: Traffic Calming

Jaywalking is a dangerous activity in America but pedestrians are safer in urban areas where jaywalking is frequent than where it is strictly prohibited. Why? Because clearly separating pedestrian spaces from traffic lanes entices drivers to speed up and overlook pedestrians and cyclists. Similarly, when small towns first introduce traffic signals, accidents tend to increase, not decrease. Green lights suck drivers into intersections with greater speed and less consideration for others. While mixing cars with pedestrians and bikes may seem like an emergency-room disaster, it isn't in practice. Numerous European municipalities mix uses because the ambiguity of right-of-way slows drivers and induces drivers, cyclists, and pedestrians to remain more alert. This strategy is most effective when planners combine it with other traffic-calming techniques to make streets safer for cyclists and pedestrians.

In my experience working with community groups, the first traffic-calming suggestion to jump into people's heads is the almighty speed bump. Speed bumps do slow traffic. However, they are noisy, and worse yet they prompt cars to speed up and slow down in an alternating succession that ends up emitting more fumes, particulates, and smog than if drivers had maintained a constant speed. Other communities step up enforcement by issuing speeding tickets to violators. Switzerland links speeding tickets to personal wealth in order to make the penalties more equitable. One repeat offender, caught driving a red Ferrari Testarossa through a village at thirty-five miles per hour faster than the posted limit, was fined $290,000—a penalty held up by Swiss courts based on the man's $22.7 million net worth. Still, speeding tickets represent an after-the-fact tactic for calming traffic and like speed bumps, are a rather blunt tool for performing the job. Here are some less famous traffic-calming stars already in use throughout the world:

- *Elevated crosswalks*: Slightly elevate and clearly mark cross-walks in order to signal that pedestrians belong on the road too.
- *Expanded sidewalks*: Push sidewalks or bike lanes into streets to narrow vehicular lanes and slow traffic.
- *Four-Way stops*: Replace two-way stop signs with four-way stop signs in order to decrease accidents and offer respite for bicyclists and pedestrians.
- *Outcroppings*: Extend urban planter beds into the roadway to force cars to slow down and negotiate the outcroppings. This strategy also slows traffic by visually pinching the road, especially if the outcroppings are placed across from one another and planted with trees.
- *Takeovers*: Completely remove road lanes and/or parking to create more space for other sidewalk activities.
- *Diversions*: Interrupt long urban streets with mini-parks (right in the center of the street) that require cars to turn but allow bicycles, pedestrians, and emergency vehicles to pass freely.[62]

An excellent alternative to calming traffic is removing it. Some cities reserve an extensive grid of lanes and streets for bikes, pedestrians, and the occasional service vehicle. This motivates people to travel by bike rather than by car, making streets safer for everyone. As bicycles become more popular in a city, planners can convert more automobile lanes and entire streets to accommodate more of them. Nevertheless, even the most bike-able cities still require motor vehicle lanes for taxis, emergency vehicles, and delivery trucks. Delivery vehicles are frequently a target of animus, but they are actually an essential component to making cities greener. A tightly packed delivery truck is a far more efficient transporter of goods than several hybrids carrying a few shopping bags each. Distributing food and other goods to neighborhood vendors allows them to operate smaller outlets close to homes so that residents can walk, rather than drive, to get their groceries.

First Step: Prioritize Bicycle Roadways

America's transportation budget provides negligible support for bicycle infrastructure despite its enormously cost-effective and socially valuable returns on investment. Compared to automotive infrastructure, bike paths and the vehicles that travel on them are far less energy intensive to build, maintain, and operate. Urban and suburban municipalities can safely increase bicycling and walking rates as demonstrated and time-tested in Germany, Japan, Denmark, and even some American cities.

Davis, California, has more bicycles than people. Campus roads at the University of California–Davis are restricted to bicycles, as are several other "greenbelts" throughout the city, making Davis a safe and enjoyable city for bicycling—a pleasure that its residents passionately take up. Residents commute year-round by bicycle with the same frequency as Europeans enjoy. Fair weather and flat topography make Davis ideal for bicycling, but the most important factors for cycling success involve land-use planning and bicycle infrastructure. Since 1966 the Davis City Council has worked to build over fifty miles of bike lanes within the city's ten square miles while containing sprawl, promoting a bicycling culture, and locating services close to residential areas—all in an explicit effort to make the city more accessible and open to walking and biking rather than driving.[63]

Another success story in the United States is on display in perhaps one of the most unlikely places—Steamboat Springs, Colorado. Compared to temperate cities, this snowy winter wonderland may seem ill suited for bicycling. However, Steamboat boasts a dedicated biking and pedestrian highway that residents of sunny San Diego can only dream about. Winding through Steamboat's downtown along the picturesque rapids of the Yampa River, the roadway connects residents to the city's schools, college, library, grocery stores, post office, hot springs, and even

Illustration 10: Prioritizing bicycle traffic This Amsterdam street prioritizes bicycle and pedestrian traffic. Retractable barriers allow emergency and delivery vehicle access. (Photograph courtesy of Jvhertum)

the ski resort. During the especially harsh winters, a couple extra layers of clothing and studded tires are enough to keep this snowplowed path active year-round. The bikeway has seamlessly worked its way into the daily life of smiling pedestrians and bicycling commuters in the area. When critics tell me that bike lanes won't work in the United States, I show them photos of Steamboat. If bicycling infrastructure can have an impact deep in the blustery Rockies, it can have an impact anywhere.

First Step: Bicycling for Youth

Across the mountain range from Steamboat, Crest View Elementary School in Boulder, Colorado, instituted a system to reward students that are frequent walkers and bikers though a program named Boltage (www.Boltage.org). Students each receive a radio-frequency identification (RFID) tag, like the ones retailers use for inventory management, to place on their bike helmet or book bag. When students walk or bike to school, they automatically receive credits that they can monitor online and accumulate to earn prizes. Executive director Tim Carlin explains that "what's especially valuable about Boltage is that it provides an evaluation component"; the organization can track the number of round-trips that students make and in turn calculate the amount of gas they save and number of calories they burn.[64] Just a handful of Boltage programs already prevent hundreds of thousands of car trips annually, reducing congestion around the schools that participate in the low-cost program. If expanded, such programs could start to measurably cut into the yearly purported $76 billion in health costs related to physical inactivity and over $40 billion in costs stemming from vehicular air pollution, most of which emanates from cold engines during the first few miles of travel (as in trips between homes and schools).[65]

In addition to getting cars away from school roundabouts, where huddles of idling tailpipes form an invisible toxic cloud awaiting students as they exit school, Boltage schools can shorten and even eliminate bus routes. Furthermore, Boltage kids are more alert upon arriving at school, having already exercised before class. Residents approve of the program as well—evidenced by one supporter living near Crest View, who gushes, "I love seeing the happy, steady parade of kids going by my house, and the families all having morning bike together time."[66] Students are equally excited about the program. "I think the most important thing that Boltage did for our school was to give students the option to ride their bikes and feel cool about it," remarks Sydney Cook, a former bike club leader. "Students were actually bragging about riding their bikes to school."[67] They'll presumably feel more empowered to use their bikes as a form of transportation for nonschool activities and into later life as well, effortlessly magnifying the power of Boltage's initial nudge.

First Step: Bicycle Insurance

When a new bicycle owner has their bicycle stolen, they become far less likely to view bicycling as a viable transportation option. Even though bikes are monetarily worth less than cars, their social value is much greater. Police and courts should treat the threat of bicycle theft as a crime equal to or greater than automotive theft. Unfortunately, insurance companies don't offer comprehensive bicycle insurance to cyclists in the United States, which would cover cyclists for liability, theft, or damage. European cyclists enjoy such coverage, and the author of *Bicycling and the Law*, Bob Mionske, insists Americans should too, citing "One of the institutional biases against cyclists is the requirement that you have to own an automobile in order to be able to purchase some important types of insurance."[68]

First Step: Reform Zoning

San Francisco sits atop a gold mine of opportunity but it remains hog-tied by harmful zoning laws. In San Francisco and many other cities, the aforementioned scenario of converting empty garages into artist studios and student flats is prohibited by archaic zoning laws that actually *mandate* garage spaces, even if they are doomed to lie empty. Allowing urban homeowners to repurpose their garages into small commercial or residential units would clearly initiate several benefits. First, small efficiency rentals would enable tens of thousands of teachers, social workers, students, artists, service employees, and others to live in the city without commuting from the burbs. Small commercial units are ideal for start-up companies, galleries, nonprofits, and other valuable alternatives to big-box stores. Furthermore, by providing more eyes on the street, ground-level flats and storefronts effectively reduce neighborhood crime rates. Finally, flats and studios are more inviting to pedestrians than the rear end of a Buick rolling from a garage.

A zoning official once told me his metropolis would be better off if the city council simply eliminated his department altogether. He may have had a point, given that America's most vibrant and environmentally successful neighborhoods *least* conform to today's zoning regulations. The most locally cherished and charming historic downtowns would in most cases be illegal to build today due to zoning restrictions stipulating land use, setbacks, height, and mandated parking. Such regulations lead to low-density patterns that are difficult to negotiate by foot, bicycle, or even transit, in effect limiting or preventing nearly all modes of community that Jane Jacobs celebrated.

First Step: Retrofitting Suburbia

Reinvigorating urban cores is only one side of the equation. About three-quarters of construction in the United States occurs outside cities, in the suburbs and rural areas surrounding them. Not everyone will want to live in cities, which is why we may be well served to integrate some of the efficiencies of cities into suburbs, making them more walkable, bikeable, and ultimately more convenient and pleasant for residents. Ellen Dunham-Jones and June Williamson take on this project in their book *Retrofitting Suburbia*. They identify a process of "incremental metropolitanism," whereby municipalities can densify failed suburban strips in order to take on many of the qualities that Jacobs promoted—short blocks and mixed-use buildings with less reliance on automotive travel. Communities throughout the United States have successfully converted their vacant big-box stores and parking lots into new community assets such as churches, schools, housing, and mixed-use buildings featuring interconnected street grids and lushly planted pedestrian access. Asphalt to assets. Dunham-Jones and Williamson claim that, "by urbanizing larger suburban properties with a denser, walkable, synergistic mix of uses and housing types, more significant reductions in carbon emissions, gains in social capital, and changes to systematic growth patterns can be achieved."[69] They point to a densely built Atlanta neighborhood named Atlantic Station, where residents drive an average of just eight miles per day in a region where the average employed individual drives sixty-six.

Of all of the obstacles to retrofitting suburbs, one challenge will trump them all: funding. Bankers are particularly prickly to mixed-use and walkable plans because they don't fit the standard set of real estate "product types," whose risks and paybacks were exhaustively studied during the high reign of suburbia.

One such product type, termed a "grocery-anchored neighborhood center," contains 100,000 to 150,000 square feet of single-story chain stores built in an L-shape on a twelve- to fifteen-acre lot on the return-home side of a freeway typically carrying twenty-five thousand cars a day. A parking lot containing four parking spaces per thousand square feet of retail separates the complex from the road. Up to three chain restaurants or bank "out-parcels" sit closer to the street. A 45,000- to 55,000-square-foot grocery-store chain anchors one end of the complex. A 15,000- to 20,000-square-foot chain drug store caps the other. Plans contain no pedestrian access or crosswalks unless mandated by local law. Double-height facades wrap the boxes in local stereotypes—fancy tiles for Southern California, imitation vigas in the Southwest, and brick facing on the East Coast.[70] We know this product type by another name: strip mall. Developers clone these so-called neighborhood centers, along with another eighteen or so accepted real-estate product types, to extend ad nauseam down suburban strips. Lined up like feeding troughs, they are calculated to return profits through the same instrumental rationality used in factory farming.

It is comparatively simple to find funding for a "grocery-anchored neighborhood center" or a "big-box-anchored power center" (another one of the nineteen accepted typologies). However, mixed-use proposals, with either residential lofts or office space above retail, are challenging for American developers to fund and are therefore not popular. In an environment where bankers scoff at "nonconforming" mixed-use buildings, and zoning officials essentially mandate standard real-estate product types, low-density, unwalkable, and high-traffic neighborhoods are guaranteed. Since American-style real estate financing is so profitable, these suburban product types are gaining popularity among developers in other parts of the world. There are alternatives.

Brookings Institution fellow Christopher Leinberger claims that those who wish to retrofit suburbia have a choice: stick with building small, expensively financed mixed-use projects, or develop new product types that financiers understand and accept.[71] He offers some alternatives to the existing product types:

- *Housing/office/artist lofts over retail*: Built adjacent to the sidewalk, ground-floor retail includes living or office space above.
- *Burying the big box*: Instead of adjoining the big box with asphalt, it is placed mid-block and surrounded by separately designed buildings containing housing or offices over retail.[72]

Suburban product types force residents to reach every destination by car. Upon arrival they must walk through treeless asphalt lots teeming with cars that silently emit toxic volatile organic compounds as they sit parked.[73] It is a wonder that Americans are willing to go anywhere by foot given that this is one of their primary experiences with walking. Yet in survey after survey across the United States, when asked to select their preference between images of walkable communities and typical suburban sprawl, people overwhelmingly point to the walkable spaces.[74]

Sprawl isn't something most people would have chosen for themselves and their families; it's just the way things are. For many Americans, it's the only way of life they can remember ever having existed. It has become comfortably familiar. But comfort, as we shall next consider, is a remarkably slippery notion.

13. Efficiency Culture

Wisdom enough to leech us of our ill is daily spun;
but there exists no loom to weave it into fabric.
—Edna St. Vincent Millay, "Upon This Age"

If we cut our per-capita electricity consumption in half, overnight, we'd still be using more juice than those living in the Netherlands.[1] This World Bank statistic forces us to entertain the possibility that Dutch people live in dark caves without modern conveniences. But in reality, their houses are not dark at all; compared with American households, Dutch homes average 20 percent more light bulbs.[2] They even have an enormous light bulb company, Phillips. In fact, life isn't so harsh in the Netherlands these days, as I can attest; I've been fortunate enough to call the country home for part of my life. Like many Europeans, the Dutch enjoy longer life spans than Americans do, less poverty, less air pollution, lower debt, incredibly clean drinking water from the tap, and a high standard of living that somehow allows them to enjoy delicious foods like Dutch apple pie, but with a fraction of our obesity rate.[3] Those slinky little rascals! It's no wonder they consistently rank higher than U.S.

citizens in international studies on happiness.[4] How do they do it? Is it really possible to achieve well-being with just a fraction of the energy input we use in the United States? Certainly. But I'd like to pose a more provocative question: Could lower energy use in the Netherlands actually be fostering Dutch happiness?

Perhaps their cheeriness does not exist in spite of their frugal energy use but rather because of it. This isn't as counterintuitive as it might seem. Since the Dutch use less energy than Americans, they don't have to manage as many of the negative side effects, costs, and limitations associated with energy capture, distribution, and use—a list of consequences that has been growing over recent decades. These side effects, which were limited mostly to apprehensions about air pollution in the 1960s, now include concerns such as political stability, price volatility, wealth transfers, social justice, economic risks, supply interruptions, supply limits, enrichment of hostile regimes, conflict, and climate change. But how can we be expected to judge the negative side effects of the energy we personally draw upon? We don't each spend one day a month on an oil rig, pumping our monthly oil allotment. How can we make choices about energy technologies if we have so little firsthand experience with their operations?

If handed a jar of toxic chemicals, most people would not be willing to pour it into their local stream or lake. In fact, many people would actively protest such an action. Yet they may hold little issue with living in a car-centered suburb and commuting daily to and from work in a gasoline-powered car, essentially pouring the same jar of chemicals into the environment five days per week. Energy researchers John Byrne and Noah Toly observe that in "the narratives of both conventional and sustainable energy, citizens are empowered to consume the products of the energy regime while largely divesting themselves of authority to govern its operations."[5] Our vehicle's exhaust system does not leave us with a vial of pollutants to be tossed into the neighbor's yard. Our grocer does not label the beef steak to indicate how much petro-fertilizer was used to grow grain that was fed to the cow, how many gallons of water the ranch polluted, how much methane the cow released, or even if the cow was treated humanely. We can assume that the most economical process was likely employed, but would it be congruent with our own values?

Time and space displace us from power production and its numerous side effects. Since we can rarely acquire sufficient technical expertise ourselves, we can travel only to a certain point before placing our trust in others, even if they may stand to profit by keeping us in the dark. And as it turns out, we've been kept conveniently in the dark on a great many things.

So, let's turn on the lights.

The Real Energy Crisis

America has plenty of energy—more than twice as much as it needs. We just waste most of it. Established energy giants are willing to embrace alternative energy since it creates a convenient diversion from this simple reality. They understand that even if America were to, say, quadruple solar, wind, and biofuel output—a lofty project in itself—the increase would hardly impact fossil-fuel demand. I've spent enough time around these people to know this well. They're laughing at us.

On the other hand, plugging leaks in the nation's energy system *could* slash fossil-fuel use, saving extraordinary sums of money in the process. The United States has plenty of room for improvement on this front. When it comes to plugging leaks, it lags far behind most other industrialized nations in almost every category.[6] In fact, the *majority* of America's power production does nothing useful at all.

Kelly Sims Gallagher, director of the Energy Technology Innovation Policy research group at Harvard asserts,

Greater energy efficiency offers leverage against all of the major economic, security, and environmental problems faced by the

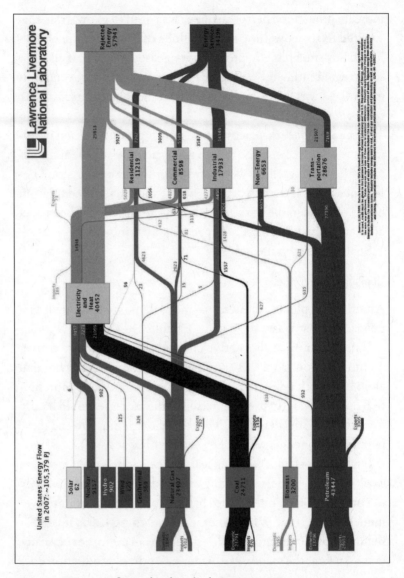

Figure 18: U.S. energy flows This chart displays primary U.S. energy sources (in petajoules) and indicates where that energy ends up. Well over half the nation's energy is lost to inefficiencies. (Image courtesy of the Livermore National Laboratory and the U.S. Department of Energy)

United States. Greater energy efficiency improves oil security and reduces emissions, the need for greater power supply capacity, and pressure on the electrical grid. It reduces the amount of money being spent on oil and gas imports, and it can improve the productivity of U.S. firms. In fact, a worthy goal would be for the United States to become the most energy-efficient economy in the world.[7]

So if energy efficiency holds such vast potential for advancing human well-being, why isn't it the center of energy policy rather than an afterthought? Why do we have a Department of Energy and no Department of Efficiency? Principally, it's because we have for so long seen energy production and well-being together, we don't know what they look like apart. This may have to change.

Designing Comfort

Energy production and well-being have walked hand in hand for some time now. Indeed, our lifestyles would never have been possible without a sizable fossil-fuel extraction industry. People think of efficiency and conservation in terms of their level of comfort, or more specifically their fear of losing it. (Efficiency means using less energy for a given service. Conservation means getting by with reduced services to save energy.) The prototypical model of American comfort—a four-bedroom, two-and-a-half-bath neocolonial on a cul-de-sac with two hybrids in the garage—was not an inevitable understanding of comfort but rather one bound to a particular history.

What exactly is comfort? Comfort cannot so easily be seen, counted, and quantified. It is an idea, perhaps even an ideal (it has even been called a "verbal invention" and a "cultural artifice"). And like most notions, it is as much a reflection of us, as it is a reflection of the spaces we have created. Historian Witold Rybczynski writes that comfort is "an idea that has meant different things at different times. . . . In the seventeenth century, comfort meant privacy, which lead to intimacy and, in turn, to domesticity. The eighteenth century shifted the emphasis to leisure and ease, the nineteenth to mechanically aided comforts—light, heat, and ventilation. The twentieth-century domestic engineers stressed efficiency and convenience."[8] The idea called *comfort* came to mean different things to different people at different times in response to a variety of economic, technological, and social forces. Contemporary American concepts of household comfort are unsurprisingly structured through a history of relatively easy access to cheap energy. If energy should become less easily accessible, our concepts of comfort may very well get caught in the crosswinds, forcing us to reconceptualize fundamental cultural ideals. As Rybczynski argues, "What is needed is a reexamination not of bourgeois styles, but of bourgeois traditions."

It is not especially surprising that homes were once much smaller than they are today, but less evident is that some people understood them as more comfortable as a result. Catherine Beecher's 1869 book, *The American Woman's Home*, which she wrote with her sister Harriet Beecher Stowe, developed an argument for small, efficient homes imagined through the eyes of the era's domestic laborers: women. Beecher's model house varied notably from the then-prevailing ideal gentleman's dominion. Beecher's model home featured a remarkable rolling cabinet large enough to define and redefine a large parlor into various configurations throughout the day and into the evening as needed. Beecher wrote, "Every room in a house adds to the expense involved in finishing and furnishing it, and to the amount of labor spent in sweeping, dusting, cleaning floors, paint, and windows, and taking care of, and repairing its furniture. Double the size of a house, and you double the labor of taking care of it, and so, vice versa."[9] Beecher saw smaller homes as more economical than larger homes, but more outstandingly, she clearly understood smaller homes as *more comfortable* as well.[10] So it appears that the rolling partitions concocted by architectural students during late-night dorm-room reengineering sessions were anticipated a century and half ago by Beecher. Indeed they were anticipated centuries before her as well.

Seventeenth-century Dutch homes, with their simple materials, functional furniture, and small footprints have something to teach us about the comforts and practicalities derived from living in small spaces. Amsterdam building plots were small and narrow since they had to be squeezed between canals and supported from below by expensive underground pilings. Builders pushed the walls of buildings out to the edges of the lot, often sharing common walls with neighbors. They topped off the narrow buildings with gables outfitted with large hooks, over which residents threw ropes to elevate furniture up and through the windows. When the Dutch wanted more space, they built in the only direction they could—up—and often several stories. Most of the buildings met the ground with a commercial space or a stoop where families sat in the evenings to socialize casually with neighbors.

These homes still stand as a seamless mixture of culture, community, and economy rendered in brick and mortar. Dutch homes were the physical embodiment of simplicity and thrift. They contained various innovations, for sure—double-hung windows for instance—but they were perhaps most remarkable for being unremarkable. Historian Steen Eiler Rasmussen once observed that while the rest of Europe had erected magnificent palaces, the Dutch had created supremely livable cities.[11] Dutch homes were reflections of a basic sense of community and domesticity that developed in other European cities as well—it's just that in the Netherlands, it occurred a century earlier. Why?

It is common for architectural historians to explain compact Dutch cities in terms of the country's diminutive size, but this bit of historicism neglects to consider that the size of the modern-day Netherlands was not perceptible to planners and designers of the time. The region was more a collection of cities and provinces, not a contiguous country with attenuated borders. The citizenry didn't feel confined. If the "size argument," which architects hold up to explain the efficiencies of Dutch urbanism, is part of an explanatory model, it certainly isn't the foundation. Nevertheless, seventeenth-century Dutch builders constructed their society upon *something*. And so, at the risk of being labeled an architectural Freudian, I argue we can discover what that something was by traveling back to the womb of Dutch architecture, if just for a paragraph.

As with all historical expeditions, it's difficult to know how far back to travel; it seems one can always go further. But a good point for our purposes is the year 1421, on St. Elizabeth's Day, when floodwaters marched in to occupy the Dutch lowlands in force, threatening the livelihood of all of the region's residents. As a result, Dutch farmers banded together into *heemraadschappen*, groups responsible for building and maintaining dykes—a project that drew upon community, over individual, action. While their obvious function was to protect against encroaching waters, perhaps more crucial were the dormant functions of these social organizations. The *heemraadschappen* offered political and social alternatives to fiefdom, leading to early democratic institutions and a "polder model" method of consensus building (which incidentally still provides utility today). As a result, the lowlands evolved as a collection of small, egalitarian landowners, without the landless peasantry of England or the aristocracy of France (the Dutch royals were comparatively powerless and poor, thanks to the wars for independence). Lowland architecture developed within the region's bourgeois inclinations, distinctively assembled in numerous medium-sized towns, occupied by a mostly city-dwelling populace of merchants and traders, many with Calvinist leanings, and of course, not to be glossed over, ready access to tradable commodities plundered from abroad. And so the story of Dutch homes is not simply a story about architecture, design, and technical development but more centrally a story about the social, economic, and political conditions of a people.[12]

By the same token, improving the way Americans build buildings and arrange neighborhoods today has very little to do with technological development (we already have a bag chockfull of ancient interventions gathering dust) and much more to do with the social and democratic fundamentals for fostering well-being.

Beyond Insulation

Buildings consume nearly half of America's power, more than all air, land, and water transportation sectors combined. If we have efficiency standards for cars, why don't we have them for buildings? This might be a good place start.

States create their own building efficiency standards and most have meager minimum requirements; others don't even have these. In 1998 a nonprofit organization called the U.S. Green Building Council developed a set of measures, called the Leadership in Energy and Environmental Design (LEED), to rank the environmental performance of buildings. Builders and architects can apply for LEED certification for new and existing buildings. Advocates credit this certification process for stepping up construction waste recycling, increasing public awareness about green design, and spurring building departments to update municipal building codes. It has had a measurable impact, but LEED has a worrisome side too.

LEED specifications reward architects and designers for outfitting buildings with expensive add-ons, such as solar panels and urban wind turbines, while it undervalues age-old strategies for green construction that cost less and are many times more effective, such as small, efficient spaces located in walkable neighborhoods. "I think people have the idea that sustainability is just a collection of exciting ideas that you can peel and stick onto your building," laments David White, a climate engineer. "Unfortunately, the exuberant creative stuff—the expensive buzz words such as 'geothermal,' 'photovoltaic,' 'double facade,' and 'absorption chiller'—only makes sense when the basic requirements, such as a well-insulated, airtight facade with good [passive] solar control, are satisfied."[13] Unfortunately, architects frequently trample wise design principles in a fanatical rush to gather LEED or public relations points. The "green" building industry may be changing the greater construction industry but it is becoming fundamentally more like it, an economically productive model inclined toward adding features rather than subtracting them. Complexity over common sense, with an eye to profit.

The first building to be certified LEED Platinum, the highest rating obtainable, was the Merrill Environmental Center, a suburban headquarters for the Chesapeake Bay Foundation. The building's main façade is almost all glass, a material that not only assures that the interior will bake during the hot summer months, but also allows precious heat to escape during the winter. In an attempt to block some of the sun's heat, a massive framework of wooden slats (recycled, of course) were erected in front of the windows, but these cast shade on the solar panels. The center received LEED credits for maximizing the lot's open space—but that open space is exactly what makes it a sprawl bomb. The foundation's former headquarters were inherently much greener because they were located in downtown Annapolis and did not force its dozens of employees to drive beyond the outer reaches of the suburbs to get to their desks.[14]

A building in Boulder, Colorado, won points for an electric-vehicle recharging station, even though there's no evidence that the planet is any better for it. Even if drivers use the charger (which they haven't been, according to one report), they would only act to reinforce the suburban model of driving rather than the urban model of walking or biking. LEED undervalues walkable communities and overvalues suburban technological fetishes.[15]

A growing number of architects, developers, and builders view LEED certification as nothing more than a fancy game of bullshit bingo where any truly green outcomes arise merely from chance. Due to LEED's highly bureaucratic, expensive, and beleaguered application procedure, only a tiny fraction of the nation's buildings are even submitted for LEED approval. In concern for the future of the program, a pair of LEED specialists admits: "The dirty little secret is that you can certify a building without doing much at all (other than mountains of paperwork) to make it green. . . . If you know how to scam LEED points, you can get the PR benefits without doing much of anything for the environment. A system developed to address greenwashing runs the risk of becoming greenwash itself!"[16]

Programs such as LEED also fool people into thinking that low-energy buildings are more expensive than regular buildings and require all sorts of technological gadgetry. We've all seen the news segments featuring people who have fallen into a trap of green conspicuous consumption, believing that if they move to the country, buy a hybrid car, and build a home with expensive, recycled glass countertops, sustainably harvested kitchen cabinetry, and fancy alternative-energy mechanisms, that they deserve to be rewarded for having somehow added to the welfare of the planet. They may indeed receive a LEED Platinum award for doing just that, but the sobering reality may be less enthralling. Truly green homes aren't extraordinary at all. As *New York Times* columnist Thomas Friedman has insisted, "If it isn't boring, it isn't green."[17]

The most efficient homes populate older, mixed-use downtown neighborhoods and occupy small lots, close to shops, restaurants, neighbors, and public transit. They aren't too large, which minimizes construction materials, decreases heating and cooling requirements, and prevents them from doubling as storage units for runaway material accumulation. They have windows with adjustable shades, plenty of roof and wall insulation, adequate weather stripping, energy-efficient appliances, and kitchens with linoleum floors. These homes are unremarkable indeed—hardly the fodder of green-eyed journalists—but they are also unremarkably green. And it may be worth noting that there is a correlation between living in these walkable neighborhoods of unremarkable homes and personal satisfaction. Unremarkably green homes make people happy—high-tech homes make them poor.[18]

Cultures of Waste

A large willow tree stands at the southern bend of Amsterdam's Muider Canal, lazily guarding the entrance to an international apartment building filled with Americans, Germans, Norwegians, and people of other nationalities. As in neighboring buildings, the basement houses a common laundry room, but unique to this building is a handwritten note taped above the washing machines.

"Use hot water setting for cotton only—it will melt other fabrics!!"

The sign shocks North Americans, often becoming the subject of conversation and speculation. Is the double exclamation point scratched at the end a product of firsthand knowledge? Some residents argue that the machines require repair. Some insist it cannot be true—a washing machine can't really melt fabric. Others claim it's a prank. They're all mistaken.

A clue to this conundrum is the note's language: it is written only in English. That's because European residents don't need the warning. They consider it common knowledge that the hot setting of a washing machine will superheat water to near-boiling temperatures appropriate for making coffee, disinfecting medical equipment, cooking spaghetti, or yes, even destroying a pair of polyester trousers. After all, how else could one clean cotton towels? Silly Americans!

American washing machines typically employ hot water from the tap, while European machines contain internal heaters to bring the water up to very high temperatures through a process that is especially energy intensive. In fact, the German setting translates as "cooking wash." Conversely, Japanese households simply wash their clothes using cold water and soap. Often they use "gray water"—water that has already been used for bathing—a process that consumes far less energy than the one employed in Europe or the United States. All of these methods produce the same result—clean clothes.

The difference between these methods is not functional as much as it is cultural. Americans typically use hot water to clean their whites but are perfectly happy using cold for colors. The Japanese, who widely value hygiene, do not cognitively associate hot water with cleanliness in the same way northern Europeans do. Nevertheless, while washing dishes, the Japanese customarily allow hot water to run in a seemingly wasteful fashion because they consider it an acceptable way to warm the person washing the dishes. Water usage peculiarities in Japan, the United States, and northern Europe are culturally informed but might not be so significant that they couldn't be changed. Other energy traditions are less flexible—bound to economic, technological, and social frameworks that limit flexibility of individuals to have an impact on their energy consumption. If energy is such a vital component of our society, why is society so frequently left out of debates about energy?[19]

Political, environmental, and media attention frequently focuses on personal energy consumption even though personal choice is only one variable in a much larger socially determined framework of energy production, distribution, and use. Many consumption practices do not necessarily contain strong strains of intentionality—they may simply arise from habit, custom, or ritual. Energy practices are encrusted with the residue of previous intentions, either our own or those of previous generations. Any efforts to remove these calcifications must permeate beyond individual action to impact the social, economic, and technological spheres they occupy.[20]

Consider transportation systems, which entrench interlocking sets of industries, products, social practices, and institutional supports. Residents can opt to take public transportation only if a transit system exists in their community. Most Americans live in areas surrounded by networks of roads and highways, not streetcars and bike lanes. Americans are also likely to associate a degree of status with owning a vehicle.[21] Individual motorized transportation in the United States corresponds with broader trends in rich nations toward private, instead of collective, use of technology, higher prioritization of convenience, and the development of desires satisfied through consumer-oriented behavior instead of nonpurchase strategies. A hard nut to crack. It will involve a range of participants, including corporations, consumers, communities, and social organizations as well as more rigorous democratic and economic institutions. Simple technological development won't do.[22]

And in this spirit, we shall move on. But before we get too far, there's another piece that will lock all of the others together. And this piece, it turns out, is in France.

The Acting Network

During the political vibrations of the 1960s, a group of French sociologists founded the Center for the Sociology of Innovation under the auspices of the elite academic institutions in France. Michel Callon, Pierre Laffitte, Bruno Latour and others would likely neither have predicted how widely influential their work would become in academic circles nor anticipated how carelessly others would ignore it.

Callon and his associates followed engineers, researchers, lab technicians, and other research specialists around on their day-to-day jobs. Over the years, these sociologists embedded themselves in laboratories, moving from corporation to corporation, and comparing notes. A decade later, they published their findings.[23]

Their arguments formed a defiant rebuttal not only to the common understanding of industrial innovation as being hierarchical or top-down but also to the conception that great ideas arise from individuals. In fact, they claimed, the dichotomies between concepts of top and bottom and internal or external were useless. Innovation, according to Callon, is a product of *networks* of people and things; it is within this intangible network where the source of innovation lies. He writes, "The network has neither a center nor a periphery, but rather is a system of relations among problematic utterances that come, equally, from the social sphere, scientific production, technology, or consumption."[24] Callon's conception of innovation presses us to think beyond simple considerations of *just* scientists, or *just* regulations, or *just* consumers, or *just* new energy technologies, or *just* environmental constraints, or just any single part of the broader systems of production, distribution, and use. He rather argues for considering the *relationships* between these various "actors." It might follow that our energy strategies should not address exclusively one actor or set of actors but should rather treat the system as a whole. This may seem an overwhelming task. Surely addressing all of the interactions between all of the people and things involved in power generation and use, all at once, would be an enormous undertaking.[25] Nonetheless, a growing number of political theorists argue that a type of market mechanism—carbon pricing—could bring this orchestration into harmony. They are somewhat correct.

Taking Carbon Pricing to its Limit

Former vice president Al Gore popularized the concept of carbon pricing in the United States. He also politicized it (although the underpinnings of carbon pricing are actually more right than left wing). Simply stated, carbon pricing requires energy users to pay extra whenever drawing upon carbon-producing activities. So in theory, riding a bike to work or visiting a local massage therapist would carry little obligation for such a fee, but driving to work or buying your own massage chair would. But this isn't a straightforward calculation.

A bicyclist in Chicago or a walker in New York City should not have to pay the same taxes a driver pays to build highways and bail out car companies. Yet everyone still relies on the interstate system for transporting goods and services, so even bicyclists should pay something. But how much?

Instead of calculating an appropriate tax for every activity or product, which would be a daunting task to say the least, proponents of carbon pricing advocate instituting a tax at the beginning of the supply chain—right on the sources of energy themselves—allowing the effects of such a tax to wriggle down through the market and into the products and services we use. So no need to calculate whether the local strawberry or the imported strawberry has a smaller carbon footprint; the low-carbon choice will be the one that costs less—saying nothing about the taste, however. Economists predict that over time people and businesses will gravitate toward more economical (and therefore less carbon-intensive) options. But this leads us to one of the first limitations of such an economy.

A financial reshuffling of energy and product markets may not budge more rigid consumption patterns. For instance, during the rapid rise in gasoline prices over the past decade, fuel consumption dropped, but only slightly. That's because gasoline consumption in the United States is relatively inelastic. That is, you can stretch the price, but the efforts to conserve are not so flexible. It would take decades before carbon pricing could restructure the American suburban landscape. It is also unclear how effective carbon pricing alone would be in shifting certain cultural practices, such as filling the space under the Christmas tree with physical gifts rather than spiritual ones. Even economists don't live exclusively through the dictates of their wallets.

As environmentalists increasingly embrace carbon pricing as a solution to excess carbon dioxide, malcontent is arising from a seemingly unlikely source. Environmentalists. Some environmentalists criticize their compatriots for blindly supporting carbon pricing, claiming that it's a neoliberal framing of the problem—one that says that excess CO_2 arises because the value of the earth's products are not priced into the market. Sounds fair enough. But critics claim that the booby trap lies in a corresponding veiled assumption: carbon pricing will solve our problems *without regulation*. They argue that carbon-pricing schemes hand over responsibilities for ecological safeguarding to a market mechanism, which is in turn largely influenced by the interests of wealthy elites and large corporations. "This market driven mechanism subjects the planet's atmosphere to the legal emission of greenhouse gases," argues anthropologist Heidi Bachram. "The arrangement parcels up the atmosphere and establishes the routinized buying and selling of 'permits to pollute' as though they were like any other international commodity."[26] Likewise, in her book *Global Spin: The Corporate Assault on Environmentalism*, professor Sharon Beder argues: "A market system gives power to those most able to pay. Corporations and firms, rather than citizens or environmentalists, will have the choice about whether to pollute and pay the charges or buy credits to do so or clean up."[27]

Since the wealthy could more easily adapt to carbon pricing than the world's poor, certain market-based pricing mechanisms could intensify economic inequities. Firms may simply pick up and move their polluting activities to regions with lax standards or weak enforcement, a phenomenon economists call "leakage." And what of the many other side effects beyond CO_2 that we have witnessed thus far? It's difficult to price human rights, cancer, working conditions, heavy metal contamination, leaks, bribery, biodiversity risks, radiation, food shortages, water security, and conflicts into a market mechanism. With all eyes on carbon, will industry slide these other harms to the back burner and let them simmer away? Perhaps.

In fact, some energy corporations have actually come out in support of carbon pricing. They assume that shifts toward carbon pricing will occur and that those shifts can take many different forms—it's to their advantage to help shape schemes that least disrupt their businesses. They are presumably hopeful that carbon-pricing schemes may protect them from what they truly fear—regulation.

Questioning Regulation

We don't always get what we expect, but what we inspect—that's the short and sweet argument for regulation. Regulators argue that market breakdowns are certain to occur and regulations can prevent such breakdowns into escalating into full-fledged ecological catastrophes.

For instance, over the course of decades, manufacturers built and sold energy-intensive refrigerators even though the technology existed to make them far more efficient. Efficiency upgrades would not have affected performance and would have saved consumers a great deal of money, but "market forces" didn't guide manufactures down that path. Instead, manufacturers felt it unwise to risk research and development funds to improve efficiency. Anyway, customers could not compare operating costs between models at the time. So why bother? In fact, manufacturers actively resisted energy consumption labeling. But in 1974 the Federal Trade Commission mandated that refrigerator labels include an estimated cost of operation, even while corporate lobbyists accused the agency of "nannyism" and pushed Congress to defang the agency's regulatory bite.

Never underestimate the value of a sticker. Even though humans are rather poor judges of long- and short-term value, the potential year-after-year savings were so patently obvious and extreme that consumers quickly embraced the more efficient models. In response, manufacturers upgraded motor designs and increased insulation on newer models. Refrigeration energy consumption plummeted. In 1974 refrigerators annually consumed one hundred kilowatt-hours per cubic foot. Today they consume less than twenty. Manufacturers had warned that prices would increase, but they have actually dropped 50 percent since the regulatory intervention. If refrigerators still operated at 1974 efficiency levels, the United States alone would require an extra thirty high-capacity fossil-fuel or nuclear power plants running around the clock to cover the difference.[28]

Here we see a potential role for regulation. Some regulations introduce transparency or choice. Others place limits on pollutants or an undesired activity. Historically, the most successful pollutant regulations have mandated initially low limits that tighten over time. These not only enable firms to plan ahead but also allow them to brew up their own solutions.

Nevertheless, this is where things start to get murky since regulations, as any free-market capitalist can explain, are bad for business and bad for jobs. It is true that regulations can have unintended consequences of their own. They are sometimes ineffective, overly expensive, or even harmful—these are the examples corporate representatives and free-market politicians hold up as scepters against regulation. However, it's the comparatively quiet and remarkably potent regulations that we hear little about as we muddle through our lives taking their protections for granted.

For anyone who is against regulations of all kinds, I recommend a trip to the Niger Delta to see what extreme deregulation looks like firsthand. Local rainfall in the delta contains acidic toxins that stunt crops and dissolve the roofs of local homes, which forces many inhabitants to sheath their roofs in asbestos, one of the few materials resistant to the corrosive rains. Residents even collect this toxic rainwater for their families to drink since the region's groundwater is even more contaminated.[29] Recently leaked diplomatic cables indicate that all main ministries of the Nigerian government have been infiltrated surreptitiously by representatives of Royal Dutch Shell, an Anglo-Dutch oil company headquartered in a country that purports to value human rights and democracy.[30]

Former *New York Times* columnist Chris Hedges extends the need for regulation to quell such runaway capitalism:

> The commodification of American Culture—the commodification of human beings whose worth is determined by the market—as well as the commodification of the natural world whose worth is determined by the market—means that each will be exploited by corporate power until exhaustion or collapse, which is why the economic crisis is intimately linked to the environmental crisis. Societies that cannot regulate capitalist forces, as Marx understood, cannibalize themselves until they die.[31]

Still, given the myriad limitations surrounding regulation and carbon pricing, how effective would they actually be in an American context? Fortunately for us, that very socioeconomic experiment has been fizzing away for over two decades, across the Atlantic, in a test tube labeled "Germany."

Die Ökosteuer

Germany was not the first country to implement large-scale carbon taxes and energy regulations to protect the environment, health, and workers. But Germany's experience is especially pertinent since the country's industrial mix and political economy are comparable to those of the United States. The future of an America with carbon taxes and strong environmental regulations has already been documented in German history books, complete with theory, methods, and a couple decades of healthy hindsight.[32]

During the late 1990s, Germany's finance minister, Oskar Lafontaine, risked his job by launching reforms to shift taxes toward carbon-intensive energy products such as gasoline and electricity. Energy corporations stood high on soapboxes to warn the country's citizens that the plan would instigate unmitigated havoc in German industry. The *Wall Street Journal* cautioned that one company alone was prepared to pull fourteen thousand jobs from the country during a time when German unemployment was already "hovering around 11 percent."[33] Multinational elites were equally pissed and the BBC gave them credit for eventually ousting Lafontaine from his cabinet position in retaliation for the proposal, but his colleagues scooped up the tax-reform plan and hastily scheduled it for a vote in parliament. Corporate heads frantically scheduled meetings with German Chancellor Gerhard Schroeder with the hope he might amend the plans, but they were ultimately unsuccessful in frightening Schroeder or the German public enough to prevent the initiative from passing.

The impact came immediately.

In just the first year, passenger rail travel ticked up 2 percent. Freight carriers shifted from trucks to more efficient rail transport at an annual rate of 7.9 percent. Carpooling shot up 25 percent.[34] But what about the threat to jobs? Three years after the ecological tax reform had passed through parliament, the German government reported that the reforms had not increased unemployment, but rather spurred the creation of sixty thousand new jobs.[35] Over the next decade, nationwide unemployment even dropped slightly.[36] Indicators of happiness and satisfaction among Germans remained high over the period as well. Even so, the German ecotax was not without its own set of limitations.

First, all carbon taxes weigh heavily on the poor since energy costs represent a larger proportion of their income. German legislators offset this effect with income tax breaks so that the net effect for lower-income households was positive. Second, while the tax reportedly offset about seven million tons of CO_2, by 2002, according to Germany's federal environmental bureau (Umweltbundesamt), this was less than expected and far below what was necessary for Germany to meet its total CO_2 reduction targets. And even though European nations limited carbon-intensive activities within their own borders, they increasingly imported enough carbon-intensive steel, concrete, and other materials to more than offset the original benefits. Carbon pricing, at least the type implemented in 1999, proved only a half-step toward larger goals, not a solution in itself.[37]

Nevertheless, instituting any meaningful shift from an income-based tax system to an energy-based tax system in the United States may be politically problematic given the overwhelming influence that energy companies have on the American political system. Industry watchdogs accuse energy corporations of expending considerable resources to misdirect journalists, frighten citizens, and ensure that members of the Senate and House are well soaked in their profit-driven liquor. For instance, in 2009, according to internal memos, the American Petroleum Institute asked its member oil companies to provide employees to populate a group it called "Energy Citizens," which it then publicized as a genuine grassroots citizens group against carbon pricing. However, ordinary citizens were actually blocked from entering the group's Houston rally, according to the watchdog group Public Citizen.[38] Greenpeace chided the rallies as "Astroturf activism." That same summer, a lobbying firm hired by the American Coalition for Clean Coal Electricity forged letterheads from the NAACP, the American Association of University Women, senior citizen groups, and other nonprofit organizations. The counterfeit letters urged congressional members to vote against carbon pricing.[39]

Furthermore, bringing America on board for international accords will pose congressional challenges. Todd Stern, the lead U.S. negotiator for the Kyoto Protocol, once lamented that in order for any international climate deal to hold, the president would have to get it through Congress for it to become binding—a true feat for any meaningful level of carbon pricing.

First Step: Carefully Shift to Energy (Not Carbon) Taxes

Redefining regulations and shifting from income to energy taxes won't solve our energy and environmental problems, but it could help make them solvable.

Americans are regularly bombarded with arguments from free-market economists who claim that "the market is always right" and leftist pundits who claim that "government regulation is always right." These extreme standpoints sneak into media, politics, and even our own vocabulary but are actually more reflective of ideology than of practical realities. Energy taxes and regulations both carry advantages and drawbacks. We'll need to learn to use these tools together and to anticipate and address their potential negative consequences.

Narrowing to a *carbon* tax alone might allow the side effects and risks of nuclear and alternative-energy production to continue unabated. Therefore, a more broadly conceived *energy* tax might be a more appropriate tool to address the wide array of energy production side effects. Just as importantly, we'll need a comprehensive *energy* tax if we intend to curb boomerang effects. Still, given our system of corporate politics, establishing significant energy taxes will be an arduous and lengthy uphill battle—one worth pursing, but not one that we should count on anytime soon.

We'll have to work together with other nations to institute global agreements preventing polluters from simply shifting their activities to regions with lax standards.[40] Most rich nations already have some form of carbon pricing in place, but China and India, where over a third of the planet's inhabitants live, do not. Nevertheless, per-capita energy consumption in these countries is low. Their citizens did not produce the carbon dioxide in our atmosphere—that was done by rich nations—and the vast majority of them are not reaping the benefits of previous energy production. Still, a few are. Therefore, we might best establish every country's obligation based on their population of *rich inhabitants*. Wealth, after all, is ultimately a product of extractive practices. Instituting environmental obligations for poorer nations based on their number of wealthy citizens would allow them to share a modest but equitable portion of global responsibility with rich nations.

First Step: Strengthen Building Efficiency Standards

The largest greenhouse-gas perpetrators in urban areas are not vehicles but buildings. Up to 80 percent of emissions in some urban areas arise from the heating, cooling, and maintenance of buildings.[41]

Simple labeling requirements slashed the energy consumption of refrigerators by 80 percent. Could a reconceptualized LEED do the same for buildings? To start, the program could award points based on actual energy impacts rather than the quantity of trendy symbolic gadgets. Or it could take its lead from the U.S. Energy Star Program, which focuses on a few straightforward factors of energy use. However, the most effective strategy might be to change the economic landscape on which buildings are built so that LEED certifications are not necessary in the first place. Building codes could require higher overall energy performance. If energy, toxins, and socially detrimental practices come with large up-front price tags, then architects, engineers, and planners will find ways to make their structures more environmentally and socially beneficial. These strategies draw upon the inventiveness of the human spirit rather than the dictates of committees.

First Step: Rediscover Passive Solar

The California Academy of Sciences sits in San Francisco's Golden Gate Park with a LEED Platinum star on its chest. Despite its location within one of the cloudiest and foggiest urban microclimates on earth, an especially expensive collection of solar photovoltaic cells circumscribes its roof. But even when the sun is shining, it's difficult to see what benefit the panels really provide, since not a single one of them directly faces the sun, a technicality that somehow seems to have eluded the building's architects and environmental planners. During an interview with Ari Harding, the building's systems engineer, I attempted numerous times to coax the photovoltaic performance figures from him but he smoothly ducked my every prod. Then he smartly pointed to a much less visible solar strategy, one that has a monumentally greater impact on the museum's energy footprint than the solar cells on its roof.

"A typical museum of this size has a standard system for air-conditioning and heating," Harding told me. "This building has no air-handling system at all."[42] In fact, the wind and sun provide the primary heating and cooling for the museum through passive-solar techniques that builders developed and refined centuries ago but that now remain almost entirely forgotten. Today, architects can easily spec commercially developed heating, venting, and air-conditioning systems that run on cheap fossil fuels. Some building codes and financiers mandate them even where they are not necessary.[43]

In the California Academy of Sciences building, a system of simple mechanical louvers and windows captures heat during the cool months. When temperatures rise, the same system releases hot air above and draws upon surrounding air currents to cool the large interior space. Harding concludes, "The energy collected through the windows is more substantial than what you get in electricity from solar cells" (whose performance, it seems, will forever remain a mystery).[44]

Trees are another underappreciated passive-solar mechanism as they shade buildings from the hot summer sun while allowing radiant heat to pass through the bare branches during the winter, warming a building's exterior. One study claims that planting a shade tree on the west side of a home can alone decrease

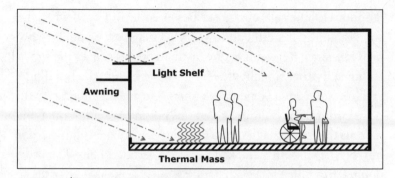

Figure 19: Passive solar strategies Light shelves, awnings, and shades can shield or reflect natural light. Awnings allow the low-angle winter sun to enter while blocking higher summer rays. Thermal masses absorb heat throughout the day and release it at night. These are just a few of many low-tech passive solar strategies to make buildings more comfortable and energy efficient.

its carbon footprint by a third.[45] Trees happen to provide the most benefit between 2:00 p.m. and 8:00 p.m., right when electricity is in greatest demand as people arrive home from work and turn on their air-conditioners. This is also when electricity is most expensive to both consumers and power companies; as utilities roll out dynamic pricing, trees stand to gain about six times more value in terms of cost savings alone.

In dense urban contexts, cities can insert tree boxes on the west edges of blocks by reclaiming on-street parking spaces and bulging out sidewalks. Incidentally, these takeovers create prime locations for green grocers or street cafés. A recent New York City "tree census" provides a replacement price tag for the city's trees—up to $90,000 for a mature oak. The census asserts that the city's trees deflect numerous costs: $28 million in air-conditioning expenditures, $5 million worth of air filtration, $36 million of storm-water absorption, and $755,000 of CO_2 absorption. According to the report, every dollar the city spends on trees returns $5.60 in benefits.[46]

First Step: Cogeneration Systems

Condominiums, apartment buildings, and row houses are generally efficient dwellings since each unit has less surface area exposed to the elements than if it were built as a separate structure. Nevertheless, developers frequently outfit each condo with its own furnace and air conditioner—a feature that may satisfy the individualist inclinations of many Americans but ends up leaving homeowners and renters with higher utility bills, more maintenance hassles, and less comfort than those residing in cogeneration districts.

In cogeneration neighborhoods, a single high-efficiency boiler heats a building or block of buildings, reducing the infrastructure costs, maintenance hassles, and energy use for everyone (it's far less expensive to maintain one central boiler than hundreds of separate furnaces). Every home retains its own heating controls and homeowners pay only for the heating or cooling they personally use. In Stockholm, a single boiler may heat an entire neighborhood. As with many other efficiency strategies in America, cogeneration bumps up against cultural barriers, not technological ones.

First Step: Monetary Reform and Decoupling

Our monetary system creates and supports the requirements for economic growth. In short, banks make loans, which must be repaid with interest. In order for businesses and individuals to repay their loans, the overall value of goods and services must rise at least in proportion to the interest. Consumption or inflation must therefore increase—the economy must expand. According to authors of a report exploring sufficiency and the rebound effect,

> Any slowing of the rate of growth has serious consequences for business (e.g., losses, bankruptcies), individuals (e.g., defaults on loans, unemployment) and politicians (e.g., loss of office). All

sectors of society have strong incentives to maximize economic growth. Set against this, any concerns about long-term environmental sustainability and quality of life can easily be overridden. Hence, the challenge is not simply to demonstrate the unsustainability of the present model of economic development and the benefits of alternative models but to propose ways in which the dependence of modern economies upon continued economic growth can be broken.[47]

Decoupling energy production from utility profits is one small monetary tool. But larger, economy-wide tools will be needed. A number of groups are investigating broader monetary reforms, including the Foundation for the Economics of Sustainability, the New Economics Foundation, and, of course, the Occupy movement.[48]

First Step: Voting Reform

Voting reforms, such as instant-runoff voting or ranked voting, could open up room to debate diverse environmental issues. Here, voters choose their first, second, and third choices, which enables them to support their favorite candidates without risk of throwing their vote away. Not only does this presumably lead to more accurate representation, but more importantly it opens up political debate to thoughtful analysis over sound bites. Voting reform, multiparty democracy, term-limit reform, and campaign finance regulations all stand to provide numerous benefits as well as some potential risks—environmentalists may realize great success by acknowledging, evaluating, and refining these democratic concepts.

First Step: Create a Department of Efficiency

Most people think the United States has a Department of Energy. It doesn't really. In essence, it simply has a clumsy bureaucratic agency called the Department of Energy (DOE). Actual energy responsibilities are dispersed among numerous other agencies, including the Environmental Protection Agency and the Departments of the Interior, Commerce, and Agriculture. As evidenced by America's dismal energy-efficiency ratings compared with other nations, the United States clearly does not have a system of energy oversight that works. The existing DOE has only a small budget for civilian energy and little capacity for developing all-encompassing energy priorities and policy recommendations, or even determining how much should be spent on energy research and development. Its budgets are held hostage in appropriation subcommittees run by politicians keen to fund local highway and water projects, many of whom question whether energy consumption, climate change, or oil dependence are even problems at all. In fact, the department's largest single expenditure goes toward weaponry.[49]

An energy department focused on efficiency rather than production could look much different. Take, for instance, a $500,000 government grant to Kettering University and the University of Washington to improve grocery store refrigeration. By simply adjusting airflow, the engineers radically increased efficiency—so much that, if extended nationally, the tweak could save $170–$200 million in energy costs, extend compressor life spans, and reduce food spoilage.[50] (Incidentally, food waste represents 40 percent of food supply, which sucks up 300 million barrels of oil annually.[51]) These are the downward energy spirals that a Department of Efficiency and Conservation could leverage into the funding spotlight.

Today corporate consultants and industry lobbyists plan, research, and write many of the DOE's reports, which leads to a net productivist leaning. Harvard energy specialist Max Bazerman insists,

Money corrupts the potential for an intelligent decisionmaking process on energy policy. Well-funded and well-organized special interest groups—concentrated constituencies intensely concerned about a particular issue—have disproportionate influence on specific policies at the expense of millions who lack a strong voice on that issue. . . . They stall reforms by calling for more thought and study or by simply donating enough money to the right politicians so that wise legislation never even comes to a vote. Their efforts effectively turn Congress and the president away from the challenge of making wise energy decisions.[52]

In response, he recommends five guiding priorities:

1. Educate the public on energy trade-offs.
2. Maximize benefits to society rather than to special-interest groups.
3. Seek energy policies that would make sense even if climate change were not an issue.
4. Identify nudges that significantly influence the behaviors of individuals and organizations in a positive direction without infringing on personal liberties.
5. Achieve buy-in on wise long-term policies that require up—front costs and consider a mild delay before policies take effect.

14. Asking Questions

What is the answer? [After a silent pause] In that
case, what is the question? —Gertrude Stein's last
words

If you're a university department chair attempting to get your environmental program successfully accredited, I am likely the last person you
should ask for input. That's because the techno-
fetishistic fruit being marketed by contemporary
environmental pornographers falls so far from
the roots of environmentalism that it's unrecognizable to me as having come from the same
tree. In fact, I recently browsed the course catalog of a prestigious university's environmental
program. It detailed a host of courses on solar
photovoltaic system design, biofuel reformation,
and wind-turbine planning. These programs are
meant to train the elite tier of future environmental experts, but it's questionable whether
such coursework will prepare students to perform environmentally meaningful work at all.

Entrusting alternative-energy technologies
with solving environmental challenges, which
at their root are social, economic, and political,
produces numerous snags. Let's begin with two

Today's Environmentalism	Future of Environmentalism
Photovoltaic installations	Passive solar
Wind turbine construction	Efficiency codes
Biofuel processing	Block rebound effects
Hybrid cars	Human rights
Electric cars	Citizen governance
Suburban geothermal systems	Walkable-community zoning
Hydrogen highways	Volunteerism
Fuel cells	Consumerism shifts
Straw-bale homes	Universal health care
Green consumerism	Social enterprise

Table 1. The Present and Future of Environmentalism

big ones. First, the technical character of the alternative-energy project greatly limits citizen involvement because most people aren't trained as technicians. Instead, environmental enthusiasts, activists, students, educators, and others are left to passively drink the green Kool-Aid—drive the green car, buy the green product, or consume the green energy. Second, prioritizing alternative energy as an environmental imperative plays into conceptions of productivism and growth that directly conflict with the stated goals of environmentalists themselves. Before mainstream environmental thinking took the technological turn, many environmentalists *criticized* growth and productivism, couching their solutions in terms of effective governance and social fundamentals. Numerous organizations still pursue these themes, and their work deserves a greater share of the spotlight.

Environmentalists of tomorrow (as Table 1 is meant to suggest) are more likely to be found studying urban sociology, public health, human rights, critical economics, ethics, international affairs, arts, hand trades, regulatory law, child welfare, nursing, and a host of other subjects infrequently recognized as environmental work.

There's nothing new about the right-hand column in Table 1. It's decidedly low-tech. And while our future will include items from the left-hand column, environmentalists will achieve their goals more quickly and compassionately by focusing on the right-hand side.

Asking Better Questions

When I criticize alternative-energy technologies, clean-energy proponents frequently grumble that I just don't get it. Each energy technology needn't be perfect, they say, because in the future we'll rely on a mix of energy sources—a little solar, a little wind, a little biofuel, and so on. They have a valid point, to be sure; I won't argue with that. But I would argue instead that "a little, plus a little, plus a little" won't get a growing consumption-based economy very far. We would need "a lot, plus a lot, plus a lot" for that. Creating meaningful quantities of any so-called clean energy certainly won't be easy or affordable. Even if we were able to pull it off, these technologies stand to intensify and entrench energy-intensive ways of life—hardly a durable formula for social or environmental prosperity.

This boomerang effect is most pronounced in economic, political, and social contexts that prioritize material growth as the sole measurement of well-being. In the United States, lavishly fueling this Wall Street model of economic expansion has led to drops in almost every quality-of-life indicator compared to other industrialized nations, including health care, happiness, equality, primary education, and trust.[1] Cheap power drives growth, expands GDP, ratchets up sprawl, and fuels surplus material consumption. Generating even more power, regardless of the means used, won't quench these factors but will rather extend their reach. Given present American demographics and consumption, an alternative-energy future doesn't look especially probable or desirable.

Even if we could afford to dramatically increase alternative-energy production, what would such a future look like? Would simply adding alternative energy to our current sociopolitical system lead to greater well-being? Or would it just leave us with another strain of fossil-fuel dependence, spinning off hyperconsumption and additional side effects?

When alternative-energy productivists do acknowledge the leaks, waste, and other consequences of "clean" energy, they quickly follow up by asserting that these effluents are less harmful than those from the exceptionally dirty fossil-fuel industries. In a very limited sense, they may be right, but they are using an inappropriate and misleading benchmark. Comparing every new energy technology to the filth of fossil fuels is hardly a reasonable yardstick for thoughtful people—especially when we have no historical experience, current data, or future backstops in place to assume that these technologies will even offset fossil-fuel use at all. Nevertheless, energy rhetoric in the United States has largely devolved into arguments pitting *production* versus *production* in manufactured pseudodebates that fool us into thinking we are making genuine energy choices. The only reason these appear to be reasonable comparisons is that we are so deeply immersed in the dirty fossil-fuel way of life that a less-dirty bad idea can seem good. (We should remember that the rise of petroleum itself was seen initially as an environmental benefit as it slowed the extermination of whales for their oil.)

Why not measure the virtues of electric vehicles against the virtues of walkable neighborhoods? Or the benefits of solar cells against the benefits of supporting comprehensive women's rights? Or the costs of nuclear energy against the costs of plugging energy leaks? These are the comparisons environmentalists should be thinking about, because in a world of limited finances and pricey resources, these are the very real trade-offs that will define our lived experience.

So where can we best invest our time, energy, resources, and research? Consider a dilemma many environmental organizations face. If you had a million dollars to reduce environmental harms, where would you spend it?

Numerous researchers have attempted to locate where you'd get the best bang for your buck. Nearly a decade ago, Robert Socolow and Stephan Pacala published an article in *Science* envisioning fifteen potential "wedges" to flatten the upward trend of CO_2 emissions (e.g., vehicle efficiency, carbon sequestration, solar power, cropping alterations, etc.), only seven of which would have to be fully implemented for success.[2] Another team at the consultancy McKinsey and Company extended this work by ranking CO_2 reduction schemes by cost and benefit.[3] Their rankings fall into three overlapping clusters: (1) energy-efficiency strategies that typically save money, (2) agriculture and forestry management that either save a little or cost a little, and (3) energy-production strategies that cost the most per ton of "avoided CO_2." Both of these prominent studies greatly influence environmental research and policy. Nevertheless, while these studies are helpful analytical tools, they are perfectly unsuitable for high-level decisionmaking. First, they draw upon the ahistorical assumption that increasing efficiency or expanding alternative-energy production will automatically displace fossil-fuel use. Further, they limit their options to trendy interventions and leave their results to be narrowly dictated by convenient cost and CO_2 abatement measurements. More fundamentally, they attend to the symptoms rather than the sources of our energy troubles. Foundational strategies such as human rights, or costs extending beyond dollars and cents, or benefits aside from CO_2 abatement are all unintelligible within such fact-making missions. Truths are as much a matter of questions as answers.

So where to spend your time, energy, or million bucks? As I hope I have convincingly argued, spending a million to deploy solar cells, wind turbines, or biofuels will do little if anything toward achieving the intended goal. It may even instigate real harms, given the present American context. And if you spend your cash on alternative-energy production, you can't spend it elsewhere. Whether you're a philanthropist, student, environmentalist, policymaker, or voter, it's helpful to keep in mind this opportunity cost of investment—spending an ounce of humanity's value on one initiative necessarily means that it cannot be used for another. It follows that as citizens in a democracy, we should orient our priorities toward the most promising funding opportunities.

If you just can't part from an alternative-energy mindset (or your boss says you need the symbolic power of alternative energy for your quarterly report), you'd likely be better off directing your resources toward research rather than immediate deployment of today's highly problematic alternative-energy schemes. Focusing on women's rights, walkable communities, or improving consumption practices would yield excellent returns. Still, if rebound effects kick in, they could negate some or all of those benefits. It seems that wherever you make your contribution, it risks great dilution. Your efforts might produce greater returns if they flowed through another context.

Environmentalists of the future will imagine and create these contexts both domestically and internationally. Domestically, environmentalists can address economic and social fundamentals such as upgrading municipal zoning, improving governance, developing voting reforms, providing health care to all citizens, and the like. Internationally, environmentalists will forge closer relationships to support context-development globally. Here, environmentalists could make a convincing argument for redirecting military spending toward supporting diplomacy, transparency, and human rights. As Bill Gates recently quipped, "We spend $80 billion a year on military R&D and we're good at shooting people. You get what you pay for."[4]

Alternative-energy technologies are only as durable as the contexts we create for them. Wind speed and turbine efficiency, for instance, only partly determine the success of a wind farm. More important are the social, economic, and political foundations supporting it. In America it's the context, not technology, which requires the most attention. In countries such as the United States, with dismal efficiency, sprawling suburbs, a growing population, and high rates of material consumption, alternative-energy technologies do the most harm as they perpetuate energy-intensive modes of living and ratchet up risks—all under the cover of a distracted mainstream environmental movement.

If we intend to decrease fossil-fuel dependence and increase the *proportion* of alternative-energy production in the future, our success will depend on the strength of our context. Below is a rough checklist showing what kind of contextual preconditions we might consider. This provisional checklist is designed for an OECD country, but you could adapt the concept to a different scale or context. (OECD is the Organisation for Economic Cooperation and Development. Essentially a rich countries' club, this international organization gathers statistics and develops international policy proposals for member nations.)

Provisional preconditions for alternative-energy deployment:

1. *Consumption*: The nation consumes less energy per capita than the average in the OECD.
2. *Human rights*: The nation has a strong human rights record including a low wealth gap between rich and poor, universal health care, and low rates of murder, teen pregnancy, crime, and incarceration.
3. *Efficiency*: The nation's cars, buildings, and industrial processes are more efficient than the OECD average. The nation has a binding long-term plan for improving efficiency into the future.
4. *Transportation*: The country is in the top fiftieth percentile of walkable and bikeable urban areas in the OECD and its legislation prioritizes these communities over car culture.
5. *Backstops*: The country has a growing energy tax and/or other backstops to stifle rebound effects.

If a region is not fulfilling or moving decisively toward every one of these very basic checkpoints, then delivering more energy to that region, whether by alternative or conventional means, may have neutral or negative consequences on long-term well-being rather than the presumed positive ones.[5] Today some OECD countries meet all of these preconditions. Others meet only a few. The United States meets none.

Preparing for Climate Change

More than a few climate scientists fear it may not matter what we do to slow climate change—it may already be too late. Others believe that harms could be avoided but hold little hope that humans are capable of mobilizing the necessary changes. Even if Americans stop burning oil, coal, and natural gas, some say, the Russians, Indians, and Chinese will burn it anyway, leading to the same global outcome either way. Keeping the world below a two-degree-Celsius global temperature rise will require every signatory nation of the Copenhagen Accord to perform within the top range of their promises according to the International Energy Agency, a goal the organization's chief economist Fatih Birol claims is "too good to be believed."[6]

Climatologists claim we'll be lucky if sea levels rise less than two feet. They expect that in forty years the probability of experiencing a summer hotter than any yet recorded will be 10–50 percent. In eighty years the chances rise to 90 percent.[7] Long before then, scientists believe that heat waves will increasingly shock crops—a single hot day can cut local agricultural yields by 7 percent. In a world with unbounded emissions, they warn, yields could decline 63–82 percent.[8]

Are these pessimistic outlooks justified? Perhaps. Does it mean we shouldn't bother implementing the first steps outlined in this book? Absolutely not, and here's why: In a world ravaged by climate change, these initial strategies will become not only valuable, but vital. Even if the first steps I have proposed are only partially realized (as they already are to varying degrees throughout the world), they should still prove advantageous.

In a world with a rapidly changing climate, we'll be better equipped to coordinate international cooperation if we've been peacefully supporting world democracies, transparency, and the rights of workers. We'll be better prepared to deal with local calamities if our neighborhoods are more accessible by walking and biking and our civic organizations are strong. If storms ravage the world's fields, it will be easier to move crop production to lesser-quality fields if there are fewer mouths to feed. If heating or cooling our homes becomes too expensive, we'll be thankful they are well insulated and designed to make the most of the sun's energy. If members of society are unequally impacted, we'll be fortunate to have a government designed for citizens, not moneyed special interests. If it comes to making difficult choices about goods and services, we'll benefit from economies with more socially based enterprises rather than those devised to consolidate profits for distant shareholders. And when the holidays arrive, we'll be thankful we've come to appreciate the many gifts of our friends and family, even if they are not the kinds that arrive wrapped in a box.

In short, the strategies we can embrace to avoid catastrophic global climate change are the same ones we'll need should the worst occur. And if those horrors don't unfurl? Well then, we'll likely be left with stronger communities, empowered women and girls, lower crime rates, cleaner air, more free time, and higher levels of happiness. Not a bad wager.

The Future Script for Clean Energy

Since this book represents a critique of alternative energy, it may seem an unlikely manual for alternative-energy *proponents*. But it is. Building alternative-energy infrastructure atop America's present economic, social, and cultural landscape is akin to building a sandcastle in a rising tide. A taller sand castle won't help.

The first steps in this book sketch a partial blueprint for making alternative-energy technologies relevant into the future. Technological development alone will do little to bring about a durable alternative-energy future. Reimagining the social conditions of energy use will.

Ultimately, we have to ask ourselves if environmentalists should be involved in the business of energy production (of any sort) while so many more important issues remain vastly underserved. Over the next several decades, it's quite likely that our power production cocktail will look very much like the mix of today, save for a few adjustments in market share. Wind and biofuel generation will become more prevalent and the stage is set for nuclear power as well, despite recent catastrophes. Nevertheless, these changes will occur over time—they will seem slow. Every power production mechanism has side effects and limitations of its own, and a global shift to new forms of power production simply means that humanity will have to deal with new side effects and limitations in the future. This simple observation seems to have gotten lost in the cheerleading for alternative-energy technologies.

The mainstream environmental movement should throw down the green energy pop-poms and pull out the bifocals. It is entirely reasonable for environmentalists to criticize fossil-fuel industries for the harms they instigate. It is, however, entirely unreasonable for environmentalists to become spokespeople for the next round of ecological disaster machines such as solar cells, ethanol, and battery-powered vehicles. Environmentalists pack the largest punch when they instead act as power production watchdogs (regardless of the production method); past environmentalist pressures have cleaned the air and made previously polluted waterways swimmable. This watchdog role will be vital in the future as biofuels, nuclear plants, alternative fossil fuels, solar cells, and other energy technologies import new harms and risks. Beyond a watchdog role, environmentalists yield the greatest progress when addressing our social fundamentals, whether by supporting human rights, cleaning up elections, imagining new economic structures, strengthening communities, revitalizing democracy, or imagining more prosperous modes of consumption.

Unsustainable energy use is a symptom of suboptimal social conditions. Energy use will come down when we improve these conditions: consumption patterns that lead to debt and depression; commercials aimed at children; lonely seniors stuck in their homes because they can no longer drive; kids left to fend for themselves when it comes to mobility or sexuality; corporate influence trumping citizen representation; measurements of the nation's health in dollars rather than well-being; a media concerned with advertising over insight, and so on. These may not seem like environmental issues, and they certainly don't seem like energy policy issues, but in reality they are the most important energy and environmental issues of our day. Addressing them won't require sacrifice or social engineering. They are congruent with the interests of many Americans, which will make them easier to initiate and fulfill. They are entirely realistic (as many are already enjoyed by other societies on the planet). They are, in a sense, boring. In fact, the only thing shocking about them is the degree to which they have been underappreciated in contemporary environmental thought, sidelined in the media, and ignored by politicians. Even though these first steps don't represent a grand solution, they are necessary preconditions if we intend to democratically design and implement more comprehensive solutions in the future.

Ultimately, clean energy is less energy. Alternative-energy alchemy has so greatly consumed the public imagination over recent decades that the most vital and durable environmental essentials remain overlooked and underfunded.

Today energy executives hiss silver-tongued fairy tales about clean-coal technologies, safe nuclear reactors, and renewable sources such as solar, wind, and biofuels to quench growing energy demands, fostering the illusion that we can maintain our expanding patterns of energy consumption without consequence. At the same time, they claim that these technologies can be made environmentally, socially, and politically sound while ignoring a history that has repeatedly shown otherwise. If we give in to accepting their conceptual frames, such as those pitting production versus production, or if we parrot their terms such as clean coal, bridge fuels, peacetime atom, smart growth, and clean energy, then we have already lost. We forfeit our right to critical democratic engagement and instead allow the powers that be to regurgitate their own terms of debate into our open upstretched mouths. Alternative-energy technologies don't clean the air. They don't clean the water. They don't protect wildlife. They don't support human rights. They don't improve neighborhoods. They don't strengthen democracy. They don't regulate themselves. They don't lower atmospheric carbon dioxide. They don't reduce consumption.

They produce power.

That power can lead to durable benefits, but only *given the appropriate context*. Ultimately, it's not a question of whether American society possesses the technological prowess to construct an alternative-energy nation. The real question is the reverse. Do we have a society capable of being powered by alternative energy? The answer today is clearly no.

But we can change that.

Future environmentalists will drop solar, wind, biofuels, nuclear, hydrogen, and hybrids to focus instead on women's rights, consumer culture, walkable neighborhoods, military spending, zoning, health care, wealth disparities, citizen governance, economic reform, and democratic institutions.

As environmentalists and global citizens, it's not enough to say that we would benefit by shifting our focus. Our very relevance depends on it.

Epilogue: A Grander Narrative?

And the people bowed and prayed / to the neon
god they made. —Paul Simon and Art Garfunkel,
"The Sound of Silence"

When approaching any large challenge, it is
sometimes difficult to know where to start, and
in our particular global crisis, nations are finding it difficult to determine *who* should start. In
response to the statistic that China is building
more than a gigawatt of new coal-fired power
plants every week, Berkeley physicist Richard
Muller asks, "Do we demand that the Chinese
stop? Do we have the right to do that? Do we
have the power to do that?" His answers are
even more straightforward, "No, no, and no."[1]

Analysts often characterize China as following
an unsustainable path compared to the United
States, but consider this: One country has a fertility rate below replacement value and an annual per-capita energy consumption equivalent to 1,500 kilograms of oil per year. The
other country's population is expanding and
its citizens annually consume over five times as
much energy, equivalent to 7,700 kilograms per
year.[2] When comparing a modestly consuming

populace whose numbers will someday be smaller with a more substantially consuming populace growing exponentially, it is not difficult to determine which is sustainable within the limits of a finite planet and which is not.[3] Americans are not in a position to preach.

As an alternative to preaching, Thomas Friedman asserts, "I would much prefer to put our energy into creating an American model so compelling that other countries would want to follow it on their own. . . . A truly green America would be more valuable than fifty Kyoto Protocols. Emulation is always more effective than compulsion."[4]

Are the first steps in this book sufficient to create such a compelling American model? No. They're simply a start. Nevertheless, I never intended to write a grand narrative. Frankly, I'm not so sure it would be worth reading if I did. But what if I had attempted to do so? What could I have drawn upon?

I might have launched directly into the gut of environmental ethics, American religion, and knowledge frameworks by quoting Lynn White, who argued in the 1960s: "More science and more technology are not going to get us out of the present ecological crisis until we find a new religion, or rethink the one we have." I could have scripted an impassioned introduction drawing upon writers such as Bill McKibben, Vandana Shiva, Joseph Stiglitz, James Lovelock, and Raj Patel.[5] And there are the many thinkers featured in the volume *Moral Ground: Ethical Action for a Planet in Peril* to consider as well.[6] I could then have tempered their optimism by quoting cantankerous theorists such as, Curtis White, author of *The Middle Mind: Why Americans Don't Think for Themselves* and *The Barbaric Heart: Faith, Money, and the Crisis of Nature*; James Howard Kunstler, author of *The Long Emergency*; Chris Hedges, author of *American Fascists*; and John Michael Greer, author of *The Ecotechnic Future*.[7] Or I could have featured the more scientific sobrieties of Elizabeth Kolbert, author of *Field Notes from a Catastrophe*, or Fen Montaigne, author of *Fraser's Penguins*.[8]

Alternately, I could have opted for an equally provocative (and ideologically charged) beginning by framing imperialism as the root of our environmental problems and quoting documentary filmmaker Philippe Diaz who claims: "The first resource we took was the land, and when you take the land away from the people, you create the slave. . . . How did these small countries like Great Britain, France, Holland, and Belgium become these huge empires with almost no resources whatsoever? Well, by taking by force, of course, resources from the South."[9] I might have drawn upon any number of historians, anthropologists, and social scientists who maintain that over the past five hundred years, we've become more efficient at performing these extractions, primarily through the economic instruments we call privatization, debt service, and free trade (as well as through good old-fashioned force and intimidation). From there, I could have quoted the activist Derrick Jensen, who observes,

> Once a people have committed (or enslaved) themselves to a growth economy, they've pretty much committed themselves to a perpetual war economy, because in order to maintain this growth, they will have to continue to colonize an ever-wider swath of the planet and exploit its inhabitants. . . . The bad news for those committed to a growth economy is that it's essentially a dead-end street: once you've overshot your home's carrying capacity, you have only two choices: keep living beyond the means of the planet until your culture collapses; or proactively elect to give up the benefits you gained from the conquest in order to save your culture.[10]

To extend his affront to growthism, I could have drawn upon insight from Daniel Quinn, Donella H. Meadows, the antics of *The Yes Men*, and solutions from the multiauthored volume

Alternatives to Economic Globalization.[11] This approach would have formed a springboard to consider inequality.

I might even have chosen to characterize America as a young dynasty system. I could have started by featuring thinkers who point out that this dynasty system has actually increased prosperity for all Americans, not just the rich. Next I could have featured social scientists who argue that while Americans have traditionally idealized storylines portraying the social and economic mobility among classes, the American socioeconomic system today is actually quite rigid—where and to whom you are born is becoming an ever more accurate predictor of future prosperity. I may have argued that unequal structures of material wealth, power, and dynastic pressure pose distinct challenges to strengthening environmental fundamentals. These themes get fleshed out in *Winner-Take-All Politics* by Paul Pierson and Jacob S. Hacker, *The Spirit Level* by Richard Wilkinson and Kate Pickett, *Griftopia* by Matt Taibbi, and by a wide range of political scientists in the book *Inequality and American Democracy*.[12]

Next I could have pointed to those theorists who argue that extreme capitalism cannot coexist with a durable environmental movement. Among them is James Gustave Speth, author of *The Bridge at the End of the World*, who admits that his conclusion, "after much searching and considerable reluctance, is that most environmental deterioration is a result of systemic failures of the capitalism that we have today and that long-term solutions must seek transformative change in the key features of this contemporary capitalism."[13] Another moderate voice I could have chosen to highlight is John Perkins, a self-described "economic hit man" who was a gear in this machinery for a decade, a position he describes in several books on the topic. In a recent interview, he insisted, "I don't think the failure is capitalism, I think it is a specific kind of capitalism that we've developed in the last thirty or forty years, particularly beginning with the time of Reagan and Milton Friedman's economic theories, which stress that the only goal of business is to maximize profits, regardless of the social and environmental costs and not to regulate businesses at all . . . and to privatize everything so that everything is run by private business."[14]

Today it is difficult to imagine that through the first hundred years of America's adolescence the government required corporations to apply for charters detailing how the company served the public good. Every ten years or so, the charters would come up for renewal and if the company's directors could not prove that the company was serving the public interest, the government revoked their charter and disbanded the company. That changed in the late 1880s, when the U.S. Supreme Court started to treat corporations more like individuals. Numerous thinkers analyze the corporation's rise from a wide array of vantage points. Among them are Joel Bakan, author of *The Corporation;* Naomi Klein, author of *The Shock Doctrine*; Carl Safina, author of *The View from Lazy Point*; as well as many volumes by David Harvey, Sam Smith, Slavoj Žižek, and of course the public intellectual, Noam Chomsky.[15] These critiques would have opened up room to imagine new forms of democracy, community, and economy, such as those envisioned in Kirkpatrick Sale's *Human Scale* and William A. Shutkin's *The Land That Could Be*.[16]

Instead of negotiating that thicket of leftist thorns, I might have chosen to avoid it by paddling my grand narrative through the varied seductions of technology itself. I could have started by discussing Aldous Huxley's *Brave New World*, cautioning how technological adoration might overcome our capacity to think.[17] Or I might have chosen E. M. Forster's novella *The Machine Stops*, written over a hundred years ago with spine-tingling premonition.[18] I could have moved on to discuss the symbolism of environmental initiatives by drawing on Yanow Dvora's *How Does a Policy Mean?*, Charles Lindblom's *Inquiry and Change*, and Neil Postman's books *Building a Bridge to the 18th Century* and *Amusing Ourselves to Death*.[19] I might have chosen to investigate the interstitial forces between society and technology more broadly, as portrayed by thinkers such as David Nye, Andrew Feenburg, Sherry Turkle, Michel Foucault, and Thomas Kuhn.[20] Many others are featured in the edited volume *Technology and Society: Building Our Sociotechnical Future*.[21] I could even have focused on the specific blend of social, personal, and technological challenges to achieving a truly sustainable energy system. But that excellent book, *Sustainable Energy Consumption and Society*, has already been written by David Goldblatt.[22]

In anticipation of those who might say that any grand narrative challenging established conceptions of capitalism, growth, inequality, consumption, and governance is but a dreamy impracticality, I might have gone so far as to quote the stalwart Milton Friedman, who observed: "Only a crisis—actual or perceived—produces real change. When that crisis occurs, the actions that are taken depend on the ideas that are lying around. That, I believe, is our basic function: to develop alternatives to existing policies, to keep them alive and available until the politically impossible becomes politically inevitable."[23] On more likely, I might simply have quoted Gar Alperovitz, who reminds us in his book *America Beyond Capitalism* that "fundamental change—indeed, radical systemic change—is as common as grass in world history."

And finally, in the epilogue, I might even have attempted to trick you into reading a "further readings" list by couching it in terms of a grand narrative.

§

Thank you for reading. If you enjoyed this book, please pass it on to a friend, write a review, or send me a note if you'd like me to give a talk at your school, community group, or other organization. You are also invited to enjoy a complimentary subscription of an ongoing series at: CriticalEnvironmentalism.org.

To contact me, visit OzzieZehner.com or GreenIllusions.org.

Resources for Future Environmentalists

If you'd like to be an environmentalist of the future, here are some websites to explore. Just select a theme that interests you and don't forget to research organizations in your own neighborhood as well—this is by no means a comprehensive listing. I will update and expand this list at GreenIllusions.org. Please contact me if you'd like to add an organization to the list. I also enjoy answering questions and talking with groups both large and small. You can find my contact information, speaking schedule, documentary films, interviews, articles, and more at OzzieZehner.com.

Human Rights: Youth Focus

Advocates for Youth (advocatesforyouth.org)
One by One (onebyone.org)
Save the Children (savethechildren.org)
U.S. Fund for UNICEF (unicefusa.org)

Human Rights: Poverty and Emergency Aid Focus

American Jewish World Service (ajws.org)
FXB USA (fxb.org)
United Methodist Committee on Relief (gbgm-umc.org/umcor)

Human Rights: Population Focus

Association for Population Libraries and Information Centers International (aplici.org)
CARE (care.org)
The Centre for Development and Population Activities (cedpa.org)
Population Action International (populationaction.org)
The Population Institute (populationinstitute.org)
Population Council (popcouncil.org)
Population Media Center (populationmedia.org)

Human Rights: Women's Issues Focus

Center for Reproductive Rights (reproductiverights.org)
Family Care International (familycareintl.org)
Global Fund for Women (globalfundforwomen.org)
Institute for Women's Policy Research (iwpr.org)
International Center for Research on Women (icrw.org)
International Women's Health Coalition (iwhc.org)
National Organization for Women Foundation (nowfoundation.org)
National Women's Law Center (nwlc.org)
Sauti Yetu Center for African Women (sautiyetu.org)
United Nations Development Fund for Women (unifem-usnc.org)
Vital Voices Global Partnership (vitalvoices.org)

Human Rights: Conflict and Violence Focus

Amnesty International (amnesty.org)
Futures Without Violence (futureswithoutviolence.org)
Human Rights Watch (hrw.org)

Human Rights: Health Focus

Doctors Without Borders (doctorswithoutborders.org)
EngenderHealth (engenderhealth.org)
Guttmacher Institute (guttmacher.org)
International Planned Parenthood Federation (ippf.org)
PATH (path.org)
Pathfinder International (pathfind.org)
Planned Parenthood Federation of America
    (plannedparenthood.org)

Volunteering

United We Serve (serve.gov)
VolunteerMatch (volunteermatch.org)
Craig's List (craigslist.org)

Commercial-Free Livelihood

Campaign for a Commercial-Free Childhood
    (commercialexploitation.org)
Rudd Center for Food Policy and Obesity at Yale
    (fastfoodmarketing.org)
Center for a New American Dream (newdream.org)
Commercial Alert (commercialalert.org)
Education and the Public Interest Center (epicpolicy.org)
Media Education Foundation (mediaed.org)
Obligation, Inc. (obligation.org)
Stuttgart Jugendhaus (jugendhaus.net)

Social Enterprise and Entrepreneurship

Wiser Earth Index (wiserearth.org)
Social Edge (socialedge.org)
New Door Ventures (newdoor.org)

Junk Mail and Packaging

Downloadable Junk Mail Kit (stopjunkmail.org)
Junk Mail Resources (newdream.org/junkmail)
Plastic Pollution Coalition (plasticpollutioncoalition.org)

GDP Alternatives

Happy Planet Index (happyplanetindex.org)
European Social Survey (europeansocialsurvey.org)
Index of Sustainable Economic Welfare (see Wikipedia entry)
Genuine Progress Indicator (see Wikipedia entry)
Gross National Happiness (gnh-movement.org)
University of Cambridge Well-being Institute
    (cambridgewellbeing.org)

Reduce Military Spending

Fund Our Communities (25percentsolution.com)
National Priorities Project (*nationalpriorities.org*)

Eating Vegan and Vegetarian

PETA's Vegetarian Starter Kit (peta.org)
Happy Cow (happycow.net)
International Vegetarian Union (ivu.org)

Safe Routes to School

Boltage (boltage.org)
National Center for Safe Routes to School (saferoutes
    info.org)
Partnership for a Walkable America (WalkableAmerica.org)

Walkable and Bikeable Neighborhoods

Carbusters Journal (carbusters.org)
Critical Mass (criticalmass.wikia.com)
PARK(ing) Day (parkingday.org)
Traffic Calming (trafficcalming.org)
World Naked Bike Ride (worldnakedbikeride.org)

Economic Reform

Foundation for the Economics of Sustainability (feasta.org)
New Economics Foundation (neweconomics.org)

Governance

American Civil Liberties Union (aclu.org)
The Center for Voting and Democracy (fairvote.org)

Charity Rankings and Philanthropy Resources

Foundation Center (foundationcenter.org)
Tides Foundation (tides.org)
American Institute of Philanthropy (charitywatch.org)
Better Business Bureau for Charities (bbb.org/us/charity)
Charity Navigator (charitynavigator.org)
American Endowment Foundation (aefonline.org)
Jumo (jumo.com)

Notes

Introduction

1. National Highway and Safety Administration, *Fatality Analysis Reporting System Encyclopedia* (Washington DC: National Highway and Safety Administration, 2011).

2. Daniel J. Benjamin et al., "Do People Seek to Maximize Happiness? Evidence from New Surveys," *Journal of Economic Literature* (forthcoming).

3. My review of media and political speech during the rise in oil prices between 2003 and 2008 in the United States shows that media coverage of alternative-energy technologies quadrupled while articles on low-tech methods of energy conservation increased just 25 percent. This is further supported by my study of the language used by politicians and the media to describe solutions during this period as well as by historical accounts of the 1970s OPEC embargo and accounts of the rise of nuclear power after World War II (see chapter 8).

1. Solar Cells and Other Fairy Tales

1. This short rendering leaves out much of the complexity the researchers uncovered, but see Monica

Leggett and Marie Finlay, "Science, Story, and Image: A New Approach to Crossing the Communication Barrier Posed by Scientific Jargon," *Public Understanding of Science* 10, no. 2 (2001).

2. Barack Obama, interview by Eric Schmidt, November 14, 2008, Google Headquarters, Mountain View CA; Thomas Markvart, *Solar Electricity*, 2nd ed. (New York: Wiley, 2000), 3–4; Greenpeace, *Global Warming Story Tour 2008*, Greenpeace, http://us.greenpeace.org; BP, *Learn More About Solar*, http://www.bp.com.

3. These are drawn from two ongoing author studies of popular science articles published in the *New York Times* and the three most widely circulated popular science magazines in the United States: *Popular Science*, *Discover*, and *WIRED*.

4. Markvart, *Solar Electricity*, 1–3.

5. There are other ways to capture the sun's energy besides solar cells. Solar thermal plants use mirrors to heat steam or molten salt, which is used to drive turbines, much like a traditional coal-fired power plant. Buildings can be designed to absorb light through their windows to heat concrete slabs or other thermal masses, which in turn radiate their heat during the night; such building techniques are labeled as passive solar.

6. Lester Brown, *Plan B 3.0* (New York: W. W. Norton, 2008), 252.

7. Bjorn Lomborg, *The Skeptical Environmentalist: Measuring the Real State of the World* (Cambridge UK: Cambridge University Press, 2001), 159.

8. The world uses about 140,000 terawatt-hours per year. International Energy Agency, *Renewables Information 2007* (Geneva: OECD/IEA, 2007). Faiman et al. calculate that a large solar collector system can produce two thousand kilowatt-hours per year for every kilowatt of installed capacity, which in turn costs $630 per kilowatt for manufacturing, $850 per kilowatt for solar collector fabrication and installation, $285 for storage batteries, and 0.5 cents per kilowatt-hour for maintenance. Faiman et al. indicate that the actual costs of large systems could turn out to be several times these amounts: D. Faiman, D. Raviv, and R. Rosenstreich, "Using Solar Energy to Arrest the Increasing Rate of Fossil-Fuel Consumption: The Southwestern States of the USA as Case Studies," *Energy Policy* 35, no. 1 (2007): 570.

9. Author's calculation using government data from: Ryan Wiser et al.,

Letting the Sun Shine on Solar Costs: An Empirical Investigation of Photovoltaic Cost Trends in California (Golden CO: National Renewable Energy Laboratory (NREL), 2006); Glen Harris and Shannon Moynahan, *The California Solar Initiative—Triumph or Train Wreck? A Year to Date Review of the California Public Utility Commissions' California Solar Initiative* (SunCentric, 2007).

10. Author's calculation using: (1) an average of 35.5 grams of carbon dioxide per kilowatt-hour of photovoltaic energy produced, based on estimations from Vasilis M. Fthenakis and Hyung C. Kim, "Greenhouse-Gas Emissions from Solar Electric and Nuclear Power: A Life-Cycle Study," *Energy Policy* 35, no. 4 (2007): 2549–57; and (2) given that the world uses about 140,000 terawatt-hours per year (see International Energy Agency, *Renewables Information 2007*); and (3) assuming a thirty-year life span for the arrays.

11. Stephanie Rosenbloom, "Giant Retailers Look to Sun for Energy Savings," *New York Times*, August 10, 2008.

12. Bernie Fischlowitz-Roberts, *Solar Cell Sales Booming*, Earth Policy Institute, January 1, 2002, http://www.earth-policy.org.

13. Janet L. Sawin, "Another Sunny Year for Solar Power," in *Energy and Climate* (Washington DC: Worldwatch Institute, 2008).

14. "Clean Technology in the Downturn: Gathering Clouds," *Economist*, November 6, 2008; Dirk Lammers, "Good Week, Bad Week for Solar Industry," *USA Today*, February 12, 2008.

15. Solarbuzz, "Photovoltaic Industry Statistics: Costs, Solar Energy Costs/Prices," Solarbuzz, http://www.solarbuzz.com.

16. Harris and Moynahan, *California Solar Initiative*, 6.

17. Ozzie Zehner, "Unintended Consequences," in *Green Technology*, ed. Paul Robbins, Dustin Mulvaney, and J. Geoffrey Golson (London: Sage, 2011).

18. Kate Galbraith, "Solar Panels Are Vanishing, Only to Reappear on the Internet," *New York Times*, September 23, 2008.

19. Save the Children, *State of the World's Mothers 2005: The Power and Promise of Girls' Education* (Westport CT: Save the Children, 2005).

20. Utilities in some parts of the world distinguish between peak and off-peak electrical pricing to reflect this varying value of electricity throughout the day. The more complex "dynamic pricing" adjusts electricity prices in real time based on supply-and-demand metrics. Smart home

wiring systems and appliances can work together to minimize electricity use during periods of high demand.

21. Severin Borenstein, *The Market Value and Cost of Solar Photovoltaic Electricity Production* (Berkeley: University of California Energy Institute Center for the Study of Energy Markets, 2008).

22. Harris and Moynahan, *California Solar Initiative*, 6.

23. R. D. Duke and D. M. Kammen, "The Economics of Energy Market Transformation Initiatives," *Energy Journal* 20, no. 4 (1999); Richard M. Swanson, "A Vision for Crystalline Silicon Solar Cells," in Department of Energy Solar Program Review Meeting, ed. DOE (Sunnyvale CA: SunPower Corporation [2004: DOE/GO–102005-2067], 2004).

24. Gregory F. Nemet, "Beyond the Learning Curve: Factors Influencing Cost Reductions in Photovoltaics," *Energy Policy* 34, no. 17 (2006): 26.

25. There are other variations of Moore's law; R. Martin Roscheisen is quoted in G. Pascal Zachary, "Silicon Valley Gets Interested in Solar Power," *New York Times*, November 7, 2008.

26. See Fthenakis and Kim, "Greenhouse-Gas Emissions from Solar Electric and Nuclear Power." Another study yielded comparable results: Kris De Decker, "De Donkere Kant Van Zonne-Energie: Hoeveel Energie Kost De Productie Van Zonnepanelen?" [The dark side of solar energy: How much energy does it take to produce solar panels?], *Low-tech Magazine*, December 15, 2008.

27. One kilowatt-hour of electricity generated by coal emits around 850 grams of CO_2 per kilowatt-hour. Natural gas produces about 450 grams of CO_2 per kilowatt-hour.

28. Borenstein, *Market Value and Cost of Solar Photovoltaic Electricity Production*, 25.

29. Borenstein, *Market Value and Cost of Solar Photovoltaic Electricity Production*, 25.

30. United Nations Intergovernmental Panel on Climate Change, "Radiative Forcing of Climate Change (6.3.3 Halocarbons)," in *Climate Change 2001: Working Group I: The Scientific Basis* (Geneva: GRID-Arendal, 2003); Richard Conniff, "The Greenhouse Gas That Nobody Knew," *Yale Environment 360*, November 13, 2008.

31. Ray F. Weiss et al., "Nitrogen Trifluoride in the Global Atmosphere," *Geophysical Research Letters* 35, no. L20821 (2008).

32. Weiss et al., "Nitrogen Trifluoride in the Global Atmosphere."

33. Ariana Eunjung Cha, "Solar Energy Firms Leave Waste Behind in China," *Washington Post*, March 9, 2008.

34. Dustin Mulvaney et al., "Toward a Just and Sustainable Solar Energy Industry," (white paper, Silicon Valley Toxics Coalition, January 14, 2009), 1.

35. David A. Taylor, "Occupational Health: On the Job with Solar PV," *Environmental Health Perspectives* 118, no. 1 (2010).

36. Mulvaney et al., "Toward a Just and Sustainable Solar Energy Industry," 25–26.

37. The DOE reports a total 0.01 percent for photovoltaics and concentrated solar power (thermal) combined. Data published in: Energy Information Administration, "Renewable and Alternative Fuels," U.S. Department of Energy, accessed November 1, 2010, http://www.eia.doe.gov/fuelrenewable.html.

38. Masdar City, "Masdar Puts Solar Power to the Test: 22 Leading International Manufacturers of PV Technology Participate in Groundbreaking Field Study in Abu Dhabi," press release, December 27, 2007, http://www.AMEinfo.com.

39. A. Kimber et al., "The Effect of Soiling on Large Grid-Connected Photovoltaic Systems in California and the Southwest Region of the United States" (paper presented at the Conference Record of the 2006 IEEE 4th World Conference on Photovoltaic Energy Conversion, Waikoloa HI, 2006), 5, 1.

40. Martin LaMonica, "A Tale of Solar Panels, Snow, and Roof Rakes," CNET, February 6, 2009, http://news.cnet.com; E. L. Meyer and E. E. van Dyk, "Assessing the Reliability and Degradation of Photovoltaic Module Performance Parameters," *IEEE Transactions on Reliability* 53, no. 1 (2004): 83–92; Kimber et al. "The Effect of Soiling on Large Grid-Connected Photovoltaic Systems."

41. C. Stanton, "Cloudy Skies for Masdar's Solar Project," *The National*, November 12, 2008.

42. Aging effects are not necessarily linear. For instance, module output may decline quickly after installation before entering a period of more stable output or degrade slowly with a sharper production drop-off later on. Some manufacturers claim their cells degrade less than 1 percent per year.

43. Mikkel Jørgensen, Kion Norrmana, and Frederik C. Krebs, "Stability/ Degradation of Polymer Solar Cells," *Solar Energy Materials and Solar Cells* 92, no. 7 (2008); Meyer and van Dyk, "Assessing the Reliability and Degradation of Photovoltaic Module Performance Parameters."

44. The more optimistic efficiency scenario is based on numbers released by a solar industry consulting firm and data clearinghouse. Solarbuzz, "Solar Cell Technologies: Thin Film Solar Cells," http://www.solar buzz.com.

45. Borenstein, *Market Value and Cost of Solar Photovoltaic Electricity Production*, 19, 20.

46. This observation is based on author interviews, media studies, and DOE reports and corroborated by Borenstein, *Market Value and Cost of Solar Photovoltaic Electricity Production*, 1, 22.

47. Peter Nieh, "The Cost of Creating Solar Cells: Green Panel Discusses the State of the Silicon Industry" (paper presented at the Always On Venture Summit, Half Moon Bay CA, December 3–4, 2008).

48. Meagan Ellis, "A 3D Solution to Solar Cell Inefficiencies," *Materials World* June 1, 2007, 5.

49. From 2008 to 2009, net solar electrical generation dropped from 864,315 thousand kilowatt-hours to 807,988 thousand kilowatt-hours in 2009, according to U.S. Department of Energy measurements. There are other factors that could have produced this drop, so more research would be required to show which part was due to decommissioning. Electric Power: U.S. Energy Information Administration, Form EIA-923, "Power Plant Operations Report," table 3, *http://www.eia.gov/ cneaf/alternate/page/renew_energy_consump*.

50. David L. Goldblatt, *Sustainable Energy Consumption and Society: Personal, Technological, or Social Change?* (Dordrecht, Netherlands: Springer, 2005), 90.

2. Wind Power's Flurry of Limitations

1. REpower Systems AG, *5M: Proven Technology in New Dimensions* (Hamburg: REpower Systems AG, 2004), 13–14.

2. "The Future of Energy: Trade Winds," *Economist* June 19, 2008, 6.

3. *Economist*, "Future of Energy: Trade Winds," 6.

4. "Case History: Wind of Change," *Economist Technology Quarterly*, December 6, 2008, 24.

5. Ozzie Zehner, "Producing Power: The Semiotization of Alternative Energy in Media and Politics," 2007, University of Amsterdam, http://www.berkeley.academia.edu.

6. K. C. Myers, "Falmouth Turbine Noise Fuels Debate," *Cape Cod Times*, May 14, 2010; Mukund R. Patel, *Wind and Solar Power Systems* (Boca Raton FL: CRC Press, 2006), 82; H. Lee Willis and Walter G. Scott, *Distributed Power Generation: Planning and Evaluation* (Boca Raton FL: CRC Press, 2000), 269.

7. Robert F. Kennedy, editorial, "An Ill Wind Off Cape Cod," *New York Times*, December 16, 2005.

8. F. A. Van der Loo, "Mediating Windpower in the Netherlands: The Task Force Windpower Implementation" (white paper, Netherlands Wind Energy Association, Utrecht, Netherlands, 2001, http://www.nwea.nl).

9. Manuela de Lucas, Guyonne F. E. Janss, and Miguel Ferrer, "The Effects of a Wind Farm on Birds in a Migration Point: The Strait of Gibraltar," *Biodiversity and Conservation* 13, no. 2 (2004); Frances Cerra Whittelsey, "The Birds and the Breeze: Making Wind Power Safe for Wildlife," *Sierra* January/February 2007, 39.

10. Whittelsey, "Birds and the Breeze," 39.

11. Patel, *Wind and Solar Power Systems*, 83.

12. K. C. Myers, "FAA: Turbines Pose Risk to Radar," *Cape Cod Times*, February 14, 2009.

13. National Research Council, *Hidden Costs of Energy: Unpriced Consequences of Energy Production and Use* (Washington DC: National Academies Press, 2010), 140.

14. Robert Booth, "Micro-Wind Turbines Often Increase CO_2, Says Study," *London Guardian*, March 26, 2007. Geoff A. Dutton, Jim A. Halliday, and Mike J. Blanch, "The Feasibility of Building-Mounted/Integrated Wind Turbines (BUWTS): Achieving Their Potential for Carbon Emission Reductions" (in the final report of the Energy Research Unit, Council for the Central Laboratory of the Research Councils, 2005, http://www.stfc.ac.uk/).

15. Dutton et al., "Feasibility of Building-Mounted/Integrated Wind Turbines," 32.

16. As quoted in Dutton et al., "Feasibility of Building-Mounted/Integrated Wind Turbines," 32.

17. Hugh Sharman, "Why Wind Power Works for Denmark," *Civil Engineering* 158 (May 2005): 72.

18. Aimee Curtright and Jay Apt, "The Character of Power Output from Utility-Scale Photovoltaic Systems," *Progress in Photovoltaics: Research and Applications* 16, no. 3 (2008): 242.

19. F. Vestergaard, "Bistand til Tyskland" [Aid for Germany], *Weekend Avisen*, November 4, 2005, 13.

20. Sharman, "Why Wind Power Works for Denmark."

21. Gerald Groenewald, "Energy Policy: Pumped Up," *Economist* January 29, 2009, 43.

22. "Building the Smart Grid," *Economist Technology Quarterly*, June 6, 2009, 16.

23. "Wiser Wires," *Economist*, October 10, 2009, 71.

24. Energy Information Administration, *Average Capacity Factors by Energy Source, 1996 through 2007*, U.S. Department of Energy, http://www.eia.doe.gov.

25. Based on 2007 wind power generation of 34,450,000 megawatt-hours divided by 16,596 megawatts wind turbine capacity times 8,765 hours per year, or 23.7 percent capacity factor. See Energy Information Administration, *Existing Capacity by Energy Source, 2007*, U.S. Department of Energy, http://www.eia.doe.gov.

26. Reliability factors can be 5 percent or lower. So to offset a traditional plant with a 90 percent reliability factor, power operators would require eighteen times the wind production.

27. Leigh Glover, "From Love-Ins to Logos: Charting the Demise of Renewable Energy as a Social Movement," in *Transforming Power: Energy, Environment, and Society in Conflict*, ed. John Byrne, Noah Toly, and Leigh Glover (London: Transaction Publishers, 2006), 256.

28. It would require quite a stretch of the figures to say that it has. Though, it is possible. If wind power is either combined with expensive energy storage technologies or distributed widely across a large grid so that when the wind lulls in one place, the grid can take up power from turbines farther away, utilities could increase wind's reliability factor to some extent. (This has been attempted in Southern Australia with limited success). Such a network will be expensive, however, and its practical feasibility will diminish as more capacity is brought online. For differing views on this issue, see Mark Diesendorf, *The Base-Load Fal-*

lacy (Sydney: University of New South Wales Institute of Environmental Studies, 2007); A. Miskelly and T. Quirk, "Wind Farming in South East Australia," *Energy and Environment* 20 (2009): 1249–55.

It's perhaps also worth noting that E.ON Netz, a German utility with 7,000 megawatts of wind farms, concluded in their *Wind Report 2005*: "Wind energy is only able to replace traditional power stations to a limited extent. Their dependence on the prevailing wind conditions means that wind power has a limited load factor even when technically available. It is not possible to guarantee its use for the continual cover of electricity consumption. Consequently, traditional power stations with capacities equal to 90 percent of the installed wind power capacity must be permanently online in order to guarantee power supply at all times." Also, a report from the Renewable Energy Foundation in the United Kingdom states: "Wind energy cannot replace conventional power stations to any significant degree." See Tom Quirk, "Australia—Where Too Much Wind Will Never Be Enough," *Science Alert* January 22, 2008; E.ON Netz, *Wind Report 2005* (Bayreuth, Germany: E.ON Netz GmbH, 2005); Miskelly and Quirk, "Wind Farming in South East Australia."

29. U.S. Department of Energy, press release, "Wind Energy Could Produce 20% of U.S. Electricity by 2030" (Washington DC: U.S. DOE, May 12, 2008); U.S. Department of Energy, "Comments on the 20 Percent Wind Report," http://www.20percentwind.org.

30. U.S. Department of Energy, *20% Wind Energy by 2030: Increasing Wind Energy's Contribution to the U.S. Electrical Supply* (Washington DC: U.S. Department of Energy, 2008), 19.

31. For information on Black and Veatch, see http://www.bv.com. For Black and Veatch contributions, see U.S. Department of Energy, *20% Wind Energy by 2030: Increasing Wind Energy's Contribution to the U.S. Electrical Supply*, front matter and appendix D.

32. U.S. Department of Energy, *2008 Wind Technologies Market Report*, (Washington DC: U.S. Department of Energy, 2009), 37.

33. Nicolas Boccard, "Capacity Factor of Wind Power: Realized Values vs. Estimates," *Energy Policy* 37 (2009): 2679.

34. American Wind Energy Association, *Wind Web Tutorial: Wind Energy Basics*, American Wind Energy Association, accessed December 2, 2009, http://www.awea.org.

35. Author interview with Liz Hartman, wind and water power technologies communication specialist, U.S. Department of Energy, Washington DC, December 14, 2009; author interview with George Minter, director of media relations and communications, Black and Veatch Corporation, Overland Park KS, December 19, 2009.

36. U.S. Department of Energy, *20% Wind Energy by 2030*.

37. Range is from 18.5 percent to 25.7 percent. This is obtained by comparing Department of Energy generation reports by the Energy Information Administration (EIA) with actual installed capacity. Capacity factors vary from year to year. Energy Information Administration, *Annual Energy Outlook 2007: With Projections to 2030*, http://eia.gov. See also Boccard, "Capacity Factor of Wind Power."

38. Energy Information Administration, *Assumptions to the Annual Energy Outlook 2009* (Washington DC: U.S. Department of Energy, 2009).

39. A factor of roughly six would probably be the most extreme case, calculated by comparing the DOE's cost projections (a 200 percent increase in costs) and capacity factors (as calculated from EIA field measurements) with Black and Veatch's projections that costs will decrease and average capacity factors will be much higher. Broken down, capacity factor differences account for roughly a factor three difference between the two extremes and cost a factor of roughly two, resulting in a compound (multiplied) factor six. This provides a rough estimation of the difference between the extreme-most values, not all value differences between the approaches.

40. U.S. Department of Energy, *20% Wind Energy by 2030*, front matter.

41. Author interview with Liz Hartman, U.S. Department of Energy, December 14, 2009.

42. William A. Sherden, *The Fortune Sellers: The Big Business of Buying and Selling Predictions* (New York: Wiley, 1998); Constance Penley, *NASA/Trek: Popular Science and Sex in America* (New York: Verso, 1997); Stephen L. Del Sesto, "Wasn't the Future of Nuclear Energy Wonderful," in *Imagining Tomorrow: History, Technology, and the American Future*, ed. Joseph J Corn (Cambridge MA: MIT Press, 1987); Daniel Sarewitz, *Frontiers of Illusion* (Temple University Press, 1996); Jerome R. Ravetz, *Scientific Knowledge and Its Social Problems* (Oxford: Oxford University Press, 1971).

43. Boccard, "Capacity Factor of Wind Power"; Nicolas Boccard, "Eco-

nomic Properties of Wind Power: A European Assessment," *Energy Policy* 38, no. 7 (2009): 3232–44.

44. Boccard, "Capacity Factor of Wind Power."
45. Anselm Waldermann, "Climate Change Paradox: Green Energy Not Cutting Europe's Carbon," *Speigel*, February 10, 2009.

3. Biofuels and the Politics of Big Corn

1. Suneeta D. Fernandes et al., "Global Biofuel Use, 1850–2000," *Global Biogeochemical Cycles* 21, no. 2 (2007).
2. Dennis Keeney, "Ethanol USA," *Environmental Science and Technology* 43, no. 1 (2009).
3. Energy Information Administration, *U.S. Energy Consumption by Energy Source* (Washington DC: U.S. Department of Energy, 2010).
4. International Energy Agency, *Biofuels for Transport: An International Perspective* (Paris: International Energy Agency, 2004); Fred Goricke and Monika Reimann, "Brasilien: Das Nationale Alkoholprogramm: Eine Verfehlte Energie-Investition" [Brazil: The national alcohol program: A missed energy investment], in *Der Fischer Oko-Almanach*, ed. Gerd Michelsen (Frankfurt: Fischer-Taschenbuch-Verlag, 1982).
5. Aditya Chakrabortty, "Secret Report: Biofuel Caused Food Crisis," *London Guardian*, July 3, 2008.
6. Quoted in Robert Zoellick, "The World This Week: Politics," *Economist*, April 19, 2008, 9.
7. Chakrabortty, "Secret Report."
8. Quoted in Joachim von Braun, "The New Face of Hunger," *Economist*, April 19, 2008, 32.
9. Christopher B. Field, J. Elliot Campbell, and David B. Lobell, "Biomass Energy: The Scale of the Potential Resource," *Trends in Ecology and Evolution* 23, no. 2 (2008): 66.
10. Keeney, "Ethanol USA."
11. "The Farm Bill: A Harvest of Disgrace," *Economist*, May 24, 2008, 46.
12. Daniel Lee Kleinman, *Science and Technology in Society* (Oxford: Blackwell, 2006), 17–21.
13. Keeney, "Ethanol USA."
14. James Bovard, "Archer Daniels Midland: A Case Study in Corporate Welfare," *Cato Policy Analysis* no. 241 (September 26, 1995).
15. Energy Information Administration, *Status and Impact of State MTBE Ban* (Washington DC: U.S. Department of Energy, 2003).

16. Energy Independence and Security Act of 2007, Public Law 110–140 (2007).

17. Data on loan guarantees, state tax breaks, farm subsidies, and water subsidies are from Doug Koplow, *Biofuels—At What Cost? Government Support for Ethanol and Biodiesel in the United States: 2007 Update* (Geneva: International Institute for Sustainable Development, 2007), 16, 17–18, 25, and 26; data on federal tax breaks are from Jesse McCurry, "Cost Segregation Fuels Depreciation Acceleration," *Ethanol Producer Magazine*, February 2006; data on research funds are from U.S. Department of Energy, "Energy Department Selects Three Bioenergy Research Centers for $375 Million in Federal Funding," press release, U.S. Department of Energy Office of Public Affairs, 2007; data on labor subsidies are from American Jobs Creation Act of 2004, Public Law 108–357; "DOE Announces up to $200 Million in Funding for Biorefineries," press release, U.S. Department of Energy, 2007; information on water subsidies is from the National Academy of Sciences, *Water Implications of Biofuels Production in the United States* (Washington DC: National Academy of Sciences Committee on Water Implications of Biofuels Production in the United States, 2007).

18. U.S. Environmental Protection Agency, *Final Changes for Certain Ethanol Production Facilities under Three Clean Air Act Permitting Programs* (Washington DC: EPA, 2007).

19. Don MacKenzie, Louise Bedsworth, and David Friedman, *Fuel Economy Fraud Closing the Loopholes That Increase U.S. Oil Dependence* (Cambridge MA: Union of Concerned Scientists, 2005).

20. Keeney, "Ethanol USA."

21. Keeney, "Ethanol USA."

22. Paul Rauber, "Corn-Fed Cars: Detroit's Phony Ethanol Solution," *Sierra*, January/February 2007, 42.

23. "Dead Water: Too Much Nitrogen Being Washed into the Sea Is Causing Dead Zones to Spread Alarmingly," *Economist*, May 19, 2008, 97.

24. National Research Council, *Hidden Costs of Energy: Unpriced Consequences of Energy Production and Use* (Washington DC: National Academies Press, 2010), 145.

25. "America's Fuel Campaign," Growth Energy, April 11, 2010, http://www.growthenergy.org.

26. *Economist*, "Dead Water," 97.

27. The following discussion is based on Matt Johnston et al., "Resetting Global Expectations from Agricultural Biofuels," *Environmental Research Letters* 4, no. 1 (2009).

28. Johnston et al., "Resetting Global Expectations."

29. Cori Hayden and Daniel Sarewitz point to conflicts here. Patent law holds that intellectual property rights must be extended to encompass the whole world. However, the international community expects that developing nations will increasingly rely on a diffusion of technology (technology that they cannot afford). As we witness today with genetic crops, HIV medications, and bioprospecting, these goals may not be compatible. See Daniel Sarewitz, *Frontiers of Illusion* (Temple University Press, 1996) and Cori Hayden, *When Nature Goes Public: The Making and Unmaking of Bioprospecting in Mexico* (Princeton: Princeton University Press, 2003).

30. David D. Zhang et al., "Global Climate Change, War, and Population Decline in Recent Human History," *Proceedings of the National Academy of Sciences* 104, no. 49 (2007).

31. Claudia Tebaldi and David B. Lobell, "Towards Probabilistic Projections of Climate Change Impacts on Global Crop Yields," *Geophysical Research Letters* 35 (2008); David B. Lobell and Christopher B. Field, "Global Scale Climate—Crop Yield Relationships and the Impacts of Recent Warming," *Environmental Research Letters* 2 (2007).

32. Mochamed Ali, David Taylor, and Kazuyuki Inubushi, "Effects of Environmental Variations on CO_2 Efflux from a Tropical Peatland in Eastern Sumatra," *Wetlands* 26, no. 2 (2007).

33. Jörn P. W. Scharlemann and William F. Laurance, "How Green Are Biofuels?" *Science* 319, no. 5859 (2008): 43.

34. Robert B. Jackson et al., "Protecting Climate with Forests," *Environmental Research Letters* 3, no. 4 (2008).

35. David Tilman, Jason Hill, and Clarence Lehman, "Carbon-Negative Biofuels from Low-Input High-Diversity Grassland Biomass," *Science* 314, no. 5805 (2006).

36. Field, Campbell, and Lobell, "Biomass Energy," 69 and 65.

37. Jim DiPeso, "Carbon Offsets: Is the Environment Getting What You Pay For?" *Environmental Quality Management* 17, no. 2 (2007).

38. Field, Campbell, and Lobell, "Biomass Energy"; Scharlemann and Laurance, "How Green Are Biofuels?"

39. "Treethanol: Woodstock Revisited," *Economist Technology Quarterly*, March 10, 2007, 16–17.
40. "Very, Very Big Corn: Ethanol and Its Consequences," *Wall Street Journal*, January 27, 2007.
41. *Wall Street Journal*, "Very, Very Big Corn," 65.
42. Cedric Briens, Jan Piskorz, and Franco Berruti, "Biomass Valorization for Fuel and Chemicals Production—A Review," *International Journal of Chemical Reactor Engineering* 6, no. R2 (2008).
43. Frank Verheijen et al., *Biochar Application to Soils: A Critical Scientific Review of Effects on Soil Properties, Processes and Functions* (Ispra, Italy: European Commission, Joint Research Centre Institute for Environment and Sustainability, 2009).
44. Kelli G. Roberts et al., "Life Cycle Assessment of Biochar Systems: Estimating the Energetic, Economic, and Climate Change Potential," *Environmental Science and Technology* 44, no. 2 (2010).
45. Field, Campbell, and Lobell, "Biomass Energy," 65.
46. Energy Information Administration, *Annual Energy Outlook 2010* (Washington DC: U.S. Department of Energy, 2009), 12.
47. Antonio Pflüger, "Potential Role of Biofuels in Future Markets," in *Biofuels Markets Congress and Exhibition* (Brussels: International Energy Agency, 2009), 11.
48. Glenn Hess, "Biofuels Are Poised to Displace Oil," *Chemical and Engineering News*, June 13, 2006.

4. The Nuclear-Military-Industrial Risk Complex

1. Blaine Harden, *A River Lost* (New York: W. W. Norton, 1997), 178.
2. Barack Obama, responding to question during the Oregon Democratic primary campaign (radio station KNDO, May 19, 2008).
3. U.S. Department of Energy, *Hanford Overview*, March 25, 2008, http://www.hanford.gov.
4. Washington State Department of Health, *The Release of Radioactive Materials from Hanford: 1944–1972* (The Hanford Health Information Network, 1997).
5. F. J. Davis et al., *An Aerial Survey of Radioactivity Associated with Atomic Energy Plants* (Oak Ridge, Tennessee: Oak Ridge National Laboratory, 1949).
6. Committee on Human Radiation Experiments, "Draft Memorandum:

The Facts and Unknowns of the Green Run" (Washington DC: National Security Archive, 1994).

7. Davis et al., *An Aerial Survey of Radioactivity Associated with Atomic Energy Plants*, 135 and 7.

8. U.S. Department of Energy, *Hanford Overview*.

9. "Science Watch: Growing Nuclear Arsenal," *New York Times*, April 28, 1987.

10. Shannon Dininny, "Hanford Plant Now $12.2 Billion: Estimated Cost of Project Keeps Growing," *Seattle Post-Intelligencer* September 8, 2006.

11. Edited for clarity and space, this list is taken directly from the U.S. Department of Energy, *Hanford Overview*.

12. Walter A. Rosenbaum, *Environmental Politics and Policies*, 7th ed. (Washington DC: Congressional Quarterly Press, 2008), 227.

13. U.S. Department of Energy, *Hanford Overview*.

14. Rosenbaum, *Environmental Politics and Policies*, 278.

15. Matthew L. Wald, "Hazards at Nuclear Plant Festering 8 Years after Warning," *New York Times*, December 24, 1992.

16. Matthew L. Wald, "Nuclear Site Is Battling a Rising Tide of Waste," *New York Times*, September 27, 1999.

17. John Stang, "New Waste Put into 'Burping Tank,'" *Hanford News*, November 15, 2002.

18. Chuck Stewart, *Hanford's Battle with Nuclear Waste Tank SY–101* (Columbus OH: Battelle Press, 2006).

19. Shannon Cram, "Escaping S–102: Waste, Illness, and the Politics of Not Knowing," *International Journal of Science in Society* 2 (2010).

20. Rosenbaum, *Environmental Politics and Policies*, 262.

21. Jim Ostroff, "Nuclear Industry Gears up for Revival," *Kiplinger Business Forecasts*, June 10, 2004.

22. Michael Grunwald and Juliet Eilperin, "Energy Bill Raises Fears About Pollution, Fraud: Critics Point to Perks for Industry," *Washington Post*, July 29, 2005.

23. Energy Policy Act of 2005, Pub. L. No. 109-58, 119 Stat. 594, 109th Cong. (2005). See also Rosenbaum, *Environmental Politics and Policies*, 259–60.

24. "Quantum Politics: India's Nuclear Deal with America," *Economist*, September 11, 2008.

25. Michael Bess, *The Light-Green Society: Ecology and Technological Modernity in France*, 1960–2000 (Chicago: University of Chicago Press, 2003), 29–30.

26. Bess, *Light-Green Society*, 30.

27. Roland E. Langford, *Introduction to Weapons of Mass Destruction* (Hoboken NJ: Wiley, 2004), 85.

28. Henry D. Sokolski, "Assessing the IAEA's Ability to Verify the NPT," in *Falling Behind: International Scrutiny of the Peaceful Atom*, ed. Henry D. Sokolski (Carlisle PA: Strategic Studies Institute, U.S. Army, 2008), 19–20, 22.

29. Henry Sokolski, ed., *Falling Behind: International Scrutiny of the Peaceful Atom* (Washington DC: Nonproliferation Policy Education Center, 2007), 14.

30. U.S. Government Accountability Office (GAO), *Nuclear Regulatory Commission: NRC Needs to Do More to Ensure That Power Plants Are Effectively Controlling Spent Nuclear Fuel*, April 8, 2005, http://www.gao.gov/products/GAO-05-339.

31. U.S. GAO, *NRC Needs to Do More*.

32. Gregory F. Nemet, "Interpreting Interim Deviations from Cost Projections for Publicly Supported Energy Technologies" (working paper series 2008-012, Robert M. La Follette School of Public Affairs at the University of Wisconsin–Madison, 2008).

33. Interview with Doug Koplow (founder of Earth Track), "Gambling on Nuclear Power: How Public Money Fuels the Industry," *Subsidy Watch*, issue 26, August 2008, http://www.globalsubsidies.org.

34. A. V. Yablokov, *Facts and Problems Related to Radioactive Waste Disposal in Seas Adjacent to the Territory of the Russian Federation* (Moscow: Office of the President of the Russian Federation, 1993).

35. U.S. Department of Energy, *Quarterly Progress Report to Congress, 2nd and 3rd Quarters FY2008* (Washington DC: Office of Civilian Radioactive Waste Management, 2008), 5.

36. U.S. Department of Energy, *Analysis of the Total System Life Cycle Cost of the Civilian Radioactive Waste Management Program, Fiscal Year 2007*, (Washington DC: U.S. Department of Energy, Office of Civilian Radioactive Waste Management [DOE/RW–0591], 2008), v.

37. Frank von Hippel, "Managing Spent Fuel in the United States: The Illogic of Reprocessing," in *Falling Behind: International Scrutiny of the*

Peaceful Atom, ed. Henry D. Sokolski (Carlisle PA: Strategic Studies Institute, U.S. Army, 2008), 166.

38. Mark Holt, CRS *Report for Congress: Civilian Nuclear Waste Disposal* (Washington DC: Congressional Research Service, 7-5700, RL33461, June 22, 2011), 8.

39. U.S. Department of Energy, *Quarterly Progress Report to Congress*, 5.

40. Richard A. Muller, *Physics for Presidents* (New York: W. W. Norton, 2008), 176.

41. National Cancer Institute, *State Cancer Profiles* (Washington DC: U.S. Centers for Disease Control and Prevention, 2010).

42. Muller, *Physics for Presidents*, 103.

43. Alexey V. Yablokov, Vassily B. Nesterenko, and Alexey V. Nesterenko, *Chernobyl: Consequences of the Catastrophe for People and the Environment* (New York: Wiley-Blackwell, 2010).

44. Other electrical generation was increasing as well, so it's impossible to determine what portion was due to the increase in nuclear capacity.

45. California decoupled its utilities, and France has a carbon tax to help prevent a return to fossil fuels.

5. The Hydrogen Zombie

1. Jerry Edgerton, "What You'll Be Driving in 2016," CBS *News*, January 14, 2010.

2. Sarah Juliet Lauro and Karen Embry, "A Zombie Manifesto: The Nonhuman Condition in the Era of Advanced Capitalism," *boundary* 35, no. 1 (2008); Vijay V. Vaitheeswaran, "Unraveling the Great Hydrogen Hoax," *Nieman Reports* (Summer 2004); Robert Zubrin, "The Hydrogen Hoax," *New Atlantis*, no. 15 (Winter 2007).

3. Arnold Schwarzenegger, address to L.A. Auto Show (State of California, 2006), *http://gov.ca.gov/news.php?id=4825*.

4. National Energy Policy Development Group, *National Energy Policy: Reliable, Affordable, and Environmentally Sound Energy for America's Future* (Washington DC: National Energy Policy Development Group, 2001).

5. *National Energy Policy*, 6–10.

6. U.S. Department of Energy, *Toward a More Secure and Cleaner Energy Future for America: A National Vision of America's Transition to a*

Hydrogen Economy—To 2030 and Beyond (Washington DC: U.S. Department of Energy, 2002).

7. U.S. Department of Energy, *Fuel Cell Report to Congress* (Washington DC: U.S. Department of Energy, 2003).

8. California Hydrogen Highway Network, *Hydrogen Production Methods and Environmental Impacts Fact Sheet* (State of California, California Hydrogen Highway Network, 2007).

9. Martin Hultman, "Back to the Future: The Dream of a Perpetuum Mobile in the Atomic Society and the Hydrogen Economy," *Futures* 41, no. 4 (2009).

10. Dave Levinthal, "Congressmen Lose Big Bucks in 2008, but Still Rank among Nation's Richest," November 4, 2009, Center for Responsive Politics, http://www.opensecrets.org.

11. *U.S. House of Representatives Committee on Government Reform, Hearing before the Subcommittee on Energy and Resources*, U.S. House of Representatives, no. 109–261 (2006), 12.

12. Zubrin, "Hydrogen Hoax."

13. Matthew Phenix, "Leno's Best Joke in Years: The BMW Hydrogen 7," *WIRED Magazine Online*, September 18, 2007.

14. Sunita Satyapal et al., "The U.S. Department of Energy's National Hydrogen Storage Project: Progress Towards Meeting Hydrogen-Powered Vehicle Requirements," *Catalysis Today* 120, nos. 3–4 (2007).

15. The fate of the Hindenburg may have been due to flammable paint, but hydrogen is shackled to the tragedy.

16. Joseph J. Romm, *The Hype about Hydrogen: Fact and Fiction in the Race to Save the Climate* (Washington DC: Island Press, 2004).

17. Mark Jaccard, *Sustainable Fossil Fuels: The Unusual Suspect in the Quest for Clean and Enduring Energy* (Cambridge UK: Cambridge University Press, 2005), 67.

18. Jeff Wise, "The Truth about Hydrogen," *Popular Mechanics*, November 1, 2006.

19. About seven hundred miles of hydrogen pipelines already crisscross the United States, but they cost about $1 million per mile to build. B. P. Somerday and C. San Marchi, "Effects of Hydrogen Gas on Steel Vessels and Pipelines," in *Materials for the Hydrogen Economy*, ed. Russell H. Jones and George J. Thomas (Boca Raton FL: CRC Press, 2008).

20. California Hydrogen Highway Network, *Renewable Hydrogen Sta-*

tions in California (State of California, California Hydrogen Highway Network, 2005), 2.

21. Zubrin, "Hydrogen Hoax."

22. Carmen Difiglio and Dolf Gielen, "Hydrogen and Transportation: Alternative Scenarios," *Mitigation and Adaptation Strategies for Global Change* 12, no. 3 (2007).

23. Proponents argued that catalytic converters employed platinum without such bubbles.

24. Romm, *Hype about Hydrogen*, 162.

25. Wise, "Truth about Hydrogen."

26. Webster as quoted in David Snow, "Fuel-Cell Stocks Not Powered Up," *WIRED Magazine Online*, November 5, 2003. For the survey, see John Webster, John DeLucchi, and Alastair Nimmons, *2003 Fuel Cell Industry Survey* (New York: PricewaterhouseCoopers, 2003).

27. As quoted in Snow, "Fuel-Cell Stocks Not Powered Up," and in John Dobosz, "Fueling Speculation," *Forbes Newsletter Watch*, October 9, 2002.

28. In 2010 the DOE relaunched a carbon sequestration plan, called Future-Gen 2.0, without the hydrogen component. David Biello, "'Clean' Coal Power Plant Canceled—Hydrogen Economy, Too," *Scientific American*, February 6, 2008.

29. Office of the Chief Financial Officer, *FY 2010 Congressional Budget Request*, vol. 3 (Washington DC: U.S. Department of Energy, 2009), 31, 62.

30. Steven Chu as quoted in David Biello, "R.I.P. Hydrogen Economy? Obama Cuts Hydrogen Car Funding," *Scientific American*, May 8, 2009.

31. Robert Zubrin, *Energy Victory: Winning the War on Terror by Breaking Free of Oil* (Amherst NY: Prometheus Books, 2007), 23.

32. Hultman, "Back to the Future."

33. Dvora Yanow states that a policy may "work symbolically through its enactment." See Dvora Yanow, *How Does a Policy Mean?: Interpreting Policy and Organizational Actions* (Washington DC: Georgetown University Press, 1996).

34. Observation from Edward Dolnick. See Edward Dolnick, *The Forger's Spell* (New York: Harper, 2008). This theoretical approach informed by: Robert L. Heilbroner, "Do Machines Make History," *Technology*

and Culture 8, no. 3 (1967); Trevor J. Pinch and Wiebe E. Bijker, "The Social Construction of Facts and Artifacts: Or How the Sociology of Science and the Sociology of Technology Might Benefit from Each Other," in *The Social Construction of Technological Systems: New Directions in the Sociology and History of Technology*, ed. Wiebe E. Bijker, T. P. Hughes, and Trevor J. Pinch (Cambridge MA: MIT Press, 1987); Thomas P. Hughes, "Technological Momentum," in *Does Technology Drive History? Dilemma of Technological Determinism*, ed. Leo Marx and Merritt Roe Smith (Cambridge MA: MIT Press, 1994); Bruno Latour, "Where Are the Missing Masses? The Sociology of a Few Mundane Artifacts," in *Shaping Technology/Building Society: Studies in Sociotechnical Change*, ed. Wiebe E. Bijker and John Law (Cambridge MA: MIT Press, 1992); Daniel Sarewitz, *Frontiers of Illusion* (Philadelphia: Temple University Press, 1996); J. M. Wetmore, "Amish Technology: Reinforcing Values and Building Community," *Technology and Society Magazine, IEEE* 26, no. 2 (Summer 2007).

6. Conjuring Clean Coal

1. Richard A. Lovett, "Coal Mining Causing Earthquakes, Study Says," *National Geographic News*, January 3, 2007; Christian D. Klose, "Geomechanical Modeling of the Nucleation Process of Australia's 1989 M5.6 Newcastle Earthquake," *Earth and Planetary Science Letters* 256, nos. 3–4 (2007); Christian Klose, "Human-Triggered Earthquakes and Their Impacts on Human Security," *Nature Precedings*, September 29, 2010; "Coal Mining in Germany: Small Earthquake in Saarland," *Economist*, March 1, 2008.

2. Energy Information Administration, *Annual Energy Outlook 2010* (Washington DC: U.S. Department of Energy, 2009).

3. Hugh Saddler, Chris Riedy, and Robert Passey, "Geosequestration: What Is It and How Much Can It Contribute to a Sustainable Energy Policy for Australia?" (Discussion Paper Number 72, Australia Institute, 2004), ix.

4. Mark Diesendorf, "Can Geosequestration Save the Coal Industry?" in *Transforming Power: Energy, Environment, and Society in Conflict*, ed. John Byrne, Noah Toly, and Leigh Glover (London: Transaction Publishers, 2006), 224.

5. Charles Duhigg, "Cleansing the Air at the Expense of Waterways," *New York Times*, October 12, 2009.

6. Diesendorf, "Can Geosequestration Save the Coal Industry?," 227.

7. Margaret A. Palmer et al., "Mountaintop Mining Consequences," *Science* 327, no. 5962 (2010).

8. As quoted in the radio news show with Amy Goodman, "Scientists Call for Ban on Mountaintop Removal," *Democracy Now!*, January 8, 2010, http://www.democracynow.org.

9. Duhigg, "Cleansing the Air at the Expense of Waterways"; Jeff Johnson, "Energy's Hidden Cost," *Chemical and Engineering News* 87, no. 44 (2009).

10. Laurie E. Winston, "Clean Coal Technology: Environmental Solution or Greenwashing?" (master's thesis, Ohio University, 2009).

11. U.S. Environmental Protection Agency, *The Federal Water Pollution Control Act as Amended by the Clean Water Act*, http://www.epa.gov.

12. Duhigg, "Cleansing the Air at the Expense of Waterways."

13. U.S. Environmental Protection Agency, *Notice of Data Availability on the Disposal of Coal Combustion Wastes in Landfills and Surface Impoundments* (Washington DC: EPA, 2007).

14. Diesendorf, "Can Geosequestration Save the Coal Industry?" 232.

15. "Dirty King Coal," *Economist*, May 31, 2007.

16. John Deutch et al., *The Future of Coal: An Interdisciplinary MIT Study* (Cambridge MA: Massachusetts Institute of Technology, 2007).

17. Klaus S. Lackner, "Climate Change: A Guide to CO_2 Sequestration," *Science* 300, no. 5626 (2003).

18. Diesendorf, "Can Geosequestration Save the Coal Industry?," 237.

19. "Dirty King Coal," *Economist*, May 31, 2007.

20. Walter Sullivan, "U.S. and French Experts Differ on Cameroon Gas Eruption," *New York Times*, January 23, 1987.

21. The industry could avoid leak risks by mixing CO_2 with widely available minerals to create bicarbonates, though this process is even more costly. See Lackner, "Climate Change: A Guide to CO_2 Sequestration."

22. Paleontologists blame ocean acidity for the mass extinction of the oceans at the end of the Permian period, 250 million years ago. Marine biologists fear history could repeat itself rather soon in geological terms.

23. "Climate Change: Sour Times," *Economist*, February 23, 2008.

24. Andrew H. Knoll et al., "Comparative Earth History and Late Permian Mass Extinction," *Science* 273, no. 5274 (1996); Saddler, Riedy, and Passey, "Geosequestration."

25. Tim Flannery, *The Weather Makers: How Man Is Changing the Climate and What It Means for Life on Earth* (New York: Grove Press, 2006).

26. Repps Hudson, "Arch Coal Chief Keeps His Focus on Carbon," *St. Louis Post-Dispatch*, September 23, 2007.

27. Steven Mufson, "Coal Industry Plugs into the Campaign," *Washington Post*, January 18, 2008.

28. Center for Responsive Politics, *Annual Lobbying on Mining* (Washington DC: Center for Responsive Politics, 2011), access date August 7, 2011, *http://www.opensecrets.org/lobby/induscode.*

29. Barack Obama and Joseph R. Biden, *Presidential Campaign Fact Sheet, Barack Obama and Joe Biden: New Energy for America* (Chicago: Obama for America, 2008).

30. U.S. Department of Energy, *Report of the Interagency Task Force on Carbon Capture and Storage* (Washington DC: Interagency Task Force on Carbon Capture and Storage, 2010).

31. Mufson, "Coal Industry Plugs into the Campaign"; Alice Ollstein, "The War against Mountaintop Mining Heats Up," *Capitol News Connection*, September 8, 2010.

7. Hydropower, Hybrids, and other Hydras

1. Bloomberg New Energy Finance, "Selected Press Coverage," Bloomberg New Energy Finance, accessed August 7, 2011, http://www.bnef .com

2. James Cameron, radio news show interview by Aaron Maté, "'Avatar' Director James Cameron Follows Box Office Success with Advocacy for Indigenous Struggles," *Democracy Now!*, April 27, 2010, http://www.democracynow.org.

3. "The Ups and Downs of Dams: Small Projects Often Give Better Returns," *Economist*, Special Report on Water, May 20, 2010.

4. Richard A. Muller, *Physics for Presidents* (New York: W. W. Norton, 2008), 340.

5. U.S. Environmental Protection Agency, *Geothermal Energy Production Wastes* (Washington DC: EPA, 2010).

6. Bloomberg New Energy Finance, *New Energy Finance Summit 2008* (London: New Energy Finance, 2008), 14.

7. Scott L. Montgomery, *The Powers That Be: Global Energy for the Twenty-First Century and Beyond* (Chicago: University of Chicago Press, 2010), 207; Daniel Clery, "Fusion Delayed: ITER Start Date Moved Again," *Science Insider*, March 11, 2010.

8. Thermal coolers employ a heated salty brine to recover vaporized refrigerant rather than use an electric compressor pump.

9. Daniel J. Soeder and William M. Kappel, *Water Resources and Natural Gas Production from the Marcellus Shale* (Washington DC: USGS Fact Sheet 2009-3032, 2009).

10. Rebecca Renner, "Pennsylvania to Regulate Salt Discharges," *Environmental Science and Technology* 43, no. 16 (2009).

11. Walter Hang, radio news show interview by Amy Goodman, "Watchdog: New York State Regulation of Natural Gas Wells Has Been 'Woefully Insufficient for Decades,'" *Democracy Now!*, November 10, 2009, http://www.democracynow.org.

12. Walter Hang, interview by Amy Goodman, *Democracy Now!*.

13. Most electric car batteries require lithium.

14. Richard Pike (Royal Society of Chemistry), "Electric Car Subsidies Do Not Serve Green Goals," *London Financial Times*, April 27, 2009. See also "Highly Charged Motoring," *Economist*, October 9, 2010.

15. General Motors, *2011 Volt*, http://www.Chevrolet.com/Volt.

16. Pacific Gas and Electric, "Electric Vehicle Charging Rate and Economics," accessed August 28, 2010, http://www.pge.com.

17. K. Parks, P. Denholm, and T. Markel, *Costs and Emissions Associated with Plug-in Hybrid Electric Vehicle Charging in the Xcel Energy Colorado Service Territory* (Technical Report NREL/TP−640-41410, National Renewable Energy Laboratory, May 2007).

18. Muller, *Physics for Presidents*, 305.

19. This observation is adapted from the author's statement in the *Economist*. See Ozzie Zehner, "On Far-Right Politicians, Guns in America, Child Benefit, Electric Cars, Swearing, Cleanliness," *Economist*, October 21, 2010.

20. National Research Council, *Hidden Costs of Energy: Unpriced Consequences of Energy Production and Use* (Washington DC: National Academies Press, 2010).

21. David Owen, *Green Metropolis: Why Living Smaller, Living Closer, and Driving Less Are the Keys to Sustainability* (New York: Riverhead, 2009), 104.

22. Ulrich Beck, *Risk Society: Towards a New Modernity* (Newbury Park CA: Sage, 1992).

8. The Alternative-Energy Fetish

1. Joe Dumit, "A Digital Image of the Category of the Person: Pet Scanning and Objective Self-Fashioning," in *Cyborgs and Citadels: Anthropological Interventions in Emerging Sciences and Technologies*, ed. Gary Lee Downey and Joseph Dumit (Santa Fe: SAR Press, 1997); Neil Postman, *Amusing Ourselves to Death: Public Discourse in the Age of Show Business* (New York: Penguin, 1985); Bruno Latour, "Morality and Technology," *Theory, Culture, and Society* 19, no. 5–6 (2002).

2. This chapter was informed by: Wiebe E. Bijker, Thomas Parke Hughes, and Trevor J. Pinch, *The Social Construction of Technological Systems: New Directions in the Sociology and History of Technology* (Cambridge MA: MIT Press, 1989); Frank W. Geels, *Technological Transitions and System Innovations: A Co-Evolutionary and Socio-Technical Analysis* (Northhampton MA: Edward Elgar, 2005); Harro Van Lente and Arie Rip, "Expectations in Technological Developments: An Example of Prospective Structures to Be Filled by Agency," in *Getting Technologies Together: Studies in Making Sociotechnical Order*, ed. Cornelis Disco and Barend Van Der Meulen (Berlin and New York: Walter de Gruyter, 1998); Harry Collins and Trevor Pinch, "Crash!: Nuclear Fuel Flasks and Anti-Misting Kerosene on Trial," in *The Golem at Large: What You Should Know About Technology* (New York: Cambridge University Press, 2002); W. Patrick McCray, "Will Small Be Beautiful? Making Policies for Our Nanotech Future," *History and Technology* 21, no. 2 (2005); L. H. Martin, H. Gutman, and P. Hutton, eds., *Technologies of the Self: A Seminar with Michel Foucault* (Amherst: University of Massachusetts Press, 1988); John Rajchman, "Foucault's Art of Seeing," *October* 44 (Spring 1988): 88–117; Michel Foucault, *Discipline and Punish: The Birth of the Prison* (Paris: Gallimard, 1977).

3. Ozzie Zehner, "The Solar Obsession: Symbolic Performance of Photovoltaic Energy in Media and Politics" (paper presented at the An-

nual Meeting of the Society for Social Studies of Science, Washington DC, October 28, 2009).

4. D. Hallman, "Climate Change: Ethics, Justice, and Sustainable Community," in *Christianity and Ecology: Seeking the Well-Being of Earth and Humans*, ed. Dieter T. Hessel and Rosemary R. Ruether (Cambridge MA: Harvard University Press, Harvard University Center for the Study of World Religions, 2000), 458; Max Weber, *The Protestant Ethic and the "Spirit" of Capitalism and Other Writings* (New York: Penguin Group USA, 2002); Aaron Norton and Ozzie Zehner, "Which Half Is Mommy?: Tetragametic Chimerism and Trans-Subjectivity," *Women's Studies Quarterly* (Winter 2008).

5. D. Cowdin, "The Moral Status of Otherkind in Christian Ethics," in *Christianity and Ecology: Seeking the Well-Being of Earth and Humans*, ed. Dieter T. Hessel and Rosemary R. Ruether (Cambridge MA: Harvard University Press, Harvard University Center for the Study of World Religions, 2000); Curtis White, *The Spirit of Disobedience: Resisting the Charms of Fake Politics, Mindless Consumption, and the Culture of Total Work* (Sausalito CA: Polipoint Press, 2006); Douglas Ezzy, "Old Traditions and New Ages: Religions and Environments," in *Controversies in Environmental Sociology*, ed. Rob White (Cambridge UK: Cambridge University Press, 2004).

6. Karl Marx observed: "Social relations are closely bound up with productive forces. In acquiring new productive forces men change their mode of production, and in changing their mode of production, in changing their way of earning a living, they change all their social relations. The hand-mill gives you society with the feudal lord; the steam-mill, society with the industrial capitalist." Karl Marx, *Misère de la Philosophie* (Paris and Brussels, 1847).

7. Timothy Weiskel, "Some Notes from Belshaz'zar's Feast," in *The Greening of Faith: God, the Environment, and the Good Life*, ed. John E. Carroll, Paul T. Brockelman, and Mary Westfall (Hanover NH: University Press of New England, 1997).

8. An excellent point made not by me but by Bill McKibben in his introduction to *The Greening of Faith: God, the Environment, and the Good Life*, ed. John E Carroll, Paul T. Brockelman, and Mary Westfall (Hanover NH: University Press of New England, 1997).

9. Martin Buber, *I and Thou* (London: Continuum International Publishing Group, 2004), 11.

10. Traci Watson, "Senate Climate Bill Would Speed Emissions Reductions," *USA Today*, September 30, 2009.

11. Daniel M. Cook et al., "Journalists and Conflicts of Interest in Science: Beliefs and Practices," *Ethics in Science and Environmental Politics* 9, no. 1 (2009).

12. Sharon Beder, "Moulding and Manipulating the News," in *Controversies in Environmental Sociology*, ed. Rob White (Cambridge UK: Cambridge University Press, 2004), 210.

13. Naomi Oreskes and Erik M. Conway, *Merchants of Doubt: How a Handful of Scientists Obscured the Truth on Issues from Tobacco Smoke to Global Warming* (New York: Bloomsbury, 2010); Naomi Oreskes, "The Scientific Consensus on Climate Change: How Do We Know We're Not Wrong?" in *Climate Change*, ed. Joseph F. DiMento and Pamela Doughman (Cambridge MA: MIT Press, 2007); Naomi Oreskes, "You Can Argue with the Facts: A Political History of Climate Change" (paper presented at the Dissent in Science: Origins and Outcomes Workshop, University of California San Diego, March 3–4, 2008).

14. Intergovernmental Panel on Climate Change, *Climate Change 1995: A Report of the Intergovernmental Panel on Climate Change* (Geneva, Switzerland: IPCC, 1995).

15. Oreskes, "You Can Argue with the Facts."

16. "Poll: Americans See a Climate Problem," *Time Magazine*, March 26, 2006.

17. Paul Farhi, "Liberal Media Watchdog: Fox News E-Mail Shows Network's Slant on Climate Change," *Washington Post* December 15, 2010.

18. This became an interest of pollsters starting around 2010. See Gallup Poll, March 4–7, 2010. The sample size was 1,014 adults nationwide.

19. Sharon Beder, *Global Spin: The Corporate Assault on Environmentalism* (Totnes, Devon, UK: Green Books, 2002).

20. Andrew Kohut et al., *The Web: Alarming, Appealing, and a Challenge to Journalistic Values: Financial Woes Now Overshadow All Other Concerns for Journalists* (Washington DC: The Pew Research Center for the People and the Press, Project for Excellence in Journalism, 2008), 5.

21. Kohut et al., *The Web*, 1.

22. Justin Lewis, Andrew Williams, and Bob Franklin, "A Compromised

Fourth Estate?" *Journalism Studies* 9, no. 1 (2008); Steven Woloshin et al., "Press Releases by Academic Medical Centers: Not So Academic?" *Annals of Internal Medicine* 150, no. 9 (2009).

23. "Save Energy, Save Money, Save the Planet," produced by Karla Murthy, interviews by David Brancaccio, *NOW*, March 28, 2008, http://www.pbs.org.

24. M. R. Levy, "Learning from Television News," in *The Future of News: Television-Newspapers-Wire Services-News Magazines*, ed. P. S. Cook, D. Gomery, and L. Lichty (Washington DC: Woodrow Wilson Center Press, 1992), 70.

25. Beder, "Moulding and Manipulating the News," 215.

26. Beder, "Moulding and Manipulating the News," 215.

27. David Domingo, "Interactivity in the Daily Routines of Online Newsrooms: Dealing with an Uncomfortable Myth," *Journal of Computer-Mediated Communication* 13, no. 3 (2008); Pablo J. Boczkowski, *News at Work: Imitation in an Age of Information Abundance* (Chicago: University Of Chicago Press, 2010).

28. Kohut et al., *The Web*, 8.

29. Kohut et al., *The Web*, 15–16.

30. Nikki Usher, "Goodbye to the News: How Out-of-Work Journalists Assess Enduring News Values and the New Media Landscape," *New Media and Society*, May 4, 2010, http://nms.sagepub.com.

31. Simon Schama, *The American Future: A History* (New York: Harper-Collins, 2009), 307–8.

32. Aidan Davison, "Sustainable Technology: Beyond Fix and Fixation," in *Controversies in Environmental Sociology*, ed. Rob White (Cambridge UK: Cambridge University Press, 2004), 136–37. Contains his longer chronology, which I have summarized in the next few paragraphs of the chapter.

33. World Commission on Environment and Development, *Our Common Future* (Oxford: Oxford University Press, 1987), 217, as quoted in Davison, "Sustainable Technology."

34. Schama, *American Future*, 311.

35. George Heaton, Robert Repetto, and Rodney Sobin, *Transforming Technology: An Agenda for Environmentally Sustainable Growth in the 21st Century* (Washington DC: World Resources Institute, 1991), vii, ix.

36. Paul Hawken, Amory Lovins, and L. Hunter Lovins, *Natural Capitalism* (New York: Little, Brown, 1999); William McDonough and Michael Braungart, *Cradle to Cradle: Remaking the Way We Make Things* (New York: North Point Press, 2002); Andrew W. Savitz and Karl Weber, *The Triple Bottom Line: How Today's Best-Run Companies Are Achieving Economic, Social, and Environmental Success—And How You Can Too* (San Francisco: Jossey-Bass, 2006).

37. Davison, "Sustainable Technology," referring to United Nations, "Plan of Implementation," in *Report of the World Summit on Sustainable Development* (Geneva: United Nations Document A/CONF.199/20*, June 18, 1998; reissued 2002), 13–20.

38. Davison, "Sustainable Technology," 136.

39. Susan Leigh Star and James R Griesemer, "Institutional Ecology, Translations' and Boundary Objects: Amateurs and Professionals in Berkeley's Museum of Vertebrate Zoology, 1907–39," *Social Studies of Science* 19, no. 3 (1989): 393.

40. See, for instance, Loet Leydesdorff and Iina Hellsten, "Measuring the Meaning of Words in Contexts: An Automated Analysis of Controversies About 'Monarch Butterflies,' 'Frankenfoods,' and 'Stem Cells,'" *Scientometrics* 67, no. 2 (2006).

41. Perhaps my threshold for being convinced by the cornucopians is too high, but I will come back to this in chapter 10. For one famous counterargument, see Bjorn Lomborg, *The Skeptical Environmentalist: Measuring the Real State of the World* (Cambridge UK: Cambridge University Press, 2001).

42. For example, as automotive CAFE regulations tightened, drivers opted for larger vehicles, greater acceleration, and more features. National Research Council, *Hidden Costs of Energy: Unpriced Consequences of Energy Production and Use* (Washington DC: National Academies Press, 2010).

43. I use "displaced externality" or "displaced side effects" to describe effects occurring at a different time or location (and therefore can frequently go unseen for some time).

44. Bruce Kochis, "On Lenses and Filters: The Role of Metaphor in Policy Theory," *Administrative Theory and Praxis* 27, no. 1 (2005); George Lakoff and Mark Johnson, *Metaphors We Live By* (Chicago: University of Chicago Press, 2003); J. Patrick Dobel, "The Rhetorical Pos-

sibilities of 'Home' in Homeland Security," *Administration and Society* 42, no. 5 (September 2010): 479–503.

45. National Research Council, *Hidden Costs of Energy*.

46. Lawrence Lessig, *Code: And Other Laws of Cyberspace* (New York: Basic Books, 1999).

47. Some scholars attribute the term "trained incapacity" to Thorstein Veblen, and while he does evoke the concept, he does not seem to use the phrase itself, which may have come from Randolph Bourne. It is similar to John Dewey's concept of "occupational psychosis." Robert K. Merton, "Bureaucratic Structure and Personality," *Social Forces* 18, no. 4 (1940); Kenneth Burke, *Permanence and Change* (Los Altos CA: Hermes Publications, 1954); Thorstein Veblen, *The Higher Learning in America: A Memorandum on the Conduct of Universities by Business Men* (New York: Huebsch, 1918).

48. This is a play on Daniel Moynihan's "There are some mistakes it takes a PhD to make."

49. Deborah G. Johnson and Jameson M. Wetmore, *Technology and Society: Building Our Sociotechnical Future* (Cambridge MA: MIT Press, 2009); Robert L. Heilbroner, "Do Machines Make History?" *Technology and Culture* 8, no. 3 (1967).

9. The First Step

1. Leigh Glover, "From Love-Ins to Logos: Charting the Demise of Renewable Energy as a Social Movement," in *Transforming Power: Energy, Environment, and Society in Conflict*, ed. John Byrne, Noah Toly, and Leigh Glover (London: Transaction Publishers, 2006), 261.

2. See Mike Davis, *Ecology of Fear: Los Angeles and the Imagination of Disaster* (London: Picador, 1999).

3. Fossil-fuel prices are exogenous, meaning that fossil-fuel prices will drop (all else being equal) if nations transfer demand to renewables through subsidies, regulations, and so on. In a context without taxes or other measures in place to artificially increase the price of fossil fuels, their demand should grow due to the decreased cost. In this case, alternative energy can spur fossil-fuel use rather than displace it. See also Michael Hoel, "Bush Meets Hotelling: Effects of Improved Renewable Energy Technology on Greenhouse Gas Emissions" (Working Paper 262, Fondazione Eni Enrico Mattei, 2009; this working

paper site is hosted by the Berkeley Electronic Press at http://www
.bepress.com); Hans-Werner Sinn, "Public Policies against Global
Warming: A Supply Side Approach," *International Tax and Public
Finance* 15, no. 4 (2008); Horace Herring, "Is Energy Efficiency Environmentally Friendly?" *Energy and Environment* 11, no. 3 (2000);
Hans-Werner Sinn, "Das Grüne Paradoxon: Warum Man Das An-
gebot Bei Der Klimapolitik Nicht Vergessen Darf" [The green paradox: Why we can't forget the supply side of climate policy], *Perspek-
tiven der Wirtschaftspolitik* (Special Issue) 9 (2008): 109–42; Reyer
Gerlagh, "Too Much Oil," *CESifo Economic Studies* 57, no. 1 (2011):
79–102; Julien Daubanes and André Grimaud, "Taxation of a Polluting Non-Renewable Resource in the Heterogeneous World," *Environmental and Resource Economics* 47 (2010): 567–88.

4. William S. Jevons, *The Coal Question: An Inquiry Concerning the Progress of the Nation, and the Probable Exhaustion of Our Coal-Mines* (1865;
repr., New York: Macmillan, 1906).

5. Rocky Mountain Institute, "Beating the Energy Efficiency Paradox,"
Treehugger, July 10, 2008.

6. Amory B. Lovins and L. Hunter Lovins, *Climate: Making Sense and
Making Money* (Old Snowmass CO: Rocky Mountain Institute, 1997).

7. Labor costs are typically twenty-five times higher than energy costs,
and labor cost savings can end up dwarfing energy cost savings. Steve
Sorrell and Horace Herring, *Energy Efficiency and Sustainable Consumption: The Rebound Effect* (New York: Palgrave Macmillan, 2008),
240; Ozzie Zehner, "Light-Emitting Diodes," in *Green Technology*, ed.
Paul Robbins, Dustin Mulvaney, and J. Geoffrey Golson (London:
Sage, 2011).

8. David Elliott, series editor preface to *Energy Efficiency and Sustainable Consumption: The Rebound Effect*, ed. David Elliott (New York:
Palgrave Macmillan, 2008), xi.

9. Simon Schama, *The American Future: A History* (New York: Harper-
Collins, 2009), 359.

10. Sam Smith once said he was a member of the search party. The idea
has stuck with me. Sam Smith, *Why Bother?: Getting a Life in a Locked-
Down Land* (Los Angeles: Feral House, 2001).

11. Nuclear power and its risks may be a part of our mix for a long time.

The resources are very large, especially if processes to extract uranium from seawater proceed as expected.

12. Glover, "From Love-Ins to Logos," 263.

13. Sohbet Karbuz, "The U.S. Military Oil Consumption," *Energy Bulletin*, May 20, 2007.

14. John Grin, "Reflexive Modernisation as a Governance Issue, Or: Designing and Shaping Re-Structuration," in *Reflexive Governance for Sustainable Development*, ed. Jan-Peter Voss, Dierk Bauknecht, and René Kemp (Northhampton MA: Edward Elgar, 2006); Charles E. Lindblom, Inquiry and Change: *The Troubled Attempt to Understand and Shape Society* (New Haven and London: Yale University Press, 1990); John Grin and Henk Graaf, "Implementation as Communicative Action," *Policy Sciences* 29, no. 4 (1996).

15. U.S. Energy Information Administration, *Consumption, Price, and Expenditure Estimates: State Energy Data System* (Washington DC: U.S. Department of Energy, 2010), http://205.254.135.24/state/seds/.

16. An admittedly narrow assessment. See Gallup-Healthways, "State and Congressional District Well-Being Reports," accessed July 15, 2011, http://www.well-beingindex.com.

17. David L. Goldblatt, *Sustainable Energy Consumption and Society: Personal, Technological, or Social Change?* (Dordrecht, Netherlands: Springer, 2005).

18. Some will disagree, saying my proposals do require sacrifice. And that's fine (I don't expect everyone to endorse every idea I propose). For example, zoning changes can spur walkable neighborhoods. Some might consider living in such neighborhoods as a sacrifice, whereas others see it as a benefit. Everyone needn't move to one to achieve an impact.

10. Women's Rights

1. Ozzie Zehner, "Population and Overpopulation," in *Green Technology*, ed. Paul Robbins, Dustin Mulvaney, and J. Geoffrey Golson (London: Sage, 2011); World Bank and United Nations Population Fund, "Family Planning and Reproductive Health Have Fallen Off Global Development Radar—World Bank, UNFPA," press release, June 30, 2009, http://www.unfpa.org.

2. The data for infant mortality are similar. U.S. Central Intelligence

Agency, "The World Factbook: United States," in *The World Factbook* (Washington DC: U.S. Central Intelligence Agency, 2008).

3. Jared Diamond, "Easter's End," *Discover*, August 1995.

4. Jared Diamond, *Collapse: How Societies Choose to Fail or Succeed* (New York: Penguin, 2005); Ronald Wright, *A Short History of Progress* (New York: Carroll and Graf, 2004).

5. Jared Diamond, "The Ends of the World as We Know Them," *New York Times*, January 1, 2005.

6. Albert A. Bartlett, *The Essential Exponential! For the Future of Our Planet* (Lincoln: University of Nebraska Press, 2004).

7. From 680 to 2,872 years at present consumption levels to just 205 to 339 years if the population grows at 1 percent. See Albert A. Bartlett, "Arithmetic, Population and Energy" (sixty-four-minute lecture, University of Nebraska, 2002), DVD.

8. David Montgomery, *Dirt: The Erosion of Civilizations* (Berkeley: University of California Press, 2007); Tim Forsyth, *Critical Political Ecology: The Politics of Environmental Science* (London: Routledge, 2003).

9. David. S. Battisti and Rosamond L. Naylor, "Historical Warnings of Future Food Insecurity with Unprecedented Seasonal Heat," *Science* 323, no. 5911 (2009).

10. N. V. Fedoroff et al., "Radically Rethinking Agriculture for the 21st Century," *Science* 327, no. 5967 (2010).

11. Wolfram Schlenkera and Michael J. Roberts, "Nonlinear Temperature Effects Indicate Severe Damages to U.S. Crop Yields under Climate Change," *Proceedings of the National Academy of Sciences* 106, no. 37 (2009).

12. Wright, *Short History of Progress*, 131–32.

13. San Francisco population: 809,000. World population is growing by about eighty-two million per year. See Population Reference Bureau, "World Population Highlights: Key Findings from PRB's 2008 World Population Data Sheet," in *Population Bulletin* (Washington DC: Population Reference Bureau, 2008).

14. U.S. Census Bureau, *International Data Base: World Population: 1950–2050* (Washington DC: U.S. Census Bureau, Population Division, 2009).

15. "Zogby Poll: Most Believe Humans Will Someday Colonize the Moon," May 3, 2007, http://www.zogby.com.

16. As quoted in "The End of Retirement," *Economist*, June 27, 2009.

17. This point is made by Bill McKibben, "Waste Not Want Not," *Mother Jones*, May/June 2009, 50.

18. Matthew Connelly, *Fatal Misconception: The Struggle to Control World Population* (Cambridge MA: Harvard University Press, 2008).

19. David Willey, "An Optimum World Population," *Medicine, Conflict, and Survival* 16, no. 1 (2000).

20. Jim Merkel, *Radical Simplicity: Small Footprints on a Finite Earth* (Gabriola Island BC: New Society Publishers, 2003), 183.

21. Joel E. Cohen, "Human Population Grows Up," in *A Pivotal Moment: Population, Justice and the Environmental Challenge*, ed. Laurie Mazur (Washington DC: Island Press, 2009).

22. Joel E. Cohen, *How Many People Can the Earth Support?* (New York: W. W. Norton, 1996).

23. Cohen, "Human Population Grows Up," 34.

24. David Pimentel and Marcia Pimentel, "The Real Perils of Human Population Growth," *Free Inquiry*, April/May 2009; U.S. Central Intelligence Agency, "The World Factbook: World," in *The World Factbook* (Washington DC: U.S. Central Intelligence Agency, 2009).

25. Angus Maddison, *Contours of the World Economy, 1–2030 AD: Essays in Macro-Economic History* (New York: Oxford University Press, USA, 2007); U.S. Central Intelligence Agency, "World Factbook: World."

26. Jim Oeppen and James W. Vaupel, "Broken Limits to Life Expectancy," *Science* 296, no. 5570 (2002).

27. *Economist*, "End of Retirement."

28. Ray Kurzweil, *The Age of Spiritual Machines: When Computers Exceed Human Intelligence* (New York: Viking, 1999); Jonathan Weiner, *Long for This World: The Strange Science of Immortality* (New York: Ecco, 2010).

29. Robert Engelman, *More: Population, Nature, and What Women Want* (Washington DC: Island Press, 2008), 234.

30. Richard A. Melanson, *American Foreign Policy since the Vietnam War* (Armonk NY: M. E. Sharpe, 2005), 195.

31. United Nations, *Social Aspects of Sustainable Development in the United States of America* (Geneva: United Nations, 1997). This information is based on the United States of America's submission to the 5th Session of the Commission on Sustainable Development.

32. Mona L. Hymel, "The Population Crisis: The Stork, the Plow, and the IRS," *North Carolina Law Review* 77, no. 13 (1998).

33. "A Special Report on Aging Populations," *Economist*, June 27, 2009, 4–5.

34. Paul Ehrlich, *The Population Bomb* (New York: Sierra Club–Ballantine Books, 1968); Paul R. Ehrlich and Anne H. Ehrlich, "Is the Population Bomb Finally Exploding?" *Free Inquiry* 29, no. 3 (2009): 32.

35. David Brooks, "The Power of Posterity," *New York Times*, July 27, 2009.

36. David Benatar, *Better Never to Have Been: The Harm of Coming into Existence* (Oxford: Oxford University Press, 2006), 2.

37. Henning Bohn and Charles Stuart, "Population under a Cap on Greenhouse Gas Emissions" (CESifo Working Paper no. 3046, 2010); Walter Laqueur, *The Last Days of Europe: Epitaph for an Old Continent* (New York: St. Martin's Griffin, 2009); Paul Demeny, "Population Policy and the Demographic Transition: Performance, Prospects, and Options," *Population and Development Review* 37, issue supplement 1 (2011): 249–74.

38. Henrik Urdal, "A Clash of Generations? Youth Bulges and Political Violence," *International Studies Quarterly* 50, no. 3 (2006); Henrik Urdal, "Population, Resources, and Political Violence: A Subnational Study of India, 1956–2002," *Journal of Conflict Resolution* 52, no. 4 (2008).

39. Richard Jackson and Neil Howe, *The Graying of the Great Powers: Demography and Geopolitics in the 21st Century* (Washington DC: Center for Strategic and International Studies, 2008); *Economist*, "A Special Report on Aging Populations."

40. At one time even the White House was open for citizens to walk in and share their concerns with the president or his staff directly.

41. As quoted in Albert A. Bartlett, "Democracy Cannot Survive Overpopulation," *Population and Environment* 22, no. 1 (2000): 65.

42. Malcolm Potts and Martha Campbell, "Sex Matters," *Foreign Policy* July/August 2009, 30.

43. Steven R. Machlin and Frederick Rohde (Agency for Healthcare Research and Quality), *Health Care Expenditures for Uncomplicated Pregnancies* (Washington DC: U.S. Department of Health and Human Services, 2007); U.S. Department of Agriculture, *Expenditures on Children*

by Families, report released August 4, 2009, available at http://www
.cnpp.usda.org.

44. Troy Onink, "The Financial Aid Game," *Forbes Magazine*, March 10,
2009.

45. Bartlett, "Democracy Cannot Survive Overpopulation," 70.

46. Connelly, *Fatal Misconception*, 378.

47. Statistical Center of Iran, *Estimation of Fertility Level and Pattern in
Iran: Using the Own-Children Method, 1972–1996* [in Persian] (Tehran: Statistical Center of Iran, 2000).

48. These statistics include unsafe abortion procedures. Mary Kimani,
"Investing in the Health of Africa's Mothers," *Africa Renewal* 21, no.
4 (2008).

49. John Guillebaud and Pip Hayes, "Population Growth and Climate
Change," *Free Inquiry*, April/May 2009.

50. Betsy Hartmann and Elizabeth Barajas-Román, "The Population Bomb
Is Back—With a Global Warming Twist," *Women in Action* no. 2 (August 1, 2009).

51. Elinor Ostrom, "A Polycentric Approach for Coping with Climate
Change" (World Bank Policy Research Working Paper WPS 5095, October 2009); Emily Boyd, "Governing the Clean Development Mechanism: Global Rhetoric versus Local Realities in Carbon Sequestration Projects," *Environment and Planning* 41, no. 10 (2009).

52. Betsy Hartmann, "10 Reasons Why Population Control Is Not the Solution to Global Warming," *Different Takes* 57 (2009); Amara Perez,
The Revolution Will Not Be Funded (Cambridge MA: South End Press,
2007).

53. Betsy Hartmann, *Reproductive Rights and Wrongs: The Global Politics
of Population Control* (Joanna Cotler Books, 1987), 301.

54. Diarmid Campbell-Lendrum and Manjula Lusti-Narasimhan, "Taking the Heat Out of the Population and Climate Debate," *Bulletin of
the World Health Organization* 87 (2009): 807.

55. Laurie Mazur and Ian Angus, "Women's Rights, Population, and Climate Change: The Debate Continues," *Climate and Capitalism*, March
7, 2010.

56. Hartmann, *Reproductive Rights and Wrongs*, 303.

57. Frances Kissling, "Reconciling Differences," in *A Pivotal Moment:*

Population, Justice, and the Environmental Challenge, ed. Laurie Mazur (Washington DC: Island Press, 2009). Acronyms edited for clarity.

58. B. Segal, "Food as a Weapon: Bucharest, Rome, and the Politics of Starvation," *Concerned Demography* 4, no. 2 (1974); "A Special Report on Water," *Economist*, May 22, 2010, 17.

59. United Nations, "Afghanistan," UNICEF's online database, Information by Country and Programme, accessed March 2, 2010, http://www.unicef.org/infobycountry/afghanistan_statistics.html.

60. Population Council, *Transitions to Adulthood: Child Marriage/Married Adolescents* (New York: Population Council, 2009).

61. Population Council, *State Department Reauthorization Bill to Prevent Child Marriage, Expand Opportunities for Girls* (New York: Population Council, 2009).

62. Barry Bearak, "The Bride Price," *New York Times Magazine*, July 9, 2006.

63. The lifetime risk of maternal death in Afghanistan is one in eight. See United Nations, "Afghanistan," http://www.unicef.org.

64. William Easterly, *The White Man's Burden: Why the West's Efforts to Aid the Rest Have Done So Much Ill and So Little Good* (New York: Penguin, 2007); Leo Bryant et al., "Climate Change and Family Planning: Least-Developed Countries Define the Agenda," *Bulletin of the World Health Organization* 87, no. 11 (2009).

65. United Nations Population Fund, "Gender Equality: A Cornerstone of Development," United Nations Population Fund, accessed July 15, 2011, http://www.unfpa.org.

66. Ehrlich and Ehrlich, "Is the Population Bomb Finally Exploding?"

67. U.S. Census Bureau's revised estimate is 420 million. See also "A Ponzi Scheme That Works," *Economist*, December 19, 2009.

68. Guttmacher Institute, *Teenagers' Sexual and Reproductive Health: Developed Countries* (New York: Alan Guttmacher Institute, 2001).

69. Amy O. Tsui, Raegan McDonald-Mosley, and Anne E. Burke, "Family Planning and the Burden of Unintended Pregnancies," *Epidemiologic Reviews* 32, no. 1 (2010).

70. Jonathan D. Klein, "Adolescent Pregnancy: Current Trends and Issues," *Pediatrics* 116, no. 1 (2005). On depression, see John Guillebaud, *Youth Quake: Population, Fertility and Environment in the 21st Century* (Manchester UK: Optimum Population Trust, 2007).

71. Klein, "Adolescent Pregnancy."

72. The statistics indicate 6,396 births to ten- to fourteen-year-old girls, 138,943 births to fifteen-to-seventeen-year-olds, and 296,493 births to eighteen- to nineteen-year-old girls. Joyce A. Martin et al., "Births: Final Data for 2006," in *National Vital Statistics Reports* (Washington DC: U.S. Department of Health and Human Services, Centers for Disease Control and Prevention, 2009).

73. Guttmacher Institute, *Teenagers' Sexual and Reproductive Health: Developed Countries*.

74. Guttmacher Institute, *Teenagers' Sexual and Reproductive Health: Developed Countries*.

75. James Trussell, "The Cost of Unintended Pregnancy in the United States," *Contraception* 75, no. 3 (2007).

76. Unintended pregnancies lead to roughly forty-two million induced abortions and about thirty-four million unintended births annually. Chicago population: 2.8 million. J. Joseph Speidel, Cynthia C. Harper, and Wayne C. Shields, "The Potential of Long-Acting Reversible Contraception to Decrease Unintended Pregnancy," *Contraception* 78, no. 3 (2008): 197–200.

77. The Kaiser Family Foundation and ABC Television, *The Kaiser Family Foundation and ABC Television 1998 National Survey of Americans on Sex and Sexual Health* (Menlo Park CA: The Kaiser Family Foundation and ABC Television, 1998).

78. Guillebaud, "Youth Quake," 14–15.

79. James Trussell, "Birth Control Is Cheaper Than Unintended Births," *New Mexico Daily Lobo*, March 30, 2007.

80. Whitty, "Last Taboo"; Murtaugh and Schlax, "Reproduction and the Carbon Legacies of Individuals"; David Wheeler and Dan Hammer, *The Economics of Population Policy for Carbon Emissions Reduction in Developing Countries* (Washington DC: Center for Global Development, 2010).

81. I don't condone this frame, especially the emphasis on "mothers" rather than "couples." Paul A. Murtaugh and Michael G. Schlax, "Reproduction and the Carbon Legacies of Individuals," *Global Environmental Change* 19, no. 1 (2009).

82. Julia Whitty, "The Last Taboo," *Mother Jones*, May/June 2010.

83. Assumes preventing 400,000 births annually at 327 million BTU/person/year. Energy Information Administration, *Energy Consumption, Expenditures, and Emissions Indicators, 1949–2008* (Washington DC: U.S. Department of Energy, 2009).

84. Assumes $4,340/installed kilowatt, annual production of one thousand kilowatt-hours output per year per installed kilowatt of capacity (typical New York State output), with panels replaced three times in a human lifespan. Does not include aging degradation, soiling losses, maintenance, repairs, and resource scarcity effects. New York State Energy Research and Development Authority, "Power Naturally," http://www.powernaturally.org; "Solar Module Price Highlights: November 2009," Solarbuzz, http://www.solarbuzz.com.

85. Judith S. Musick, *Young, Poor, and Pregnant: The Psychology of Teenage Motherhood* (New Haven and London: Yale University Press, 1995).

86. Guttmacher Institute, *Teenagers' Sexual and Reproductive Health: Developed Countries.*

87. Amy Schalet, "Sexual Subjectivity Revisited: The Significance of Relationships in Dutch and American Girls' Experiences of Sexuality," *Gender and Society* 24, no. 3 (2010): 310–11, 325.

88. Save the Children, *State of the World's Mothers 2009: Investing in the Early Years* (Westport CT: Save the Children, 2009), 6–7.

11. Improving Consumption

1. Annika Carlsson-Kanyama, Marianne P. Ekström, and Helena Shanahan, "Food and Life Cycle Energy Inputs: Consequences of Diet and Ways to Increase Efficiency," *Ecological Economics* 44, nos. 2–3 (2003).

2. Ozzie Zehner, "Unintended Consequences," in *Green Technology*, ed. Paul Robbins, Dustin Mulvaney, and J. Geoffrey Golson (London: Sage, 2011); David Owen, *Green Metropolis: Why Living Smaller, Living Closer, and Driving Less Are the Keys to Sustainability* (New York: Riverhead, 2009), 38.

3. Michael Bess, *The Light-Green Society: Ecology and Technological Modernity in France, 1960–2000* (Chicago: University of Chicago Press, 2003), 186.

4. Bess, *Light-Green Society*, 187.

5. John Elkington, Julia Hailes, and Joel Makower, *The Green Consumer* (New York: Viking, 1990).

6. Joel Makower, "Industrial Strength Solution," *Mother Jones*, May/June 2009, 60.

7. Scot Case, *It's Too Easy Being Green: Defining Fair Green Marketing Practices* (Washington DC: United States Congress: Energy and Commerce Panel, 2009).

8. E. F. Schumacher, *Small Is Beautiful: Economics as if People Mattered* (London: Blond and Briggs, 1973).

9. Clive Hamilton and Richard Denniss, *Affluenza: When Too Much Is Never Enough* (Crows Nest, Australia: Allen and Unwin, 2005), 36.

10. Dale Kunkel, "Children and Television Advertising," in *Handbook of Children and the Media*, ed. Dorothy G. Singer and Jerome L. Singer (London: Sage, 2001); Victor C. Strasburger, "Children and TV Advertising: Nowhere to Run, Nowhere to Hide," *Journal of Developmental and Behavioral Pediatrics* 22, no. 3 (2001).

11. Thomas N. Robinson et al., "Effects of Fast Food Branding on Young Children's Taste Preferences," *Archives of Pediatrics and Adolescent Medicine* 161, no. 8 (2007). See also Juliet B. Schor, *Born to Buy: The Commercialized Child and the New Consumer Culture* (New York: Scribner, 2004).

12. Robinson et al., "Effects of Fast Food Branding."

13. Aaron Norton, interview with the author, February 10, 2011.

14. David A. Kessler, *The End of Overeating: Taking Control of the Insatiable American Appetite* (New York: Rodale, 2009); Michelle M. Mello, Eric B. Rimm, and David M. Studdert, "The Mclawsuit: The Fast-Food Industry and Legal Accountability for Obesity," *Health Affairs* 22, no. 6 (2003); William E. Kovacic et al., *Marketing Food to Children and Adolescents* (Washington DC: Federal Trade Commission, 2008).

15. Neil Postman, *The Disappearance of Childhood* (New York: Vintage, 1994); Neil Postman, *Building a Bridge to the Eighteenth Century* (New York: Knopf, 1999).

16. Schor, *Born to Buy*, 83.

17. Valley Lake Girl Scouts, "Free and Fun Field Trips," www.valleylakegirlscouts.org.

18. Schor, *Born to Buy*.

19. Douglas Holt as quoted in Schor, *Born to Buy*, 48–49.

20. Schor, *Born to Buy*, 100–101.

21. Schor, *Born to Buy*, 117.

22. Adam Koval as quoted in Schor, *Born to Buy*, 110.

23. Adam Koval as quoted in Schor, *Born to Buy*, 186.

24. Author interview used with permission, February 5, 2010. Name withheld by request.

25. In 2009, twelve thousand Botox injections were administered to kids between the ages of thirteen and nineteen. Peggy Orenstein, *Cinderella Ate My Daughter: Dispatches from the Front Lines of the New Girlie-Girl Culture* (New York: HarperCollins, 2011).

26. Keith Suter, *Global Order and Global Disorder: Globalization and the Nation-State* (Westport CT: Praeger, 2003), 73.

27. Schor, *Born to Buy*, 167.

28. Thorstein Veblen, *The Theory of the Leisure Class: An Economic Study of Institutions* (New York: Macmillan, 1899).

29. Hamilton and Denniss, *Affluenza*, 178.

30. Tim Kasser, *The High Price of Materialism* (Cambridge MA: MIT Press, 2002), 59.

31. From the foreword by Richard M. Ryan, in Kasser, *High Price of Materialism*, xi.

32. Robert H. Frank, "Post-Consumer Prosperity: Finding New Opportunities Amid the Economic Wreckage," *American Prospect*, April 2009.

33. U.S. Census Bureau, *Median and Average Square Feet of Floor Area in New One-Family Houses Completed by Location* (Washington DC: U.S. Census Bureau, 2010).

34. Jon Mooallem, "The Self-Storage Self," *New York Times Magazine*, September 20, 2009; Gail Steketee and Randy Frost, *Stuff: Compulsive Hoarding and the Meaning of Things* (New York: Mariner Books, 2011).

35. Frank, "Post-Consumer Prosperity."

36. Hamilton and Denniss, *Affluenza*, 90.

37. To find out where you rank, visit: http://www.globalrichlist.com.

38. Barry Schwartz, *The Paradox of Choice: Why More Is Less* (New York: Ecco, 2003); Daniel Gilbert, *Stumbling on Happiness* (New York: Vintage, 2007).

39. Laura Novak, "For-Profit Crusade against Junk Mail," *New York Times*, September 6, 2007.

40. Jim Ford, *Junk Mail's Impact on Global Warming* (San Francisco: Forest Ethics, 2009).

41. U.S. Environmental Protection Agency, *Municipal Solid Waste in the United States* (Washington DC: EPA Office of Solid Waste, EPA530-R–08-010, 2008), 84.

42. Jonathan Levine and Lawrence D. Frank, "Transportation and Land-Use Preferences and Residents' Neighborhood Choices: The Sufficiency of Compact Development in the Atlanta Region," *Transportation* 34, no. 2 (2007); Daniel J. Benjamin et al., "Do People Seek to Maximize Happiness? Evidence from New Surveys," *Journal of Economic Literature* (forthcoming); Gary Pivo and Jeffrey Fisher, *The Walkability Premium in Commercial Real Estate Investments* (University of Arizona Responsible Property Investing Center and Indiana University Benecki Center for Real Estate Studies, 2010); Michael Duncan, "The Impact of Transit-Oriented Development on Housing Prices in San Diego, California," *Urban Studies* (forthcoming).

43. Henry Samuel, "Millionaire Gives Away Fortune That Made Him Miserable," *Daily Telegraph*, February 8, 2010.

44. John Browne, *Beyond Business: An Inspirational Memoir from a Visionary Leader* (London: Weidenfeld and Nicolson, 2010).

45. Michelle R. Nelson, Mark A. Rademacher, and Hye-Jin Paek, "Downshifting Consumer = Upshifting Citizen? An Examination of a Local Freecycle Community," *Annals of the American Academy of Political and Social Science* 611, no. 1 (2007); Hannah Salwen and Kevin Salwen, *The Power of Half: One Family's Decision to Stop Taking and Start Giving Back* (New York: Mariner Books, 2011).

46. Pink argues that traditional systems of monetary reward are harmful motivation killers. Paying blood donors ends up reducing the pool of people willing to give blood. Firms paying large bonuses are prone to reckless behavior and risk taking by managers. Daniel Pink, *Drive: The Surprising Truth About What Motivates Us* (New York: Riverhead, 2009).

47. Hamilton and Denniss, *Affluenza*, 188.

48. Dale Kunkel et al., *Report of the APA Task Force on Advertising and Children* (Washington DC: American Psychological Association, 2004).

49. Martin Caraher, Jane Landon, and Kath Dalmeny, "Television Advertising and Children: Lessons from Policy Development," *Public Health*

Nutrition 9, no. 5 (2007); Mary Story and Simone French, "Food Advertising and Marketing Directed at Children and Adolescents in the U.S.," *International Journal of Behavioral Nutrition and Physical Activity* 1, no. 1 (2004).

50. Karen Pine and Avril Nash, "Dear Santa: The Effects of Television Advertising on Young Children," *International Journal of Behavioral Development* 26, no. 6 (2002): 529.

51. Kunkel et al., "Report of the APA Task Force."

52. Kunkel et al., "Report of the APA Task Force," 20, 22–23.

53. C. A. Adebamowo et al., "Milk Consumption and Acne in Teenaged Boys," *Journal of the American Academy of Dermatology* 58, no. 5 (2008); F. W. Danby, "Acne and Milk, the Diet Myth, and Beyond," *Journal of the American Academy of Dermatology* 52, no. 2 (2005).

54. Commercial Alert, "Parent's Bill of Rights," http://www.commercialalert.org; Schor, *Born to Buy*.

55. Lawrence J. Schweinhart, "How the Highscope Perry Preschool Study Grew: A Researcher's Tale," HighScope, http://www.highscope.org.

56. See http://www.edibleschoolyard.org.

57. Laurence S. Seidman, *The USA Tax: A Progressive Consumption Tax* (Cambridge MA: MIT Press, 1997).

58. This would require a payment system to help poorer people. Economist Robert Frank recommends linking consumption taxes to economic activity. Frank, "Post-Consumer Prosperity." See also Jonathan Barry Forman, *Making America Work* (Washington DC: The Urban Institute, 2006), 146.

59. Virginia Spitler et al., *Hazard Screening Report: Packaging and Containers for Household Products* (Washington DC: U.S. Consumer Product Safety Commission, 2005).

60. PV probably does not offset coal-fired power production in the United States (see chapter 9 regarding the boomerang effect). But if it did, the nation's 1,256 megawatts of PV capacity in 2009 (a solar industry estimate) with a capacity factor of 14 percent would produce about a third the output of a single five-hundred-megawatt coal-fired power plant. If we accept Ford's (2009) estimate that junk mail yields the CO_2 of eleven five-hundred-megawatt coal-fired plants with a capacity factor of 73.6 percent (see chapter 2, figure 5), the carbon difference would be greater than twenty-three. The energy impact of "no

thanks" stickers would be less than twenty-three but still many times larger than the nation's photovoltaic impact. See Ford, *Junk Mail's Impact on Global Warming*; Energy Information Administration, *Average Capacity Factors by Energy Source, 1996 through 2007*, U.S. Department of Energy, http://www.eia.doe.gov.

61. Simon Kuznets, *National Income, 1929–32*, 73rd Cong., 2nd sess., Senate document no. 124, 1934, 7.

62. Breakdown: 29.4 cents for military operations, 7.9 cents to service military debt, and 3.8 cents for veteran's benefits. National Priorities Project, *Interactive Tax Chart*, National Priorities Project, accessed August 8, 2011, http://www.nationalpriorities.org.

63. Rep. Dennis Kucinich, radio news show interview with Amy Goodman on Afghanistan War, "We're Acting Like a Latter Day Version of the Roman Empire," *Democracy Now!*, December 2, 2009, http://www.democracynow.org.

64. Author interview, name withheld by request, July 29, 2010.

65. Most would pay for themselves in energy and cash savings.

66. See, for instance, Peter Dauvergne, *The Shadows of Consumption: Consequences for the Global Environment* (Cambridge MA: MIT Press, 2008), 133–68.

67. Michael Brower and Warron Leon, *The Consumer's Guide to Effective Environmental Choices: Practical Advice from the Union of Concerned Scientists* (New York: Three Rivers Press, 1999).

68. Worldwide, about a third of greenhouse-gas emissions arise from land use changes made for ranching and crop production. Susan Subak, "Global Environmental Costs of Beef Production," *Ecological Economics* 30, no. 1 (1999); Alex Avery and Dennis Avery, "Beef Production and Greenhouse Gas Emissions," *Environmental Health Perspectives* 116, no. 9 (2008); Intergovernmental Panel on Climate Change, *Climate Change 2007: Synthesis Report* ed. R. K. Pachauri and A. Reisinger (Geneva: IPCC, 2007).

69. H. Steinfeld et al., *Livestock's Long Shadow: Environmental Issues and Options* (Rome: United Nations FAO/LEAD, 2006).

70. Timothy J. Key, Gwyneth K. Davey, and Paul N. Appleby, "Health Benefits of a Vegetarian Diet," *Proceedings of the Nutrition Society* 58 (1999); Fu et al., "Effects of Long-Term Vegetarian Diets on Cardiovascular Autonomic Functions in Healthy Postmenopausal Women,"

American Journal of Cardiology 97, no. 3 (2006); Jane Hart, "The Health Benefits of a Vegetarian Diet," *Alternative and Complementary Therapies* 15, no. 2 (2009): 64.

71. Ubuntu is ranked one of the best vegetarian restaurants in the United States. It has had two head chefs, neither vegetarian. Annet C. Hoek, *Will Novel Protein Foods Beat Meat?: Consumer Acceptance of Meat Substitutes—A Multidisciplinary Research Approach* (Wageningen Netherlands: Wageningen University, 2010); G. E. Varner, "In Defense of the Vegan Ideal: Rhetoric and Bias in the Nutrition Literature," *Journal of Agricultural and Environmental Ethics* 7, no. 1 (1994); C. R. Gale et al., "IQ in Childhood and Vegetarianism in Adulthood: 1970 British Cohort Study," *BMJ* 334, no. 7587 (2007).

12. The Architecture of Community

1. Clifford Geertz, *Agricultural Involution: The Process of Ecological Change in Indonesia* (Berkeley: University of California Press, 1963), 622.
2. James Howard Kunstler, *The Geography of Nowhere: The Rise and Decline of America's Man-Made Landscape* (New York: Touchstone, 1994); Robert D. Putnam, *Bowling Alone: The Collapse and Revival of American Community* (New York: Simon and Schuster, 2000); Scott W. Allard and Benjamin Roth, *Strained Suburbs: The Social Service Challenges of Rising Suburban Poverty*, in the Metropolitan Opportunity Series (Washington DC: Brookings Institution, 2010).
3. Allard and Roth, *Strained Suburbs*; Arthur C. Nelson et al., *A Guide to Impact Fees and Housing Affordability* (Island Press: Washington DC, 2008).
4. Mark C. Childs, *Parking Spaces: A Design, Implementation, and Use Manual for Architects, Planners, and Engineers* (Boston: McGraw-Hill, 1999).
5. C. McShane, *Down the Asphalt Path: The Automobile and the American City* (New York: Columbia University Press, 1994).
6. Childs, *Parking Spaces*.
7. M. McClintock, *Street Traffic Control* (New York: McGraw-Hill, 1925).
8. "The Connected Car," *Economist Technology Quarterly*, June 6, 2009, 18.
9. South Coast Air Quality Management District, *AQMD Fact Sheet: Study*

of *Air Pollution Levels inside Vehicles* (Research Triangle Park NC: California Air Resources Board, 1999).

10. Alan Pisarski, *Commuting in America III: The Third National Report on Commuting Patterns and Trends* (Washington DC: Transportation Research Board of the National Academies, 2006).

11. Kalle Lasn, "Culture Jamming," in *The Consumer Society Reader*, ed. Juliet B. Schor and Douglas B. Holt (New York: W. W. Norton, 2000).

12. Cristina Milesi et al., "Mapping and Modeling the Biogeochemical Cycling of Turf Grasses in the United States," *Environmental Management* 36, no. 3 (2005).

13. Richard V. Pouyat, Ian D. Yesilonis, and Nancy E. Golubiewski, "A Comparison of Soil Organic Carbon Stocks between Residential Turf Grass and Native Soil," *Urban Ecosystems* 12, no. 1 (2009): 46; Elizabeth Kolbert, "Turf War," *New Yorker*, July 21, 2008.

14. See June Fletcher, "Giving up on the Outdoors," *Wall Street Journal*, June 8, 2007.

15. David Owen, *Green Metropolis: Why Living Smaller, Living Closer, and Driving Less Are the Keys to Sustainability* (New York: Riverhead, 2009), 191.

16. See Michael Leccese, *Charter of the New Urbanism*, ed. Kathleen McCormick (New York: McGraw-Hill, 2000).

17. Thomas Sieverts, *Cities without Cities: An Interpretation of the Zwischenstadt* (London: Spon Press, 2003), x.

18. Alexis de Tocqueville, *Democracy in America*, vol. 2, trans. Henry Reeve (London: Saunders and Otley, 1840), 622.

19. Modified "lets" to "rents," from de Tocqueville, *Democracy in America*.

20. De Tocqueville, *Democracy in America*, 622.

21. Simon Schama, *The American Future: A History* (New York: HarperCollins, 2009), 304.

22. See Robert W. Burchell, *Sprawl Costs: Economic Impacts of Unchecked Development* (Washington DC: Island Press, 2005).

23. M. J. Lindstrom and H. Bartling, *Suburban Sprawl: Culture, Theory, and Politics* (Lanham MD: Rowman and Littlefield, 2003); Susan Okie, *Fed Up!: Winning the War against Childhood Obesity* (Washington DC: Joseph Henry Press, 2005); Roger Silverstone, *Visions of Suburbia* (London: Routledge, 1997); Putnam, *Bowling Alone*.

24. Allard and Roth, *Strained Suburbs*; Steven Raphael and Michael Stoll,

Job Sprawl and the Suburbanization of Poverty, in the Metropolitan Opportunity Series (Washington DC: Brookings Institution, 2010).

25. Richard M. Haughey, *Higher-Density Development: Myth and Fact* (Washington DC: Urban Land Institute, 2005); Brett Hulsey, *Sprawl Costs Us All* (Madison WI: Sierra Club, 1996).

26. Robert Lewis, ed. *Manufacturing Suburbs: Building Work and the Home on the Metropolitan Fringe* (Philadelphia: Temple University Press, 2004).

27. Judson Welliver, "Detroit, the Motor-Car Metropolis," *Munsey's* (1919), 655.

28. Hardy Green, *The Company Town: The Industrial Edens and Satanic Mills That Shaped the American Economy* (New York: Basic Books, 2010); Edward Glaeser, *Triumph of the City: How Our Greatest Invention Makes Us Richer, Smarter, Greener, Healthier, and Happier* (New York: Penguin Press, 2011).

29. Arthur C. Nelson, Julian C. Juergensmeyer, James C. Nicholas, and Liza K. Bowles, *A Guide to Impact Fees and Housing Affordability* (Washington, DC: Island Press, 2008).

30. Lindstrom and Bartling, *Suburban Sprawl*, 110.

31. Robert Bruegmann, *Sprawl: A Compact History* (Chicago: University of Chicago Press, 2005); Kiril Stanitov and Brenda Case Scheer, *Suburban Form* (New York: Routledge, 2004).

32. Curtis White, *The Middle Mind: Why Americans Don't Think for Themselves* (San Francisco: Harper, 2004).

33. Owen, *Green Metropolis*, 1.

34. Owen, *Green Metropolis*, 3–4.

35. Daniel J. Benjamin et al., "Do People Seek to Maximize Happiness? Evidence from New Surveys," *Journal of Economic Literature* (forthcoming).

36. Jane Jacobs, *The Death and Life of Great American Cities* (New York: Random House, 1961).

37. Owen, *Green Metropolis*, 29.

38. Why stairs and not elevators? James Howard Kunstler has said, "Oddly, the main reason we're done with skyscrapers is not because of the electric issues or heating-cooling issues per se, but because they will never be renovated! They are one-generation buildings. We will not have the capital to renovate them—and all buildings eventually require renovation! We likely won't have the fabricated modular materials they

require, either—everything from the manufactured sheet-rock to the silicon gaskets and sealers needed to keep the glass curtain walls attached." James Howard Kunstler as quoted in Kerry Trueman, "James Howard Kunstler: The Old American Dream Is a Nightmare," *Grist*, March 9, 2011.

39. The 2008 and 2009 automotive bailouts could have provided every American with a bicycle and a comprehensive network of bike paths throughout the entire nation extensive enough to cover 90 percent of all trips. Ozzie Zehner, "Three Visions for Detroit," *Forbes*, December 4, 2008.

40. Frances Willard, *A Wheel within a Wheel: How I Learned to Ride the Bicycle, with Some Reflections by the Way* (Chicago: Woman's Temperance Publishing Association, 1895), 11.

41. John Pucher and John L. Renne, "Socioeconomics of Urban Travel: Evidence from the 2001 NHTS," *Transportation Quarterly* 57, no. 3 (2003).

42. John Pucher and Lewis Dijkstra, "Promoting Safe Walking and Cycling to Improve Public Health: Lessons from the Netherlands and Germany," *American Journal of Public Health* 93, no. 9 (2003): 1511.

43. Steven Miller, "Why Health Care Reform Should Be a Transportation Issue (and visa [*sic*] versa)," *The Public Way* (blog), September 14, 2009, http://blog.livablestreets.info/. Countries with more bikes on their streets also have lower obesity levels. David R. Bassett et al., "Walking, Cycling, and Obesity Rates in Europe, North America, and Australia," *Journal of Physical Activity and Health* 5, no. 6 (2008).

44. Colin D. Mathers et al., "Estimates of Healthy Life Expectancy for 191 Countries in the Year 2000: Methods and Results" (discussion paper as part of the Global Programme on Evidence for Health Policy, World Health Organization, 2001); National Institute on Aging, *Exercise and Physical Activity: Your Everyday Guide from the National Institute on Aging* (Washington DC: U.S. National Institutes of Health, 2009).

45. Pucher and Dijkstra, "Promoting Safe Walking and Cycling," 1510.

46. There were 1,338 cyclist deaths in Germany in 1980 and 425 in 2007. Bundesanstalt für Strassenwesen, *International Traffic and Accident Data* (Bergisch Gladbach, Germany: Bundesanstalt für Strassenwesen [Federal Highway Research Institute], 2009).

47. There were 1,003 cyclist deaths in the United States in 1975 and 716 in

2008. National Highway Traffic and Safety Administration, "Traffic Safety Facts," in DOT HS 811 156 (Washington DC: United States Department of Transportation, National Highway Traffic and Safety Administration, 2009); Pucher and Dijkstra, "Promoting Safe Walking and Cycling."

48. Deborah A. Hubsmith, *Safe Routes to School National Partnership* (Washington DC: Testimony for the House Subcommittee on Highways and Transit, concerning transportation bill SAFETEA-LU, October 7, 2007, 110th Cong.).

49. Hubsmith, *Safe Routes*.

50. List quoted directly from Pucher and Dijkstra, "Promoting Safe Walking and Cycling," 1509.

51. Jeffrey Zupan, as quoted in Owen, *Green Metropolis*, 120.

52. Owen, *Green Metropolis*, 136–37. For Curitiba, see Richard Register, *Ecocities: Rebuilding Cities in Balance with Nature* (New Society, 2006).

53. Daniel Eran Dilger, "Big Disasters: The San Francisco Freeways, a Disaster of Design, Engineering and Planning," *Roughly Drafted Magazine*, August 20, 2001.

54. Edward Epstein, "Ceremony Opens an Era of Optimism for S.F. Embarcadero," *SF Gate*, March 5, 2008.

55. The Dutch government distributes federal contracts and support evenly across cities rather than favoring just a few large metropolitan areas. This supports thriving mid-size cities, assuring that everyone can live in a walkable village. Even with a national population of seventeen million, the largest city has fewer than one million residents.

56. Jim Conley and Arlene Tigar McLaren, *Car Troubles: Critical Studies of Automobility and Auto-Mobility* (Farnham, Surrey, UK: Ashgate, 2009).

57. *Economist Technology Quarterly*, "The Connected Car"; Frost and Sullivan, *Sustainable and Innovative Personal Transport Solutions—Strategic Analysis of Carsharing Market in North America (N748-18)*, January 18, 2010, http://www.frost.com.

58. David Zhao (research analyst, Frost and Sullivan Automotive Practice), "Carsharing: A Sustainable and Innovative Personal Transport Solution with Great Potential and Huge Opportunities," January 28, 2010, http://www.frost.com.

59. Zipcar, *Zipcar Media Kit*, ed. Nancy Scott, accessed July 18, 2010, http://www.zipcar.mediaroom.com.

60. Owen, *Green Metropolis*, 138–39.

61. Charles Komanoff, *Free Transit Plan* (New York: Nurture New York's Nature, 2009).

62. Reid Ewing and Steven Brown, *U.S. Traffic Calming Manual* (Washington DC: American Planning Association, 2009).

63. Yan Xing, Susan L. Handy, and Theodore J. Buehler, "Factors Associated with Bicycle Ownership and Use: A Study of Six Small U.S. Cities" (presented at the Annual Meeting of the Committee on Bicycle Transportation, Institute of Transportation Studies and Department of Environmental Science and Policy, University of California–Davis, 2008).

64. Quote updated to reflect name change from Freiker to Boltage. See Nicole Resnick, "Commuter School," *Bicycling*, October 2009, 17.

65. Centers for Disease Control and Prevention, *Preventing Chronic Diseases: Investing Wisely in Health* (U.S. Department of Health and Human Services, CDC: Chronic Disease Prevention, 2008); U.S. Department of Health and Human Services, *Healthy People 2010: Understanding and Improving Health* (Washington DC: U.S. Department of Health and Human Services, 2010), http://www.healthpeople.gov.

66. "Schools," Freiker, accessed September 25, 2009, http://www.freiker.org.

67. Quote updated to reflect name change from Freiker to Boltage. As quoted in Resnick, "Commuter School."

68. Bob Mionske, "Legally Speaking—Insurance for Your Bike," *VeloNews*, January 17, 2008, http://www.velonews.competitor.com.

69. Ellen Dunham-Jones and June Williamson, *Retrofitting Suburbia: Urban Design Solutions for Redesigning Suburbs* (New York: Wiley, 2008), 3.

70. Christopher B. Leinberger, "Retrofitting Real Estate Finance: Alternatives to the Nineteen Standard Product Types," *Places: A Forum for Environmental Design* 17, no. 2 (2005).

71. Christopher B. Leinberger, "Building for the Long-Term," *Urban Land*, December 2003.

72. Leinberger, "Retrofitting Real Estate Finance."

73. Terry Tamminen, *Lives Per Gallon: The True Cost of Our Oil Addiction* (Washington DC: Island Press, 2006).

74. Leinberger, "Retrofitting Real Estate Finance"; Anton Clarence

Nelessen, *Visions for a New American Dream: Process, Principles, and Ordinances to Plan and Design Small Communities* (Chicago: American Planning Association, 1994).

13. Efficiency Culture

1. World Bank, *World Development Indicators* (Washington DC: World Bank, 2005).

2. International Energy Agency, *Cool Appliances: Policy Strategies for Energy-Efficient Homes, Energy Efficiency Policy Profiles* (Paris: OECD, 2003).

3. Life span: J. Komlos and M. Baur, "From the Tallest to (One of) the Fattest: The Enigmatic Fate of the American Population in the 20th Century," *Economics and Human Biology* 2, no. 1 (2004). Poverty: T. M. Smeeding et al., *United States Poverty in a Cross-National Context* (Syracuse NY: Center for Policy Research, and Luxembourg Income Study, Syracuse University, 2001). Air pollution: A. P. Verhoeff et al., "Air Pollution and Daily Mortality in Amsterdam," *Epidemiology* 7, no. 3 (1996). Debt: "The Netherlands Country Briefing," the Economist Intelligence Unit, 2008, accessed March 14, 2011. Obesity: John Pucher and Lewis Dijkstra, "Promoting Safe Walking and Cycling to Improve Public Health: Lessons from the Netherlands and Germany," *American Journal of Public Health* 93, no. 9 (2003).

4. For a review, see "Polls, Wealth and Happiness: Where Money Seems to Talk," *Economist* July 12, 2007.

5. John Byrne and Noah Toly, "Energy as a Social Project: Recovering a Discourse," in *Transforming Power: Energy, Environment, and Society in Conflict*, ed. John Byrne, Noah Toly, and Leigh Glover (London: Transaction Publishers, 2006), 21.

6. Kelly Sims Gallagher et al., *Acting in Time on Energy Policy*, ed. Kelly Sims Gallagher (Washington DC: Brookings Institution Press, 2009), 11.

7. Gallagher, *Acting in Time*, 11.

8. Witold Rybczynski, *Home: A Short History of an Idea* (New York: Viking Penguin, 1986), 231.

9. Catherine E. Beecher, *A Treatise on Domestic Economy: For the Use of Young Ladies at Home and at School* (New York: Harper, 1856), 259.

10. Rybczynski, *Home*, 162.

11. See Steen Eiler Rasmussen, "The Dutch Contribution," *Town Planning Review* 24, no. 3 (1953); Steen Eiler Rasmussen, *Towns and Buildings Described in Drawings and Words* (Cambridge MA: The MIT Press, 1969).

12. Aaron Betsky and A. Eeuwens, *False Flat: Why Dutch Design Is So Good* (London: Phaidon, 2004); Rybczynski, *Home*, 52.

13. As quoted in Anya Kamenetz, "The Green Standard?" *Fast Company*, October 2007.

14. David Owen, *Green Metropolis: Why Living Smaller, Living Closer, and Driving Less Are the Keys to Sustainability* (New York: Riverhead, 2009), 227–29.

15. Electric vehicles and grid hybrids yield higher damages than other technologies, even given expected technological advancements to 2030, according to: National Research Council, *Hidden Costs of Energy: Unpriced Consequences of Energy Production and Use* (Washington DC: National Academies Press, 2010). Owen, *Green Metropolis*, 223.

16. Auden Schendler and Randy Udall, "LEED Is Broken . . . Let's Fix It," August 9, 2005, iGreenBuild, accessed August 8, 2011, http://iGreen Build.com.

17. Thomas L. Friedman, *Hot, Flat, and Crowded: Why We Need a Green Revolution—And How It Can Renew America* (New York: Farrar, Straus and Giroux, 2008), 269.

18. Anton Clarence Nelessen, *Visions for a New American Dream: Process, Principles, and Ordinances to Plan and Design Small Communities* (Chicago: American Planning Association, 1994); Robert D. Putnam, *Bowling Alone: The Collapse and Revival of American Community* (New York: Simon and Schuster, 2000); Owen, *Green Metropolis*.

19. Harold Wilhite et al., "Cross-Cultural Analysis of Household Energy Use Behaviour in Japan and Norway," *Energy Policy* 24, no. 9 (1996); David L. Goldblatt, *Sustainable Energy Consumption and Society: Personal, Technological, or Social Change?* (Dordrecht, Netherlands: Springer, 2005).

20. R. R. Dholakia, N. Dholakia, and A. F. Firat, "From Social Psychology to Political Economy: A Model of Energy Use Behavior," *Journal of Economic Psychology* 3 (1983); Lawrence Lessig, *Code: And Other Laws of Cyberspace* (New York: Basic Books, 1999).

21. Jim Conley and Arlene Tigar McLaren, *Car Troubles: Critical Studies*

of Automobility and Auto-Mobility (Farnham, Surrey, UK: Ashgate, 2009).

22. Inge Røpke, "The Dynamics of Willingness to Consume," *Ecological Economics* 28, no. 3 (1999); Gert Spaargaren and Bas A. S. Van Vliet, "Lifestyles, Consumption, and the Environment: The Ecological Modernisation of Domestic Consumption," *Environmental Politics* 9, no. 1 (2000).

23. Michel Callon, "The Sociology of an Actor-Network: The Case of the Electric Vehicle," in *Mapping the Dynamics of Science and Technology*, ed. John Law and Arie Rip (London: McMillan, 1986); Michel Callon, "Some Elements of a Sociology of Translation: Domestication of the Scallops and the Fishermen of St. Brieuc Bay," in *The Science Studies Reader*, ed. Mario Biagioli (1986; repr., New York: Routledge, 1998); Bruno Latour, *The Pasteurization of France* (Cambridge MA: Harvard University Press, 1988); Bruno Latour and Steve Woolgar, *Laboratory Life: The Social Construction of Scientific Facts* (Beverly Hills and London: Sage, 1979); John Law, "On the Methods of Long-Distance Control: Vessels, Navigation, and the Portuguese Route to India," in *Power, Action and Belief: A New Sociology of Knowledge*, ed. John Law (London: Routledge and Kegan Paul, 1986).

24. For detailed analysis, see François Dosse, *Empire of Meaning: The Humanization of the Social Sciences* (Minneapolis: University of Minnesota Press, 1998), 11. See also Callon, "The Sociology of an Actor-Network"; Callon, "Some Elements of a Sociology of Translation."

25. Latour and Callon include physical artifacts as actors. Callon, "Some Elements of a Sociology of Translation"; Olga Amsterdamska, "Surely You Are Joking, Monsieur Latour!" *Science, Technology, and Human Values* 15, no. 4 (1990).

26. As quoted in Michael Specter, "Big Foot," *New Yorker*, February 25, 2008.

27. Sharon Beder, *Global Spin: The Corporate Assault on Environmentalism* (White River Junction VT: Chelsea Green, 2002), 104.

28. Richard A. Muller, *Physics for Presidents* (New York: W. W. Norton, 2008), 316.

29. Kingsley Kubeyinje and Tony Nezianya, "Delta Communities Protest Neglect: Press for Government Aid, Oil Company Compensation," in *Africa Recovery* (Lagos: United Nations Africa Recovery, 1999);

Ofeibea Quist-Arcton, "Gas Flaring Disrupts Life in Oil-Producing Niger Delta," *Morning Edition*, National Public Radio, July 24, 2007.

30. Shell's leadership informed U.S. diplomats that the Nigerian government had "forgotten that Shell had seconded people to all the relevant ministries and that Shell consequently had access to everything that was being done in those ministries." Abuja Embassy, "Shell MD Discusses the Status of the Proposed Petroleum Industry Bill," Wikileaks, October 20, 2009, http://www.saharareports.com.

31. Book talk on October 17, 2010, at the Sanctuary for Independent Media in North Troy, New York; Chris Hedges, *Death of the Liberal Class* (New York: Nation Books, 2010).

32. The Heinrich Böll Foundation fosters political conversations between Germany and other nations (http://www.boell.org).

33. Vanessa Fuhrmans, "Energy, Insurance Giants Threaten to Cut Investments over Tax Plan," *Wall Street Journal*, March 1, 1999.

34. Hans Diefenbacher, Volker Teichert, and Stefan Welhelmy, "How Have Ecotaxes Worked in Germany?" in *Growth: The Celtic Cancer*, ed. R. Douthwaite and J. Jopling (Devon, UK: Green Books, 2004), 134.

35. Umwelt, "Positive Effekte Der Ökologischen Steuerreform," [Positive effects of the eco-tax reform], *Umwelt* 2 (2002): 94–97.

36. *The World Factbook: Germany* (Washington DC: U.S. Central Intelligence Agency, 2009).

37. Umwelt, "Positive Effekte Der Ökologischen Steuerreform"; Stefan Bach, "Wirkungen Der Ökologischen Steuerreform in Deutschland," [Effects of the eco-tax reform in Germany] in *Wochenbericht Des DIW* (Berlin: Deutsches Institut für Wirtschaftsforschung, 2001), 222.

38. Kert Davies, radio news show interview by Amy Goodman, "Astroturf Activism: Leaked Memo Reveals Oil Industry Effort to Stage Rallies against Climate Legislation," *Democracy Now!*, August 21, 2009, http://www.democracynow.org.

39. Brian McNeill, "In Wake of Fake Letters to Congress, Groups Call for Ban on Such Tactics," *Charlottesville Daily Progress*, September 4, 2009; Stephanie Strom, "Coal Group Is Linked to Fake Letters on Climate Bill," *New York Times*, August 4, 2009.

40. Julien Daubanes and André Grimaud, "Taxation of a Polluting Non-Renewable Resource in the Heterogeneous World," *Environmental and Resource Economics* 47 (December 2010): 567–88.

41. Buildings use more energy than any other sector of the economy including the industrial and transportation sectors. U.S. Department of Energy, *Building Technologies Program* (Washington DC: U.S. Department of Energy, Energy Efficiency and Renewable Energy, 2008).

42. Ari Harding, interview by the author, San Francisco, November 23, 2009.

43. Steve Sorrell, "Barrier Busting: Overcoming Barriers to Energy Efficiency," in *The Economics of Energy Efficiency: Barriers to Cost-Effective Investment*, ed. Steve Sorrell et al. (Northampton MA: Edward Elgar, 2004).

44. Ari Harding, interview by the author. For reference see Ken Haggard, David Bainbridge, and Rachel Aljilani, *Passive Solar Architecture Pocket Reference* (London: Earthscan, 2009).

45. Half is due to trunk growth that releases upon death. Geoffrey H. Donovan and David T. Butry, "The Value of Shade: Estimating the Effect of Urban Trees on Summertime Electricity Use," *Energy and Buildings* 41, no. 6 (2009): 667.

46. New York City Department of Parks and Recreation, *NYC Street Tree Census* (New York, 2007).

47. Steve Sorrell and Horace Herring, *Energy Efficiency and Sustainable Consumption: The Rebound Effect* (New York: Palgrave Macmillan, 2008), 257

48. For the Foundation for the Economics of Sustainability, see http://www.feasta.org. For the New Economics Foundation, see http://www.neweconomics.org.

49. Totaling over $10 billion. Robert Alvarez in Matthew L. Wald, "U.S. Drops Research into Fuel Cells for Cars," *New York Times*, May 7, 2009. Gallagher et al., *Acting in Time*, 91–92.

50. Dawn Hibbard, "The Big Chill," *Kettering Perspective*, Summer 2009.

51. "A Hill of Beans," *Economist*, November 28, 2009.

52. Gallagher et al., *Acting in Time*, 173.

14. Asking Questions

1. Curtis White, *The Spirit of Disobedience: Resisting the Charms of Fake Politics, Mindless Consumption, and the Culture of Total Work* (Sausalito CA: Polipoint Press, 2006), 132; Robert D. Putnam, *Bowling Alone: The*

Collapse and Revival of American Community (New York: Simon and Schuster 2000); Richard Wilkinson and Kate Pickett, *The Spirit Level: Why Greater Equality Makes Societies Stronger* (New York: Bloomsbury, 2009).

2. Robert Socolow and Stephan Pacala, "Wedges: Solving the Climate Problem for the Next 50 Years with Current Technologies," *Science* 305, no. 5686 (August 13, 2004): 968–72.

3. Tomas Nauclér and Per-Anders Enkvist, *Pathways to a Low Carbon Economy: Version 2 of the Global Greenhouse Gas Abatement Cost Curve*, (New York: McKinsey and Company, 2009).

4. Bill Gates quoted in Steven Mufson, "Gates Coming to White House to Appeal for More Energy Research Dollars," *Washington Post*, June 9, 2010.

5. This concept could use critique and refinement. I present it to spur thought and look forward to discussing, debating, and discovering more rigorous preconditions. This list measures states-of-being for simplicity but the more important analysis may involve "rates." A region may fail a top 50 percent test but has instituted changes to move toward it. Here, alternative energy could induce benefits. Eventually, a 50 percent benchmark won't be useful as more countries fulfill these preconditions.

6. Fatih Birol as quoted in: Economist, "Facing the Consequences," The Economist November 27(2010).

7. Synthesis of twenty climate models: David. S. Battisti and Rosamond L. Naylor, "Historical Warnings of Future Food Insecurity with Unprecedented Seasonal Heat," *Science* 323, no. 5911 (2009).

8. Wolfram Schlenkera and Michael J. Roberts, "Nonlinear Temperature Effects Indicate Severe Damages to U.S. Crop Yields under Climate Change," *Proceedings of the National Academy of Sciences* 106, no. 37 (2009).

Epilogue

1. Richard A. Muller, *Physics for Presidents* (New York: W. W. Norton, 2008), 300–301.

2. World Bank, "Energy Use (Kilograms of Oil Equivalent Per Capita)," in *World Development Indicators* (Washington DC: World Bank, 2010).

3. To be clear, I do not support China's one-child policy or other democratic subversions. Also, China's per-capita consumption is growing, but is unlikely to reach American levels. Much of China's energy footprint comes from energy-intensive industries relocated to their shores from Europe and the United States.

4. Thomas L. Friedman, *Hot, Flat, and Crowded: Why We Need a Green Revolution—And How It Can Renew America* (New York: Farrar, Straus and Giroux, 2008), 176.

5. Bill McKibben's works include *The End of Nature* (New York: Random House, 1989); *Deep Economy: The Wealth of Communities and the Durable Future* (New York: Holt, 2007); and *Eaarth: Making a Life on a Tough New Planet* (New York: Times Books, 2010). Vandana Shiva's works include *Soil Not Oil: Environmental Justice in an Age of Climate Crisis* (New York: South End Press, 2008); *Water Wars: Privatization, Pollution, and Profit* (New York: South End Press, 2002); and *Staying Alive: Women, Ecology, and Development* (New York: South End Press Classics, 2010). Books by Joseph E. Stiglitz include *Making Globalization Work* (New York: W. W. Norton, 2006) and *Mismeasuring Our Lives: Why GDP Doesn't Add Up* (New York: The New Press, 2010). James Lovelock's works include *Gaia: A New Look at Life on Earth* (New York: Oxford University Press, 2000); *The Revenge of Gaia: Earth's Climate Crisis and the Fate of Humanity* (New York: Basic Books, 2007); and *The Vanishing Face of Gaia: A Final Warning* (New York: Basic Books, 2010). Books by Raj Patel include *The Value of Nothing: How to Reshape Market Society and Redefine Democracy* (New York: Picador, 2010) and *Stuffed and Starved: The Hidden Battle for the World Food System* (London: Portobello Books, 2007).

6. These thinkers include: Fred W. Allendorf, Ecumenical Patriarch Bartholomew I, Mary Catherine Bateson, His Holiness Pope Benedict XVI, Thomas Berry, Wendell Berry, Marcus J. Borg, J. Baird Callicott, Courtney S. Campbell, F. Stuart Chapin III, Robin Morris Collin, Michael M. Crow, the Dalai Lama, Avner de-Shalit, Alison Hawthorne Deming, Brian Doyle, David James Duncan, Massoumeh Ebtekar, Jesse M. Fink, James D. Forbes, Dave Foreman, Thomas L. Friedman, James Garvey, Norman Habel, Thich Nhat Hanh, Paul Hawken, Bernd Heinrich, Linda Hogan, bell hooks, Dale Jamieson, Derrick Jensen, Martin S. Kaplan, Angayuqaq Oscar Kawagley, Ste-

phen R. Kellert, Robin W. Kimmerer, Barbara Kingsolver, Shepard Krech III, Ursula K. Le Guin, Hank Lentfer, Carly Lettero, Oren Lyons, Wangari Muta Maathai, Sallie McFague, Bill McKibben, Katie McShane, Curt Meine, Stephanie Mills, N. Scott Momaday, Kathleen Dean Moore, Hylton Murray-Philipson, Gary Paul Nabhan, Seyyed Hossein Nasr, Michael P. Nelson, Nel Noddings, Barack Obama, David Orr, Ernest Partridge, Pope John Paul II, John Perry, Rolf O. Peterson, Edwin P. Pister, Carl Pope, Robert Michael Pyle, David Quammen, Daniel Quinn, Kate Rawles, Tri Robinson, Libby Roderick, Holmes Rolston III, Deborah Bird Rose, Jonathan F. P. Rose, Rosmarin, Carl Safina, Scott Russell Sanders, Lauret Savoy, Nirmal Selvamony, Ismail Serageldin, Peter Singer, Sulak Sivaraksa, Fred Small, Gary Snyder, James Gustave Speth, Alana Summers, Brian Swimme, Bron Taylor, Paul B. Thompson, George Tinker, Joerg Chet Tremmel, Quincy Troupe, Mary Evelyn Tucker, Jose Galizia Tundisi, Beth Turner, Brian Turner, Kaylynn Sullivan TwoTrees, Steve Vanderheiden, John A. Vucetich, Kimberly A. Wade-Benzoni, Sheila Watt-Cloutier, Xin Wei, Alan Weisman, Jack E. Williams, Cindy Deacon Williams, Terry Tempest Williams, E. O. Wilson, and Ming Xu. See Kathleen Dean Moore and Michael P. Nelson, *Moral Ground: Ethical Action for a Planet in Peril* (San Antonio: Trinity University Press, 2010).

7. Curtis White's books include *The Middle Mind: Why Americans Don't Think for Themselves* (San Francisco: Harper, 2004); *The Spirit of Disobedience: Resisting the Charms of Fake Politics, Mindless Consumption, and the Culture of Total Work* (Sausalito CA: Polipoint Press, 2006); and *The Barbaric Heart: Faith, Money, and the Crisis of Nature* (Sausalito CA: Polipoint Press, 2009). The work of James Howard Kunstler includes *The Geography of Nowhere: The Rise and Decline of America's Man-Made Landscape* (New York: Touchstone, 1994); *The Long Emergency: Surviving the Converging Catastrophes of the Twenty-First Century* (New York: Atlantic Monthly Press, 2005); and *World Made by Hand: A Novel* (New York: Atlantic Monthly Press, 2008). Chris Hedges is the author of *Death of the Liberal Class* (New York: Nation Books, 2010); *American Fascists: The Christian Right and the War on America* (New York: Free Press, 2007); *War Is a Force That Gives Us Meaning* (New York: PublicAffairs, 2002); and *Empire of Illusion:*

The End of Literacy and the Triumph of Spectacle (New York: Nation Books, 2009). John Michael Greer's work includes *The Ecotechnic Future: Envisioning a Post-Peak World* (Gabriola Island BC: New Society Publishers, 2009) and *The Long Descent: A User's Guide to the End of the Industrial Age* (Gabriola Island BC: New Society Publishers, 2008).

8. Elizabeth Kolbert, *Field Notes from a Catastrophe: Man, Nature, and Climate Change* (New York: Bloomsbury, 2006); Fen Montaigne, *Fraser's Penguins: A Journey to the Future in Antarctica* (New York: Henry Holt, 2010).

9. Taken from a spoken interview and edited for readability. See Philippe Diaz, radio news show interview by Amy Goodman, "Filmmaker Philippe Diaz on 'The End of Poverty?'," *Democracy Now!*, November 10, 2009, http://democracynow.org.

10. Derrick Jensen, "The Tyranny of Entitlement," *Orion*, January/February 2011.

11. Daniel Quinn's works include *Ishmael: An Adventure of the Mind and Spirit* (New York: Bantam, 1995); *The Story of B* (New York: Bantam, 1997); and *Beyond Civilization: Humanity's Next Great Adventure* (New York: Broadway, 2000). Donella H. Meadows, *Thinking in Systems: A Primer* (New York: Chelsea Green, 2008); Donella H. Meadows, Jorgen Randers, and Dennis L. Meadows, *Limits to Growth: The 30-Year Update* (New York: Chelsea Green, 2004). The film *The Yes Men* is by Andy Bichlbaum, Mike Bonanno, Chris Smith, and Dan Ollman. John Cavanagh et al., *Alternatives to Economic Globalization* (San Francisco: Berrett-Koehler Publishers, 2002).

12. Paul Pierson and Jacob S. Hacker, *Winner-Take-All Politics: How Washington Made the Rich Richer—and Turned Its Back on the Middle Class* (New York: Simon and Schuster, 2010). Richard Wilkinson and Kate Pickett, *The Spirit Level: Why Greater Equality Makes Societies Stronger* (New York: Bloomsbury, 2009). Matt Taibbi, *Griftopia: Bubble Machines, Vampire Squids, and the Long Con That Is Breaking America* (New York: Spiegel and Grau, 2010). These thinkers include: Kay Lehman Scholzman, Benjamin Page, Sidney Verba, Morris Fiorina, Larry Bartels, Hugh Heclo, Rodney Hero, Lawrence Jacobs, Jacob Hacker, Suzanne Mettler, and Diane Pinderhughes. See Lawrence R. Jacobs and Theda Skocpol, eds., *Inequality and American Democracy:*

What We Know and What We Need to Learn (New York: Russell Sage Foundation Publications, 2007).

13. James Gustave Speth, *The Bridge at the Edge of the World: Capitalism, the Environment, and Crossing from Crisis to Sustainability* (New Haven and London: Yale University Press, 2009).

14. John Perkins, radio news show interview by Amy Goodman, "Hoodwinked: Former Economic Hit Man John Perkins Reveals Why the World Financial Markets Imploded—and How to Remake Them," *Democracy Now!*, November 10, 2009, http://democracynow.org.

15. Joel Bakan, *The Corporation: The Pathological Pursuit of Profit and Power* (New York: The Free Press, 2005); Naomi Klein, *The Shock Doctrine: The Rise of Disaster Capitalism* (New York: Picador, 2008); Carl Safina, *The View from Lazy Point: A Natural Year in an Unnatural World* (New York: Henry Holt, 2011); David Harvey, *The Enigma of Capital: And the Crises of Capitalism* (New York: Oxford University Press, 2010); Sam Smith, *Why Bother?: Getting a Life in a Locked-Down Land* (Los Angeles: Feral House, 2001); Slavoj Žižek, *First as Tragedy, Then as Farce* (London: Verso, 2009); Noam Chomsky, *Understanding Power: The Indispensable Chomsky* (New York: New Press, 2002).

16. Kirkpatrick Sale, *Human Scale* (New York: Coward, McCann and Geoghegan, 1980); William A. Shutkin, *The Land That Could Be: Environmentalism and Democracy in the Twenty-First Century* (Cambridge: MIT Press, 2001).

17. Aldous Huxley, *Brave New World* (London: Chatto and Windus, 1932).

18. E. M. Forster, *The Machine Stops* (Oxford and Cambridge Review, 1909).

19. Dvora Yanow, *How Does a Policy Mean?: Interpreting Policy and Organizational Actions* (Washington DC: Georgetown University Press, 1996); Charles E. Lindblom, *Inquiry and Change: The Troubled Attempt to Understand and Shape Society* (New Haven and London: Yale University Press, 1990); Neil Postman, *Amusing Ourselves to Death: Public Discourse in the Age of Show Business* (New York: Penguin, 1985) and *Building a Bridge to the 18th Century* (New York: Knopf, 1999).

20. David E. Nye's books include *Electrifying America: Social Meanings of a New Technology, 1880–1940* (Cambridge MA: MIT Press, 1990); *Consuming Power: A Social History of American Energies* (Cambridge

MA: MIT Press, 1999); and *Technology Matters: Questions to Live With* (Cambridge MA: MIT Press, 2007). Andrew Feenburg, *Between Reason and Experience: Essays in Technology and Modernity* (Cambridge: MIT Press, 2010). Sherry Turkle, *Alone Together: Why We Expect More from Technology and Less from Each Other* (New York: Basic Books, 2011). Michel Foucault, *Power/Knowledge: Selected Interviews and Other Writings, 1972–1977*, ed. C. Gordon (New York: Pantheon, 1980); L. H. Martin, H. Gutman, and P. Hutton, eds., *Technologies of the Self: A Seminar with Michel Foucault* (Amherst: University of Massachusetts Press, 1988). Thomas S. Kuhn, *The Structure of Scientific Revolutions*, 3rd ed. (Chicago: University Of Chicago Press, 1996).

21. These thinkers include: Freeman Dyson, Francis Fukuyama, Stellan Welin, Bill Joy, Robert L. Heilbroner, Trevor J. Pinch, Wiebe Bijker, Thomas P. Hughes, Lawrence Lessig, Bruno Latour, Langdon Winner, George Ritzer, Richard Dyer, Rachel N. Weber, Daniel Sarewitz, Jameson M. Wetmore, Harry Collins, Gary Chapman, Judy Wajcman, Noela Invernizzi, Guillermo Foladori, Roopali Phadke, Bruce Schneider, Torin Monahan, David Elliott, Kristin S. Shrader-Frechette, Michael Bess, and others. See Deborah G. Johnson and Jameson M. Wetmore, *Technology and Society: Building Our Sociotechnical Future* (Cambridge MA: MIT Press, 2009).

22. David L. Goldblatt, *Sustainable Energy Consumption and Society: Personal, Technological, or Social Change?* (Dordrecht, Netherlands: Springer, 2005).

23. Milton Friedman, introduction to *Capitalism and Freedom* (Chicago: University of Chicago Press, 1962).

Index

Brown, Lester, 8
Browne, John, 244–45
Brundtland Commission, 163
Buber, Martin, 152
buildings: Dutch example of,
    307; efficiency of cogeneration
    systems in, 327; and LEED
    critique, 309–12; virtue of
    simplicity in, 307–8. *See also*
    passive solar
Bush, George W.: hydrogen
    betrayal by, 116; hydrogen
    support by, 107; wind energy
    support by, 51
Byrne, John, 302

CAFE biofuel exemption, 69
café culture, 288
California Academy of Sciences,
    324
California decoupling, 180–82
California Division of Highways,
    286
California per-capita energy use,
    182
California Solar Initiative, 15
Callon, Michel, 314
Calvinism, 308
Cameron, David, 41
Cameron, James, 134
campaign finance reform, 322,
    329–30
Campbell, Martha, 207
cancer coal-related risks, 124
capacity factor: description of,
    49–50; routine exaggerations of,
    56–57
Cape Wind Project, 39
capitalism: critiques of, 196; support
    for, 346–47; well-being and, 333.

See also neoliberalism; Occupy
    Wall Street
carbon capture and sequestration:
    challenges, 126–31; cost of, 126–
    27; description of, 126; energy
    required, 126; impact of, 128;
    leaks, 127; Obama assessment of,
    130; risks, 127–28; storage issues,
    126–28
carbon credits, 317–18. *See also*
    carbon dioxide; carbon pricing;
    carbon trading
carbon dioxide: oceanic uptake
    of, 127–28; responsibilities to
    reduce, 176; wedges to reduce,
    335. *See also* carbon pricing;
    greenhouse gasses
carbon footprint of childbirth, 219
carbon pricing, 210, 315–18, 322
carbon trading, 210
car culture, xvi, 264; emergence
    of, 265–67; and freeway
    construction, 286
carrying capacity, 243
car sharing, 289–90
Carson, Johnny, 82
Carter, Jimmy 66, 162
The Cato Institute, 38
Cavanagh, John, 346
cellulosic ethanol: bioprospecting
    for, 77; description of, 76;
    expense of, 77; *Wall Street
    Journal* analysis of, 77. *See also*
    biofuels
centenarians, 200
Center for the Sociology of
    Innovation, 314
The Centre for Policy Studies, 38
Channel One, 251
charcoal, 78–79

fertility: global average, 199; in
Iran, 209; Japan, 205; reductions,
208–9; replacement rate, 199
fertilizers, 192
fetishes: cheerleading for
energy-related, 340; in LEED
critique, 311; of suburbia, 274;
technofetishism, 331–33. *See also*
semiotics
First steps: characteristics of,
183–84; flexibility of, 184; goals
of, 183–84
First steps list: addressing complex
population challenges, 220;
approaching population concerns
of poor regions, 214–16;
approaching women's welfare in
America, *216–20*; bicycling for
youth, 295–96; bicycle insurance,
297; carefully shift to energy
(not carbon) taxes, 323–24; car
sharing, 289–90; cogeneration
systems, 327; congestion pricing,
290–91; create a department of
efficiency, 328–30; ditch the GDP,
253–56; eliminate advertising
to kids, 246–50; enable
downshifting, 244–46; from
cars to cafés, 288; from parking
to parks, *288–89*; introduce
junk-mail choice, 253; monetary
reform and decoupling, 327;
prioritize bicycle roadways,
294–95; promote volunteering,
246; rediscover passive solar,
324–26; reform zoning, 297–98;
retrofitting suburbia, 298–300;
shift military investment to
real energy security, 256–57;
shift taxes from income to

consumption, 251–52; smart
packaging, 252; social enterprises
for youth, 250; strengthen
building efficiency standards,
324; traffic calming, 292–93;
vegetarianism, 258–61; voting
reform, 328
Fleck, Ludwik, 31
flex-fuel vehicles, 69
Fonda, Jane, 81
food export bans, 213
food security: and fertilizer use, 192;
climate change effects on, 193,
339; population issues involving,
192–93, 213
Ford Motor Company, 14, 108
Forster, E. M., 347
Foucauldian power flows, 162,
180–81. *See also* codevelopment,
hydrogen economy example of
Foucault, Michel, 348
FOX news, 158
fracking, 140–42
framing: of "accidents," 82; of fossil
fuels as a renewable resource,
107–8; of the hydrogen dream,
107–10, 119–20; of technological
solutions, 171. *See also* semiotics;
productivism
France: abortion prevalence in, 218;
nuclear ambitions in, 92–93
Frank, Robert, 237, 239
freeways: construction of, 286;
effects of, 286–87. *See also* car
culture
Friedman, Milton, 346–47, 348
Friedman, Thomas, 311, 344
fuel cells: automotive programs for,
107; lifespan of, 114; limitations
of, 114; platinum usage in, 115

Fukushima meltdown, 91, 101, 102–3
fusion: explanation of, 137; ongoing projects for, 138; promise of, 138
future of environmentalism, 287–88, *331–33*, 342

Ganal, Michael, 107
gangsta, 231. *See also* semiotics
Gans, Herbert, 237
Gates, Bill, 336
GDH, 255
GDP reform, *252–56*
Geertz, Clifford, 263
General Motors Volt electric car, 143
Genuine Progress Indicator, 256
George Marshall Institute, 157
geothermal: degradation of sources for, 137; earthquakes from, 137; limitations of, 136–37; output potential of, 137; source of, 136; types of systems for, 136
Germany: bicycling in, 281–82; birthrate in, 205; eco taxes in, 320–22; resistance to nuclear power, 91
Gini Coefficient of Wealth Disparity, 256
Girl Scouts, 230
Global Climate Coalition, 157
Global Spin (Beder), 317
Glover, Leigh, 50
Goldblatt, David, 348
Goodall, Jane, 198, 251
Gore, Al, 315
governance: online resources for, 353
government secrecy: document ORNL-341, *84–85*; Green Run,

83–85; Hanford Site, *83–90*; leaked diplomatic cables on Nigeria, 319; radioactive dispersions, *83–85*
Green Building Council, 309
The Green Consumer (Makower), 226
greenhouse gasses: from biofuels 74–76; from buildings, 324; from coal, 122; from electric vehicles, 143; from meat production, 260; from natural gas, 143; oceanic uptake of, 127–28; responsibilities to reduce, 176; from solar cells, 9, 17–18; wedges to reduce, 335
green lifestyle. *See* ecoconsumerism
Green Metropolis, 275–76
Greenpeace, 5, 322
Green Run, 82–85
greenwashing: Astroturf activism, 322; BP, 160; clean coal, 124–26; ecoconsumerism, 224–7; LEED critique, 311; solar, 27. *See also* ecoconsumerism; fetishes; semiotics
Greer, John Michael, 344
Griesemer, Jim, 164
Griftopia (Taibbi), 346
gross domestic health, 255
gross domestic product, *253–56*
Gross National Happiness, 256
groundwater contamination: biofuel impact on, 66; in Niger Delta, 319; nuclear related, 83, 85; photovoltaic related, 18–20
Grove, Sir William, 106
growth: end of, 168; limits to, 163–64, 166–68. *See also* economics; growthism; neoliberalism

skeptics, 157–58; coal industry, 125, 131–32; electric vehicles, 143, 146; influence on media, 159
public space: congestion pricing, 290–91; parking takeovers, 288
public transportation: Curitiba model, 285; density of locale, 284–85; effects on children, 250–51; history of, 265; success factors for, 284–85

quality of life survey, 256
Quantum Technologies, 108
questioning goals, 178
questions about the future, 334
Quinn, Daniel, 345

Rabeder, Karl, 244
radiation, 100–103
radioactive waste: geothermal-related, 137; natural gas–related, 141. *See also* nuclear waste
ranching footprints, 259–60
Rapa Nui. *See* Easter Island
Rasmussen, Steen Eiler, 307
ratchet effect: of alternative vehicles, 146; benchmark for avoiding, 337; challenges of addressing, 171–75; of cheap power, 333; of comparative consumerism, 237; of suburbanization, 270–72
Reagan, Ronald, 346–47
REDD, 210
Rees, Amanda, 273
refrigerators, 318–19, 329
regulation: of building energy consumption, 324; outlook for, 340–42; utility of, 318–20
reliability factor, 49

Renewable Energy Consulting Services, 52
Renewable Fuels Association, 70
reproduction selfish project, 204–5
research vs. production, 336
resources, 349–53
retrofitting suburbia, *298–300*
risk society, 145, 171–72, 327–28
Robbins, Tom, 133
Roberts, Kelli, 78
Rockefeller, David, 65
Rocky Mountain Institute, 175
Romm, Joseph, 114
Roots and Shoots, 251
Rosenbaum, Walter, 90
Rostand, Edmond, 121
Royal Dutch Shell, 320
Running the Numbers (Jordan), 197
Rybczynski, Witold, 305–6

sacrificing sacrifice, 178–79
Safe Routes to School, 283
Safina, Carl, 347
Salazar, Ken, 130
Sale, Kirkpatrick, 347
Save the Children, 221
Schalet, Amy, 220–21
Schama, Simon, 162, 176, 270
Schor, Juliet, 231–35
Schroeder, Gerhard, 321
Schumacher, Fritz, 227
science and technology studies. *See* actor-network theory; boundary objects; codevelopment; congruency; Foucauldian power flows; framing; risk society; self-fashioning; semiotics; social construction of technology; trained incapacity; unintended consequences

In the Our Sustainable Future series